Birds of Prey and Wind Farms

Hermann Hötker · Oliver Krone
Georg Nehls
Editors

Birds of Prey and Wind Farms

Analysis of Problems and Possible Solutions

Springer

Editors
Hermann Hötker
Michael-Otto-Institut im NABU
Bergenhusen
Germany

Georg Nehls
BioConsult SH GmbH & CoKG
Husum, Schleswig-Holstein
Germany

Oliver Krone
Leibniz-Institut für Zoo-und
Wildtierforschung
Berlin
Germany

Translated by Katrin Knudsen from Hötker H, Krone O, Nehls G (eds) Greifvögel und Windkraftanlagen: Problemanalyse und Lösungsvorschläge. Schlussbericht für das Bundesministerium für Umwelt, Naturschutz und Reaktorsicherheit. Michael-Otto-Institut im NABU, Leibniz-Institut für Zoo- und Wildtierforschung, BioConsult SH, Bergenhusen, Berlin, Husum

ISBN 978-3-319-53401-5 ISBN 978-3-319-53402-2 (eBook)
DOI 10.1007/978-3-319-53402-2

Library of Congress Control Number: 2017931344

Printed on acid-free paper

This Springer imprint is published by Springer Nature
The registered company is Springer International Publishing AG
The registered company address is: Gewerbestrasse 11, 6330 Cham, Switzerland

Acknowledgements

We thank the Federal Ministry for the Environment, Nature Conservation, Building and Nuclear Safety for the financial support (Project FKZ: 0327684/0327684A/0327684B), and we are grateful to the Project Management Jülich, in particular Tobias Verfuß, for their patience and support.

We owe Katrin Knudsen our heartfelt thanks for translating the manuscript.

Contents

Chapter 1
Research Issues and Aims of the Study

Hermann Hötker

1.1 Project Background

The project was carried out against the backdrop of a further development of the use of renewable sources of energy which was (Federal Ministry for the Environment, Nature Conservation and Nuclear Safety (BMU) 2004) and still is (BMU 2010) the declared aim of the federal German government. This aim is related to the efforts to reduce greenhouse gas emissions and to phase out nuclear power (BMU 2011, http://www.bmu.de/energiewende/doc/47465.php, 30.6.2011). After the USA, Germany occupies a leading position in wind energy use. Further development of onshore wind energy use in Germany will mainly be carried out within areas which were already identified to be suitable for wind farms, although additional areas may be of regional importance as expected for Brandenburg and other federal states. Increases in installed capacity are planned to be mainly achieved through construction of larger wind turbines and increased hub heights.

Especially in the US but as well in Europe, concerns that wind turbines may have impacts on wildlife and birds in particular were already expressed in the early stages of wind energy use. In the US these concerns were mainly fuelled by experiences with the first large wind farm in the Altamont Pass in California, where hundreds of birds of prey were killed annually since its erection with initially nearly 5000 wind turbines. This also affects protected species like the golden eagles (*Aquila chrysaetos*) (Orloff and Flannery 1992; Smallwood and Thelander 2004). In Germany, experiences to date confirmed high sensitivity of birds of prey towards wind turbines. 1700 victims of collisions with wind turbines are currently (as of 7.3.2013) recorded in the central German database for collision victims. With 626 birds (37%), birds of prey make up a disproportionately large proportion (Dürr

H. Hötker (✉)
NABU (Nature and Biodiversity Conservation Union),
Michael-Otto-Institute, Goosstroot 1, 24861 Bergenhusen, Germany
e-mail: Hermann.Hoetker@NABU.de

H. Hötker et al. (eds.), *Birds of Prey and Wind Farms*,
DOI 10.1007/978-3-319-53402-2_1

2013). Especially affected are red kites (*Milvus milvus*) with 190 reported collisions, a species endemic to Europe, with Germany hosting more than half of the breeding population, as well as the white-tailed sea eagle (*Haliaeetus albicilla*) with 78 reported collisions. Both of these species are listed in Annex I of the EC Birds directive.

To date, possibilities to reduce the risks by means of structural measures or choice of suitable locations are limited, as little is known about the circumstances of collisions (Drewitt and Langston 2006).

Potential risks for birds of prey and other large bird species are a major controversial issue in the designation of new wind turbine sites. The Lower Saxony County Association for example recommends to observe distances of up to several kilometres from nest sites of different bird of prey species in the designation of wind farm sites (Niedersächsicher Landkreistag (Lower Saxony County Association) 2006). Apart from the above mentioned white-tailed sea eagle and red kites, especially the Montagu's harrier (*Circus pygargus*) is considered relevant with a recommended distance of up to 12.5 km. Similar distances are recommended by the Working Group of German State Bird Conservancies (Länderarbeitsgemeinschaft der Vogelschutzwarten 2007). Consistent implementation of such recommendations would considerably reduce onshore space available for wind energy use.

Several court decisions in Bavaria did not approve the construction of wind turbines, because these would have been located too close to breeding sites of the Montagu's harrier (see for example Administrative Court Würzburg, W 5 K 04.291; Administrative Court Ansbach AN 18 K 03.02016, A). In the Hellweg Börde in North Rhine-Westphalia the Montagu's harrier plays a major role in conflicts between wind energy use and nature conservation in the designation of Special Protection Areas under the Bird's Directive.

Although bird of prey collisions with wind turbines have long been reported, the reasons for the frequent collisions of these diurnal and skilful flyers were largely unknown. It was assumed that birds of prey showed little fear of solid structures like wind turbines and therefore stayed longer than other species in the rotor swept area. It could be expected that collisions with wind turbines did not occur due to restricted perception of the turbines as likely for birds migrating at night, but due to species-specific behaviours. Therefore, understanding of these behaviours was of major importance in the development of avoidance measures.

Sensory physiological investigations have shown though that fast moving monochromatic rotor blades may not be perceived as a potential threat by birds of prey in the close-up area (Hodos et al. 2001). Experimental field studies on the effectiveness of marking turbine blades were not available at the time. Assumptions were made that the perception of birds of prey is mainly focused downwards or side-wards in flight (Martin 2011).

No studies on the reasons of collisions or other conflicts of birds of prey were available from Germany. Several studies in other European countries and the US mainly dealt with the determination of collision rates and identification of problematic locations for wind turbines (overview in: Hötker et al. 2004). A comprehensive study on the avoidance of collisions in the well-known Altamont wind farm was published in the US (Smallwood and Thelander 2004) shortly before the present research project was proposed, but was specifically tailored to local species and conditions. Still, some of the suggested approaches could be evaluated regarding their transferability to conditions in Germany. A number of studies on the determination of collision rates was available mainly from Scotland (Madders and Whitfield 2006; Band et al. 2007) which were important in the development of basic methods for the present study. In the early stages of the present study, first studies on the reduction of collisions of white-tailed sea eagles were carried out in Norway with focus on the implementation of mitigation measures (Bevanger et al. 2009).

In the beginning of the project, VHF transmitters, well-proven in various studies, were used in bird of prey monitoring, which can be tracked from mobile ground stations. GPS satellite transmitters were available in the second phase of the project, which regularly transmit locations to a satellite where data can be retrieved at regular intervals.

1.2 Scope

Aim of the study was to investigate the reasons for the higher collision rates of birds of prey in comparison to other species and to analyse additional conflicts between birds of prey and a further development of wind energy use, to assess the importance of these conflicts for the conservation of endangered species and to develop solutions for a reduction or avoidance of conflicts.

The results were to be presented in form of a guideline for development of wind turbines in breeding areas of sensitive bird of prey species. Specific suggestions on how to reduce collision risk through changes in the surrounding area of wind turbines, marking of turbine blades or selection of the wind turbine location were to be drawn up. The effects of these suggestions were to be verified in field studies.

To achieve this, the risk for birds of prey within existing wind farms needed to be determined in behavioural observations. Hunting behaviour, flight height and evasive movements of three relevant species were to be investigated with the additional support of telemetric methods. The question whether the collision risk changes with larger wind turbines and increasing hub height was an important aspect. Investigations were mainly carried out for red kites, white-tailed sea eagle and Montagu's harrier. These species were chosen, because red kites and white-tailed sea eagles are among the most frequent collision victims in comparison to the size of their respective populations. The Montagu's harrier as a highly endangered species throughout Europe was proposed as a further species, as this species was frequently rated as sensitive towards wind turbines and breeding and foraging habitats of Montagu's harriers, large-scale open agricultural landscapes, are at the same time especially suitable for wind energy use. Many wind farms are located in the north-west of Schleswig-Holstein and the Hellweg Börde in North Rhine-Westphalia, the main areas of distribution of the Montagu's harrier.

Detailed questions regarding the single species can be found in the next chapter.

References

Band W, Madders M, Whitfield DP (2007) Developing field and analytical methods to assess avian collision risk at wind farms. In: de Lucas M, Janss G, Ferrer M (eds) Birds and Wind Power. Quercus, Madrid, p 259–275.
Bevanger K, Berntsen F, Clausen S, Dahl EL, Flagstad Ø, Follestad A, Halley D, Hanssen FE, Hoel PL, Johnsen L, Kvaløy P, May R, Nygård T, Pedersen HC, Reitan O, Steinheim Y, Vang R (2009) Pre- and post-construction studies of conflicts between birds and wind turbines in coastal Norway (BirdWind). Progress Report 2009, NINA Report 505, 70 pp.
Bundesministerium für Umwelt, Naturschutz und Reaktorsicherheit (BMU) (2004) Themenpapier Windenergie. Bundesministerium für Umwelt, Naturschutz und Reaktorsicherheit, Art 2122, Berlin.
Bundesministerium für Umwelt, Naturschutz und Reaktorsicherheit (BMU) (2010) Umweltbericht 2010. Bundesministerium für Umwelt, Naturschutz und Reaktorsicherheit, Berlin.
Bundesministerium für Umwelt, Naturschutz und Reaktorsicherheit (BMU) (2011) Energie für Deutschland - Eckpunktepapier der Bundesregierung. http://www.bmu.de/energiewende/doc/47465.php. Accessed 30 Jun 2011.
Drewitt AL, Langston RHW (2006) Assessing the impacts of wind farms on birds. Ibis 148, Suppl 1:29–42.
Dürr T (2013) Vogelverluste an WKA in Deutschland und Fledermausverluste weltweit. Summe der Funde seit 1989. Stand 7.3. 2013. Daten aus dem Archiv der Staatlichen Vogelschutzwarte im Landesamt für Umwelt, Gesundheit und Verbraucherschutz Brandenburg. http://www.lugv.brandenburg.de/cms/detail.php/bb1.c.321381.de. Accessed 07 Mar 2013.
Hodos W (2001) Minimization of Motion Smear: Reducing Avian Collisions with Wind Turbines. NREL/SR-500-33249, Maryland.
Hötker H, Thomsen K-M, Köster H (2004) Auswirkungen regenerativer Energiegewinnung auf die biologische Vielfalt am Beispiel der Vögel und der Fledermäuse – Fakten, Wissenslücken, Anforderungen an die Forschung, ornithologische Kriterien zum Ausbau von regenerativen Energiegewinnungsformen. Endbericht für das Bundesamt für Naturschutz, Förd. Nr. Z1.3-684 11-5/03. Michael-Otto-Institut im NABU, Bergenhusen, pp 80.
Länder-Arbeitsgemeinschaft der Vogelschutzwarten (2007) Abstandsregelungen für Windenergieanlagen zu bedeutsamen Vogellebensräumen sowie Brutplätzen ausgewählter Vogelarten. Berichte zum Vogelschutz 44:151–153.
Madders M, Whitfield DP (2006) Upland raptors and the assessment of wind farm impacts. Ibis 148, Suppl 1:43–56.
Martin GR (2011) Understanding bird collisions with man-made objects: a sensory ecology approach. Ibis 153:239–254.
Niedersächsichen Landkreistag (2006) Hinweise zur Berücksichtigung des Naturschutzes und der Landschaftspflege sowie zur Durchführung der Umweltprüfung und Umweltverträglichkeitsprüfung bei Standortplanung und Zulassung von Windenergieanlagen. Informationsdienst Naturschutz Niedersachsen 1/2006:16–37.
Orloff S, Flannery A (1992) Wind turbine effects on avian activity, habitat use and mortality in Altamont Pass and Solano County wind resources areas 1989-1991. California Energy Commission, Bio-Systems Analysis, Tiburon.
Smallwood KS, Thelander CG (2004) Developing methods to reduce bird mortality in the Altamont Pass Wind Resource Area. Final report by BioResource Consultants to the California Energy Commission, pp 363.

Chapter 2
Project Structure and Methodological Approach

Hermann Hötker

2.1 Project Structure

The study was carried in the five German federal states of Schleswig-Holstein, Mecklenburg-Western Pomerania, Brandenburg, Saxony-Anhalt and North Rhine-Westphalia in cooperation of seven institutions with long-term experience in research of birds of prey and ecological impacts of wind turbines.

Participating institutions

- NABU (Nature And Biodiversity Conservation Union) Michael-Otto-Institute, (www.bergenhusen.nabu.de), Goosstroot 1, D-24861 Bergenhusen (project lead)
- BioConsult SH GmbH & Co. KG, (www.biocosult-sh.de), Schobüller Str. 36, D-25813 Husum
- Leibniz Institute for Zoo and Wildlife Research (IZW), (www.izw-berlin.de), Alfred-Kowalke-Str. 17, D-10315 Berlin
- ÖKOTOP GbR (www.oekotop-halle.de), Willy-Brandt-Straße 44, D-06110 Halle (Saale)
- Förderverein für Ökologie und Monitoring von Greifvogel- und Eulenarten e.V. (MEROS—Monitoring European Raptors and Owls), (www.greifvogel monitoring.de), Buchenweg 14, 06132 Halle (Saale)
- Staatliche Vogelschutzwarte (State Bird Observatory and Field Centre) of Landesumweltamt Brandenburg (Brandenburg Environmental Agency), Buckower Dorfstr. 34, D-14715 Nennhausen-Buckow

H. Hötker (✉)
NABU (Nature and Biodiversity Conservation Union)
Michael-Otto-Institute, Goosstroot 1, 24861 Bergenhusen, Germany
e-mail: Hermann.Hoetker@NABU.de

H. Hötker et al. (eds.), *Birds of Prey and Wind Farms*,
DOI 10.1007/978-3-319-53402-2_2

- Arbeitsgemeinschaft Biologischer Umweltschutz im Kreis Soest e.V. (Soest district working group for nature conservation and landscape management), (www.abu-naturschutz.de), Teichstraße 19, D-59505 Bad Sassendorf-Lohne.

Three of the institutions submitted a total of three coordinated project proposals to the Federal Ministry for the Environment, Nature Conservation and Nuclear Safety (BMU) (NABU (Nature And Biodiversity Conservation Union)—Michael-Otto-Institute (coordinator); BioConsult SH; Leibniz Institute for Zoo and Wildlife Research). Cooperation is shown in Fig. 2.1. The project was subdivided into several partial projects which were carried out by one or several of the project partners (Fig. 2.1). The partial projects were of differing duration (see below).

After the end of the first phase of the project some of the partial projects were prolonged until 31.12.2010.

Over the entire project period the project was accompanied by a working group comprising the following members (participating for differing periods of time):

Mrs. Hofmann, Mrs. Radecke	BMU
Mr. Heider, Mr. Verfuß	Project developer Jülich (PTJ)
Mr. Igel	German Federal Agency for Nature Conservation (BfN)
Mrs. Findeisen	Umweltbundesamt (UBA)
Mr. Wetzig, Mrs. Lepinski	German Wind Energy Association (BWE)
Mr. Ratzbor	Schmal & Ratzbor
Mr. Schlüter	Enerplan
Mr. Boris de Wolf	Enercon
Mr. Jaehne	Working Group of German State Bird Conservancies (Länderarbeitsgemeinschaft staatl. Vogelschutzwarten)

The basic project as well as extended versions of abstracts of project events were published on the project home page (http://bergenhusen.nabu.de/forschung/greifvoegel/).

Apart from data collection and evaluation, two major project events were scheduled within the project period. The first event, an international workshop on 21 and 22 October 2008 in Berlin served to exchange expertise with colleagues from Germany and abroad. The results of this workshop are documented in detail on the internet (Hötker 2009). Also in Berlin, the most important preliminary project results were presented to an audience of experts of wind farm planners and operators. Authority representatives, lawyers, researchers and representatives of nature conservation organizations in a final event on 8 November 2010. This event is likewise documented on the internet (http://bergenhusen.nabu.de/forschung/greifvoegel/berichtevortraege/).

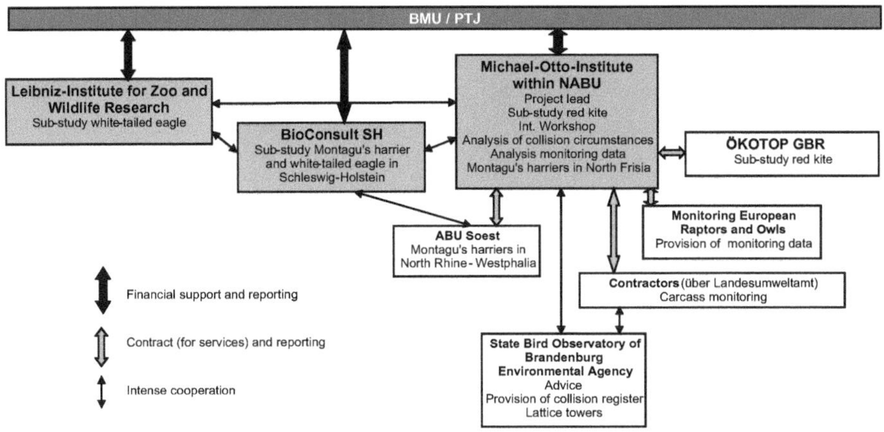

Fig. 2.1 Cooperation of the project partners of the project "Birds of prey and wind turbines: Analysis of problems and possible solutions"

2.2 Methodological Approach

It was planned to investigate the behaviour of the target species white-tailed sea eagle (*Haliaeetus albicilla*), red kite (*Milvus milvus*) and Montagu's harrier (*Circus pygargus*) within existing wind farms and to determine the threat level. To achieve this, flight paths and heights as well as reactions and behaviours within wind farms were to be observed and recorded. An investigation merely based on recorded collision victims would not have resulted in reliable results considering the short project period. Risk estimates based on behavioural observations were therefore assumed to be suitable. Direct observations were to be supported by telemetric methods. Using telemetric monitoring more and higher quality data could be collected within a shorter period of time than in observations alone. Use of telemetric monitoring allowed to track individual birds of prey within a larger area and to determine use or avoidance of wind farms within the home range. Last but not least, telemetric monitoring helped to save expenses.

Investigations were to be carried out within the main areas of distribution of the species in Germany (white-tailed sea eagle: Schleswig-Holstein, Mecklenburg-Western Pomerania, red kite: Saxony-Anhalt, Montagu's harrier: north-western Schleswig-Holstein, Hellweg Börde in North Rhine-Westphalia).

Additionally, existing data was to be reviewed and the circumstances of collisions of red kites and white-tailed sea eagles were to be researched.

2.2.1 Sub-study on Red Kite (2007–2010)

Studies of red kites pursued three objectives. Firstly, the question whether red kites show avoidance behaviour towards wind farms during the breeding season or are quite in the contrary attracted to wind farms due to specific structures (fallows around wind turbines as an attractive foraging ground) was to be answered. Secondly, methods to keep red kites away from wind turbines were to be tested. For this purpose the tower base area of some of the wind turbines were to be covered with agricultural film or gravel to prevent an establishment of rodent populations, which might attract red kites towards the dangerous rotor swept area. Some of the turbine base fallows remained uncovered for reference. Thirdly, it was to be verified, whether red kites may be kept away from wind farms through creation of attractive sources of food elsewhere.

Two methodological approaches were used in the collection of data. Standardised watches of red kites were carried out in the experimentally modified wind farms. Other bird of prey species, especially common buzzards (*Buteo buteo*) and black kites (*Milvus migrans*), were included. The behaviour of the birds (flight height, distance from wind turbines, interactions, etc.) was documented to allow for conclusions on the relatively high number of collision victims and to verify the effectiveness of measures taken to reduce the collision risk. Secondly, red kites were to be equipped with small direction-finding transmitters to gain insight in habitat use in relation to wind farms or turbines.

Red kites breeding birds were investigated in Saxony-Anhalt, where the population density of the species is highest in Germany and a large number of wind turbines were constructed and alternative sources of food were established due to compensation measures at the same time.

Investigations of red kites were to be carried out in close cooperation between ÖKOTOP GbR and the NABU Michael-Otto-Institute. A contract for services was concluded with ÖKOTOP GbR.

2.2.2 Sub-study on Telemetric Monitoring of Montagu's Harriers (2007–2009)

Investigations of Montagu's harriers breeding pairs in western part of Schleswig-Holstein aimed to complement the comprehensive approach by taking account of open landscape habitats. The North Frisian marshes of the North German plain hosted a breeding population of about 50 Montagu's harrier pairs with a large number of wind turbines within their habitats. Therefore, this region was highly suitable to investigate the potential for conflict with wind turbines. Two questions were to be answered in the investigations of Montagu's harriers. The risk of colliding with wind turbines and a possible disturbance effect of wind farms causing a displacement of breeding pairs. Possible displacement was investigated in the two sub-studies "Montagu's harriers in the Hellweg Börde" and "Montagu's harriers in North Frisia" (see below).

Investigation of the collision risk was of major importance due to extensive overlap of breeding areas and areas for wind energy use. To achieve this, flight movements of this species within existing wind farms were investigated. The collision risk was estimated in standardised watches of the behaviour of this species within wind farms, documenting flight direction and height as well as reactions towards wind turbines and distance from and duration of stay within wind farms. Telemetric monitoring by means of direction-finding transmitters was possible for both Montagu's harriers and red kites. Telemetric monitoring was important to track and observe the birds during long-range hunting flights. Telemetric monitoring of Montagu's harriers was carried out by BioConsult SH.

2.2.3 Sub-study on White-Tailed Sea Eagle (2007–2010)

Aim of the investigations was to determine which wind turbine models in the surrounding area of white-tailed sea eagle nests held particular risks for the eagles and which turbines were less problematic.

Visual observations supported by telemetric methods were carried out. Observations aimed to determine flight paths of the eagles and how far wind farms are included in the home range. The size of white-tailed sea eagles allowed using spatially high resolution GPS transmitters or data loggers which allow to record flight movements over a longer period of time to determine spatial and temporal activity patterns. Recorded coordinates can be transmitted via GSM (mobile communication) or downloaded as UHF signal. At the same time, transmitter and data logger transmit signals in the VHF band which can be tracked and traced by means of a portable receiver. Activities of single eagles may be tracked individually.

To determine the impacts of wind turbines on local breeding pairs and their breeding success, young birds as well as adult birds were to be equipped with GPS transmitters. Dr. Krone, IZW, was responsible for telemetric monitoring of white-tailed sea eagles. Observations of tagged white-tailed sea eagles were carried out by staff members of IZW and BioConsult SH.

2.2.4 Sub-studies on Montagu's Harriers in the Hellweg Börde and in North Frisia (2007–2008)

Possible disturbance effects causes by wind turbines were especially in North Rhine-Westphalia in the focus of discussions. This issue was to be investigated by comparison of long-term data on the development of the breeding population of

Montagu's harriers to data on the development of wind energy use in the Hellweg Börde near Soest and in the northern part of North Frisia. Long-term, exact data were available for both regions suitable to carry out GIS-based analyses. Evaluation of the data for the Hellweg Börde was to be performed by the ABU Soest, which has already mapped breeding sites of this species since 1993. Data for the northern part of North Frisia were available from the species protection programme and was to be evaluated by the NABU Michael-Otto-Institute.

2.2.5 Sub-study Analysis of the Circumstances of Bird of Prey Collisions (2007–2009)

The State Bird Observatory and Field Centre of the Brandenburg Environmental Agency keeps a register of birds which have collided with wind turbines. Within the framework of the sub-study, collisions of the investigated bird of prey species and other large bird species (herons, storks, crane, further bird of prey species) was to be assessed with focus on the circumstances (habitat, site conditions, wind turbine model, age of the bird, etc.) and analysis of the data.

2.2.6 Sub-study on Lattice Towers (2008)

The analysis of the circumstances of bird of prey collisions aimed to assess the impacts of different wind turbine models on the mortality of birds of prey. The central database of the Brandenburg Environmental Agency (State Bird Observatory) did not include data on wind turbines with lattice towers and a relatively small amount of data on searches of wind turbines with a total height of over 140 m. Statements made on the basis of this database were therefore only to a limited amount transferable to wind turbines with a capacity of between 2.5 and 3 MW and not at all to wind turbines with lattice towers. In 2008 targeted investigations to record mortality were carried out in two wind farms with ten wind turbines each to estimate the potential risk for birds of prey caused by these currently mainly used types of wind turbine.

2.2.7 Band Model

Detailed telemetric data and visual observations of red kites and Montagu's harriers were used to simulate the collision risk in Band models (Band et al. 2007). Even if the absolute risk of collision cannot be estimated, the method allows for comparison of relative risks.

2.2.8 Sub-study on Bird of Prey Monitoring (2007–2009)

Since 1988, data on bird of prey populations and reproduction rates was annually collected in more than 500 mostly larger monitoring areas within the framework of the MEROS (Monitoring European Raptors and Owls) research project. Over the entire project period, wind turbines erected within these monitoring areas as well as possible influences on the size of the populations or the reproductive success were recorded.

References

Band W, Madders M, Whitfield DP (2007) Developing field and analytical methods to assess avian collision risk at wind farms. In: de Lucas M, Janss G, Ferrer M (eds) Birds and Wind Power. Quercus, Madrid, p 259–275.

Hötker H (2009) Birds of Prey and Wind Farms: Analysis of Problems and Possible Solutions – A brief introduction to the project and the workshop. In: Hötker H (ed) Birds of Prey and Wind Farms: Analysis of Problems and Possible Solutions. Documentation of an international workshop in Berlin, 21st and 22nd October 2008. http://bergenhusen.nabu.de/imperia/md/images/bergenhusen/bmuwindkraftundgreifwebsite/birds_of_prey_and_windfarms_documentation_2009.pdf, Assessed 30 Jun 2013.

Chapter 3
Red Kite

Kerstin Mammen, Ubbo Mammen and Alexander Resetaritz

3.1 Introduction

In Germany, the red kites (*Milvus milvus*) takes the second position in the nationwide statistics of collision victims (see Chap. 12). More victims of collisions with wind turbines were only reported for the common buzzard (*Buteo buteo*), whose breeding population is at least nine times as large as that of the red kite (Mebs and Schmidt 2006).

Since 1991, the breeding population of the red kite is declining in Germany (Mammen and Stubbe 2009a, b). The strongest decline occurred between 1991 and 1997 (Mammen 2000). A main reason for the decline is food shortage due to intensified farming and the decrease or worsening availability of food. Currently, reasons for a decline in areas of high density of red kites in Saxony-Anhalt are loss of breeding sites due to disappearance of tree populations in the open landscape, increasing inter-specific nest site competition and increased predation (Nicolai and Mammen 2009; Nicolai et al. 2009). Poisoning and direct human persecution during migration and in wintering grounds may have played an equally important role (Villafuerte et al. 1998; Berny and Gaillet 2008). Wind energy use is named as an additional reason for the decline, but is weighed differently by different authors (Stubbe and Stubbe 2007; Aebischer 2009; Mammen 2009; Nicolai et al. 2009).

Results of studies on the behaviour of red kites towards wind farms are partly contradictory (Brauneis 1999; Bergen 2001, 2002; Stübing 2001; Traxler et al. 2004). Hötker et al. (2004) gave a summary of the current state of knowledge up to the year 2004, and indicated a need for research of the impacts of wind turbines on

---

K. Mammen (✉) · U. Mammen · A. Resetaritz
ÖKOTOP GbR, Willy-Brandt-Straße 44, 06110 Halle (Saale), Germany
e-mail: Kerstin.Mammen@oekotop-halle.de

K. Mammen · A. Resetaritz
NABU (Nature and Biodiversity Conservation Union), Michael-Otto-Institute,
Goosstroot 1, 24861 Bergenhusen, Germany

H. Hötker et al. (eds.), *Birds of Prey and Wind Farms*,
DOI 10.1007/978-3-319-53402-2_3

red kites and white-tailed sea eagles (*Haliaeetus albicilla*) in particular. Investigations of habitat use and home range sizes were presented among others by Porstendorfer (1994, 1998), Walz (2001), Hagge et al. (2003), Resetaritz (2006), Resetaritz et al. (2006), Nachtigall (1999, 2008) and Nachtigall et al. (2003) without particular focus on wind turbines and red kite.

The first study focusing on red kite and wind power was carried out in 2005 in the Querfurter Platte in Saxony-Anhalt (Strasser 2006; Mammen et al. 2009).

Between 2007 and 2010, behavioural observations of red kites within wind farms as well as telemetric monitoring to determine habitat use of red kite breeding in the close-up area of wind farms were carried out for the project reported here. Additionally, the effectiveness of covering turbine base fallows was checked and the use of additional food sources (cut alfalfa crops, composting plants) outside the wind farm site was recorded. Aim of the study was to find out, why red kites collide as often with wind turbines and to develop measures to minimise conflicts.

This chapter summarizes the results of the study. More details on several aspects of the study can be found in an annex to this report (https://bergenhusen.nabu.de/imperia/md/nabu/images/nabu/einrichtungen/bergenhusen/projekte/bmugreif/endbericht_greifvogelprojekt_anhang.pdf).

3.2 Methods

3.2.1 Study Areas

For the present project, investigations on the red kite were carried out in Saxony-Anhalt, the federal state with a state-wide density of 11 red kite breeding pairs per 100 km^2, the highest density in Germany and the world. Four wind farms, situated within the fairly featureless, large-scale intensively farmed agricultural landscape typical for this federal state, were chosen. The more closely investigated two main study areas (Querfurt in the agriculturally used part of the Querfurter Platte and Druiberg in the hilly Nordharz foothills) were compared to reference areas with similar landscape characteristics (Dessau and Speckberg). From 2009, after the end of the first phase of the project, the Speckberg study area became the second main study area. The study areas Druiberg and Dessau were discontinued after 2009. From 2010, a further wind farm near Queis (Queis study area) was included in telemetric monitoring. Figure 3.1 gives an overview of the study areas in Saxony-Anhalt.

The study areas cover an area of 3 km beyond the boundaries of the respective wind farm. Depending on the number of wind turbines installed within a wind farm, the size of the study area covered between 47.8 and 88.6 km^2. Large-scale crop fields (typical size: 30–70 ha) with only few dominating crops (wheat, oilseed rape, barley, beet, maize) and a lack of larger forest areas (with the exception of the Dessau study area which included parts of the Mosigkau Heath in the east) were

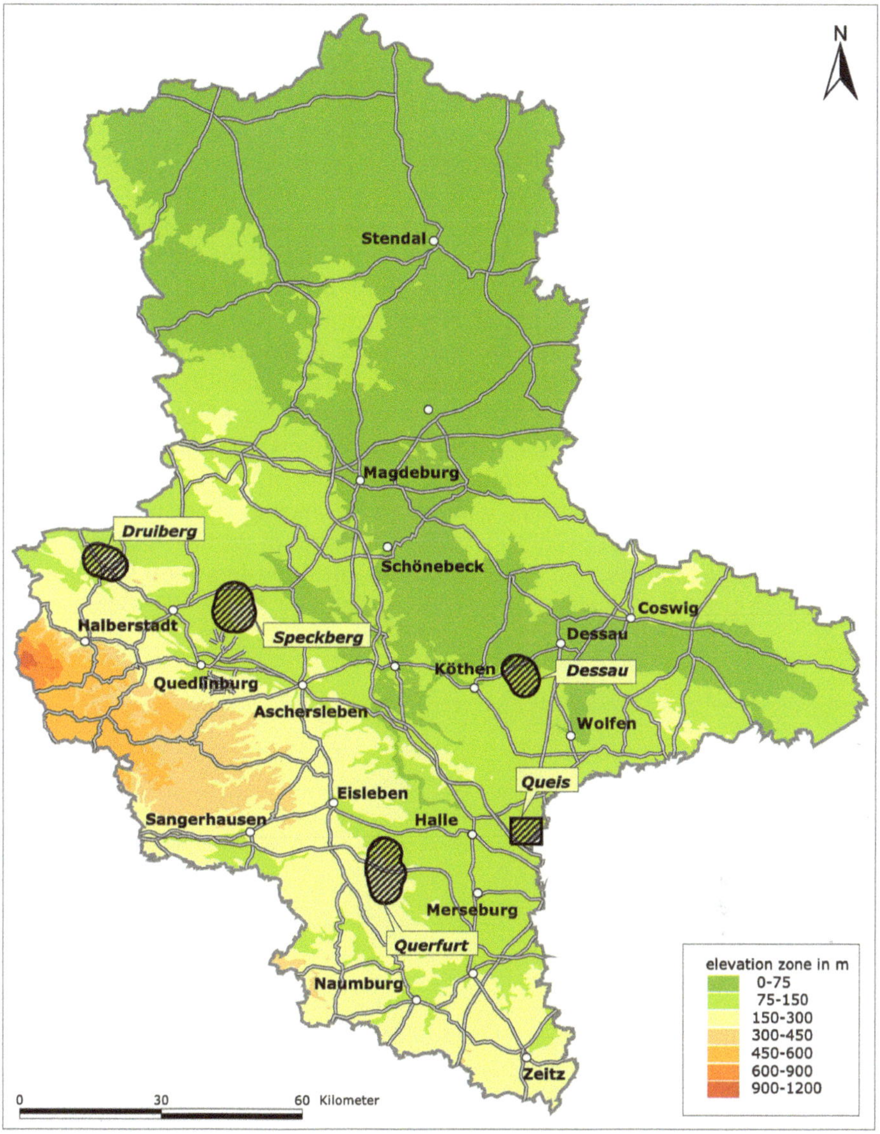

Fig. 3.1 Location of the study areas for the red kite partial project in Saxony-Anhalt

typical for the study areas. In the Querfurt study area, high growing poplar groves can be found on the afforested slopes of an open pit mining dump. In all study areas, single thickets occurred within the crop areas. Apart from these, only hedgerows planted in the 1950s–1980s (rows of poplars, windbreaks), local remains of wet woodland and riparian forests as well as interrupted rows of fruit trees lining the roads were available to birds of prey for nesting.

Querfurt

The main part of this study area (Figs. 3.2 and 3.3) is intensively farmed. Highly fertile loess and black earth soils with soil values between 82 and 97 prevail (Arndt 2002) mainly used to cultivate cereals and beet but also oilseed rape and maize. Many of the fields are bordered by hedgerows or rows of poplars and fruit trees of different ages. Due to the considerable size of the fields (mean: approximately 40 ha), the landscape is of relatively low structural density. Rows of 50 year old poplars in the east and south of the study area directly adjoining the wind farm were the most important nest sites for red kite within this study area.

The two directly adjoining wind farms Dornstedt (14 wind turbines) and Obhausen/Nemsdorf (20 wind turbines) are located in the centre of the study area. Of the total of 34 wind turbines, 30 are already in operation since 2001. In 2005 the Dornstedt wind farm was extended by four turbines. All wind turbines are Enercon E-66 turbines varying only in hub height (85 m in the Obhausen/Nemsdorf wind farm and 98 m in the Dornstedt wind farm) and colour of the tower (white or light grey). The rotor of 70 m diameter is three-bladed, having either tubular steel or prestressed concrete towers, which are painted in shades of green at the base (WKA 2005). A radius of 3 km around these 34 turbines equals an area of 61.3 km^2, which was the area used as a study area in 2007. In 2007, a further 13 wind turbines

Fig. 3.2 Wind farm Querfurt. View from the south over the entire wind farm. The southernmost four wind turbines are not in the picture (*Photo* U. Mammen, 07.05.2011)

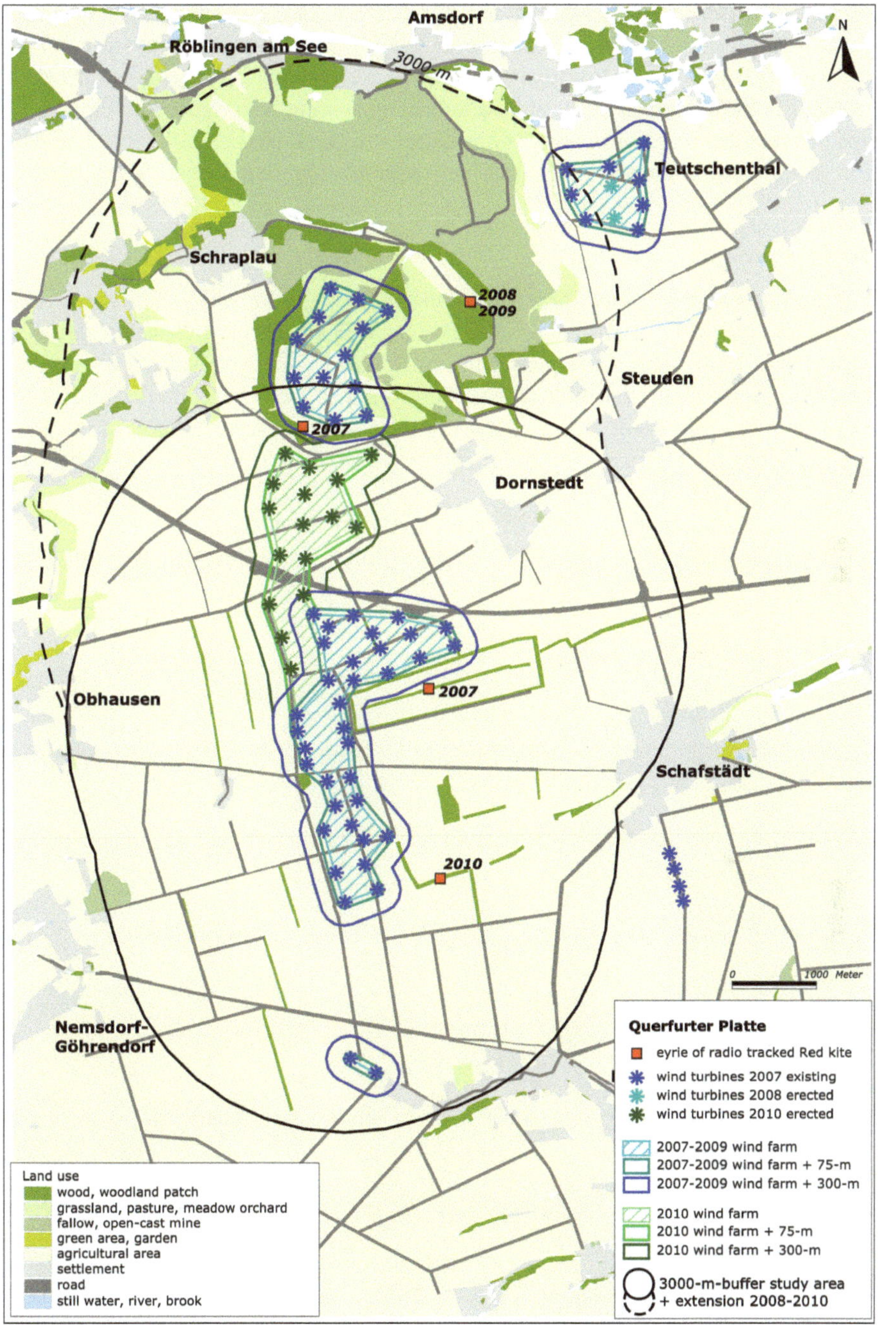

Fig. 3.3 Location and boundaries of the Querfurt study area including wind turbines with the surrounding area of 3 km (study area) as well as the wind farm polygon +75 and +300 m buffer (wind farm area and wind farm impact area)

(Vestas V90, 105 m hub height, 90 m rotor diameter) were erected on the slag heap of the open pit mine Amsdorf (Asendorfer Kippe), and the study area was extended to cover 88.6 km^2 accordingly, to include these turbines. In 2010, a further 16 turbines (Fuhrländer FL-2500, 100 m hub height) went into operation, closing the gaps between the wind farm areas in the south and north of the study area. The wind farm within the Querfurt study area subsequently stretched over 8 km north to south.

Speckberg

Deep loess soils, mainly chernozems and rendzina, with soil values between 78 and 94 prevail in the area (Töpfer 1995). In the Bodeaue ground-water gley soils occur (LAU 2000). The major part of the study area is intensively farmed (Figs. 3.4 and 3.5). Cultivation of winter wheat, barley and oilseed rape predominates followed by root crops (beet, maize, potatoes). Grassland covers only 5% of the study area and is mainly restricted to floodplain areas. The area is farmed with large-scale crops.

Fields are bordered by rows of trees, hedgerows or roads, but the structural density is generally low within the study area. Solely the Bodeaue in the south-west

Fig. 3.4 Speckberg study area. View from the north-east over the southern part of the wind farm with 12 wind turbines. In the centre, the woodland with the nest of the satellite-tagged bird "Alte Dame"

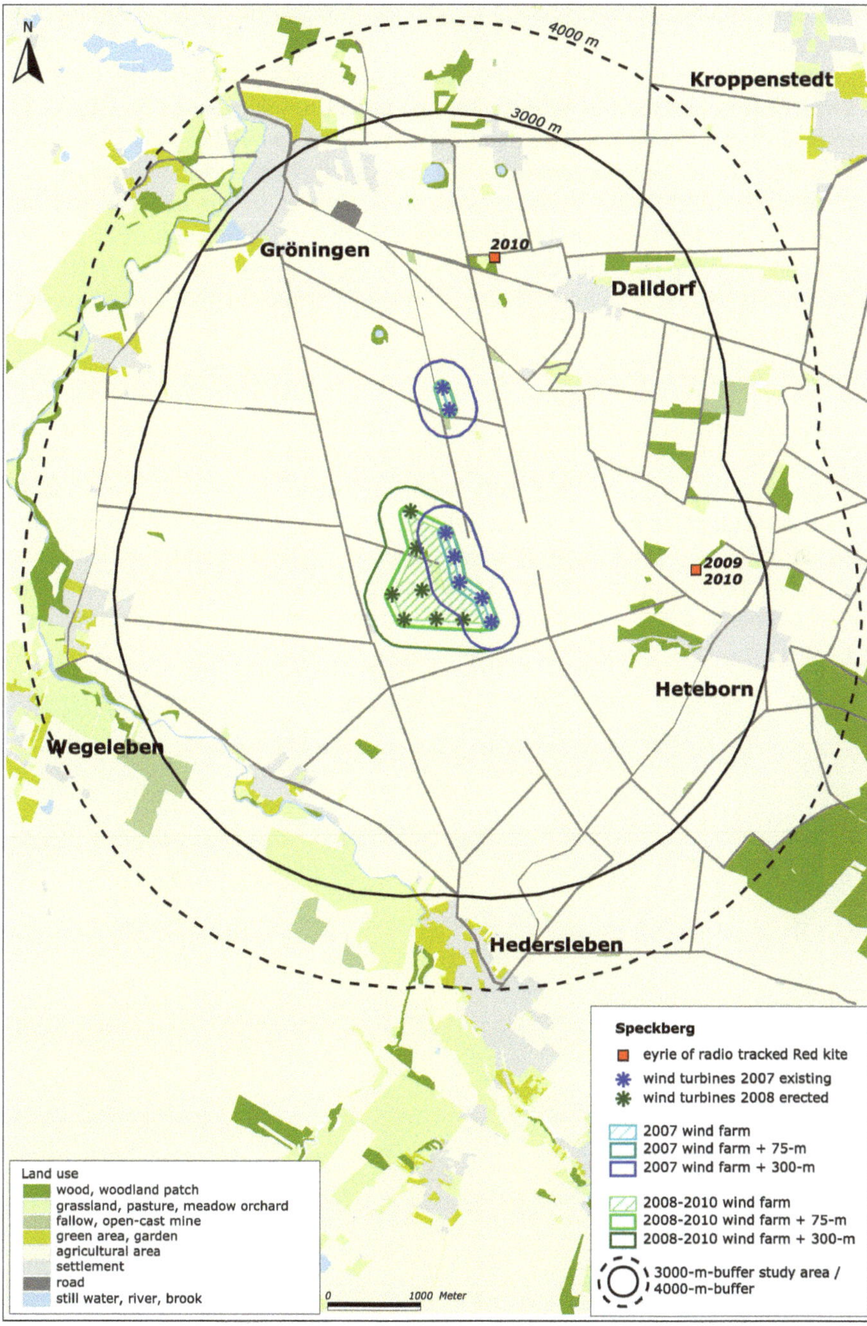

Fig. 3.5 Location and boundaries of the Speckberg study area including wind turbines with the surrounding area of 3 km (study area) as well as the wind farm polygon +75 and +300 m buffer (wind farm area and wind farm impact area)

is rich in woodland. Several smaller woods and meadow orchards can be found in this area.

14 wind turbines are located in the centre of the study area. Five Repower MM 82 turbines were installed in 2006 and in operation from the start of this study. The three-bladed turbines have a rotor diameter of 82 m and a hub height of 100 m. The conical steel towers are pure white. Until mid of 2007, two further Enercon E-70 wind turbines (2.3 MW) with the typical tower base in shades of green followed spatially detached in the north. These turbines have a rotor diameter of 71 m and a hub height of 98 m. In 2009, the wind farm in the southern part of the study area was extended by seven Enercon E 82 turbines with a rotor diameter of 82 m and a hub height of 98 m.

Druiberg

Loess over clay rendzinas or loess over loamy black soils or loam rendzinas prevail in the area (Regionale Planungsgesellschaft Harz 2007), which is characterized by intensive farming. Typically, winter cereals (winter wheat, winter barley, triticale) are cultivated but also maize, oilseed rape and beet. Within the wind farms, crops are relatively small-scale in comparison to regional conditions. Roads and agricultural roads are partly lined by rows of fruit trees and hedgerows (Figs. 3.6 and 3.7). Smaller patches of woodland can be found on the site of the former radar

Fig. 3.6 Druiberg study area. View on the wind farm form the north-west. The double row of turbines is clearly visible

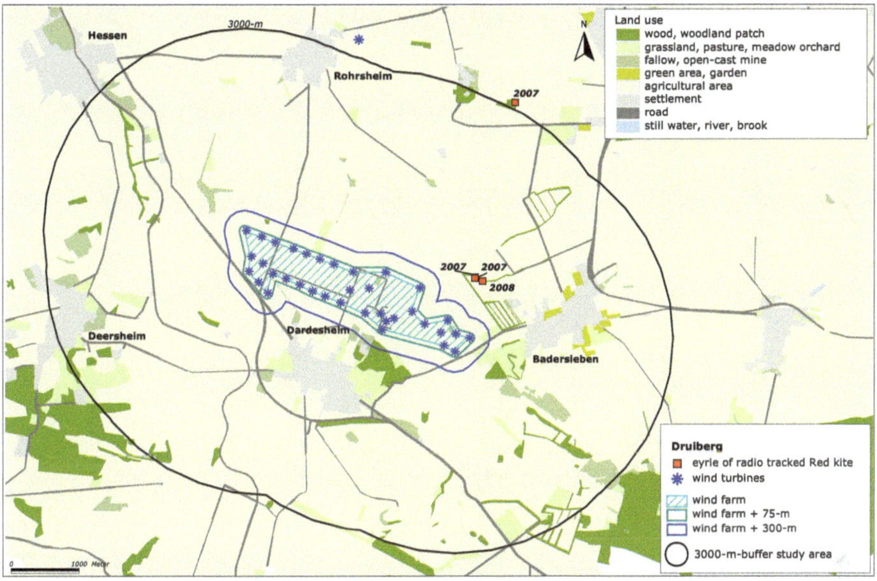

Fig. 3.7 Location and boundaries of the Druiberg study area including wind turbines with the surrounding area of 3 km (study area) as well as the wind farm polygon +75 and +300 m buffer (wind farm area and wind farm impact area)

station and in the north-western and eastern part of the wind farm. Larger wooded areas are a mixed forest area south-east of the wind farm and coherent poplar rows east and north-east of the wind farm.

The wind farm comprises a total of 36 wind turbines. The turbines were erected between 1995 and 2006. In the eastern part of the current wind farm, initially three Lagerwey LW 27/250 turbines with a nominal capacity of 250 kW each, a hub height of 40 m and a rotor diameter of 27 m were erected. In the same year, four Micon M 1500-600 turbines with a capacity of 600 kW each, a hub height of 40 m and a rotor diameter of 43 m were added in this area. East of these, 16 Enercon E-66 turbines with a capacity of 2 MW each, a hub height of 114 m and a rotor diameter of 70 m were constructed in 2004. Between 2004 and 2006, 12 Enercon E-70 turbines with mainly the same specifications as the E-66 turbines were added. Rotors were of a different design with a slightly larger rotor diameter. Also in 2006, an Enercon E-112 was erected, the turbine with the highest capacity so far. This turbine has a capacity of 6 MW and a hub height of 125 m with the same rotor diameter (Radach pers. comm. 2007).

Dessau study area

Loess and black soils prevail in the study area. The major part of the area is intensively farmed with smaller groups and rows of trees (Figs. 3.8 and 3.9). North of the B 185, on the edge of the Mosigkau Heath, and in an area between the

Fig. 3.8 View into the Dessau wind farm

villages Libbesdorf and Scheuder (headwater area of the Ziethe), pastures and larger fallows can be found. Parts of the woodland areas of the Mosigkau Heath are included in the eastern part of the study area. Further woodland areas, mainly scattered pine forests, are located in the northern part of the study area. Around former LPG (Agricultural Production Cooperative) sites, contiguous poplar rows are common throughout the study area. The ditch systems of Landgraben, Schindergraben and Fließgraben run through the entire study area.

The Quellendorf/Libbesdorf wind farm comprises 22 Enercon wind turbines, mainly Enercon E-66 with a hub height of 98 m and a rotor diameter of 66 m. Most of the turbines were erected in 2005/2006.

Queis

The study area is part of the Hallesches Ackerland, dominated by loess substrates, directly adjoining the sandy loess areas (LAU 2000). To a large extent, the area is intensively farmed (Figs. 3.10 and 3.11). Fallows and meadows or pastures make up only a small proportion of the area. Structuring elements as rows of trees, hedgerows or copses are rare. The Strengbach, lined with poplars, runs through the eastern part of the study area. Further rows of poplars suitable for nesting are mainly available in the southern part of the study area. Larger, contiguous woods or forest lack completely. With a general elevation of between 106 and 115 m above sea level, the terrain is rather flat throughout.

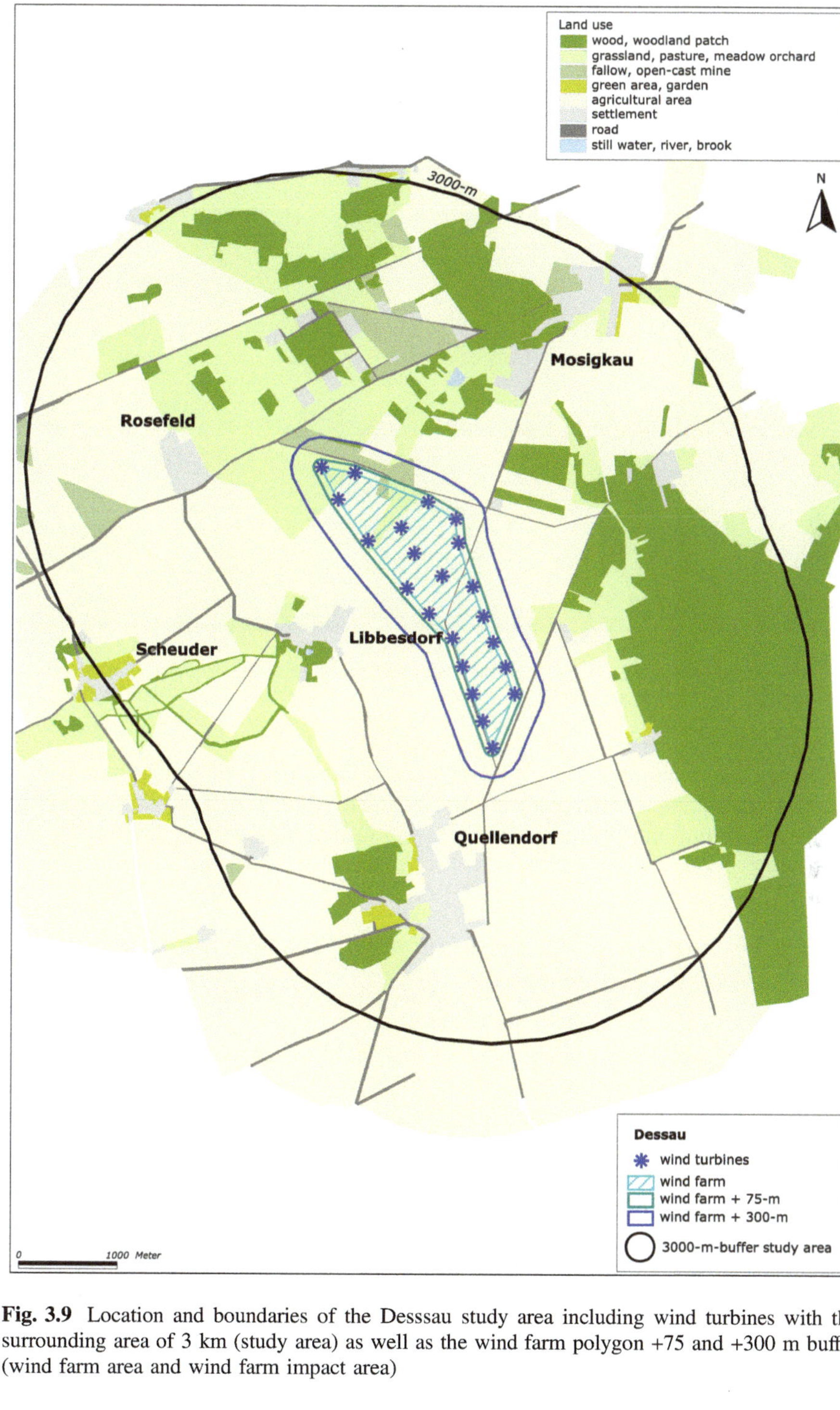

Fig. 3.9 Location and boundaries of the Desssau study area including wind turbines with the surrounding area of 3 km (study area) as well as the wind farm polygon +75 and +300 m buffer (wind farm area and wind farm impact area)

Fig. 3.10 Wind farm Queis. View from the south into the wind farm. The three Vestas turbines are located on the right-hand side

The wind farm comprises 17 wind turbines, 14 Enercon E-66 turbines with a hub height of 65 m and a rotor diameter of 70 m and three Vestas turbines. 16 of the turbines were put into operation in 2003, one turbine followed in 2006.

3.2.2 Monitoring of the Population and the Reproduction Rate of Birds of Prey

Monitoring of the bird of prey population was based on mapping of nest sites. Prior to foliage growth (March to early April), the woody structures within the study areas were systematically searched for nests suitable for medium-sized birds of prey every year. The nest sites were entered in topographical maps or recorded using GPS. End of April to early May the nests were checked for occupation while especially watching out for newly built nests.

Nests which were apparently unoccupied but showed evidence of potential use by birds of prey—fresh foliage on the nest, adult birds circling above the nest, fresh moulted feathers or fresh faeces—were revisited and checked several times. Only those nests with observations of an adult or, at a later time, of at least one young bird were rated as occupied.

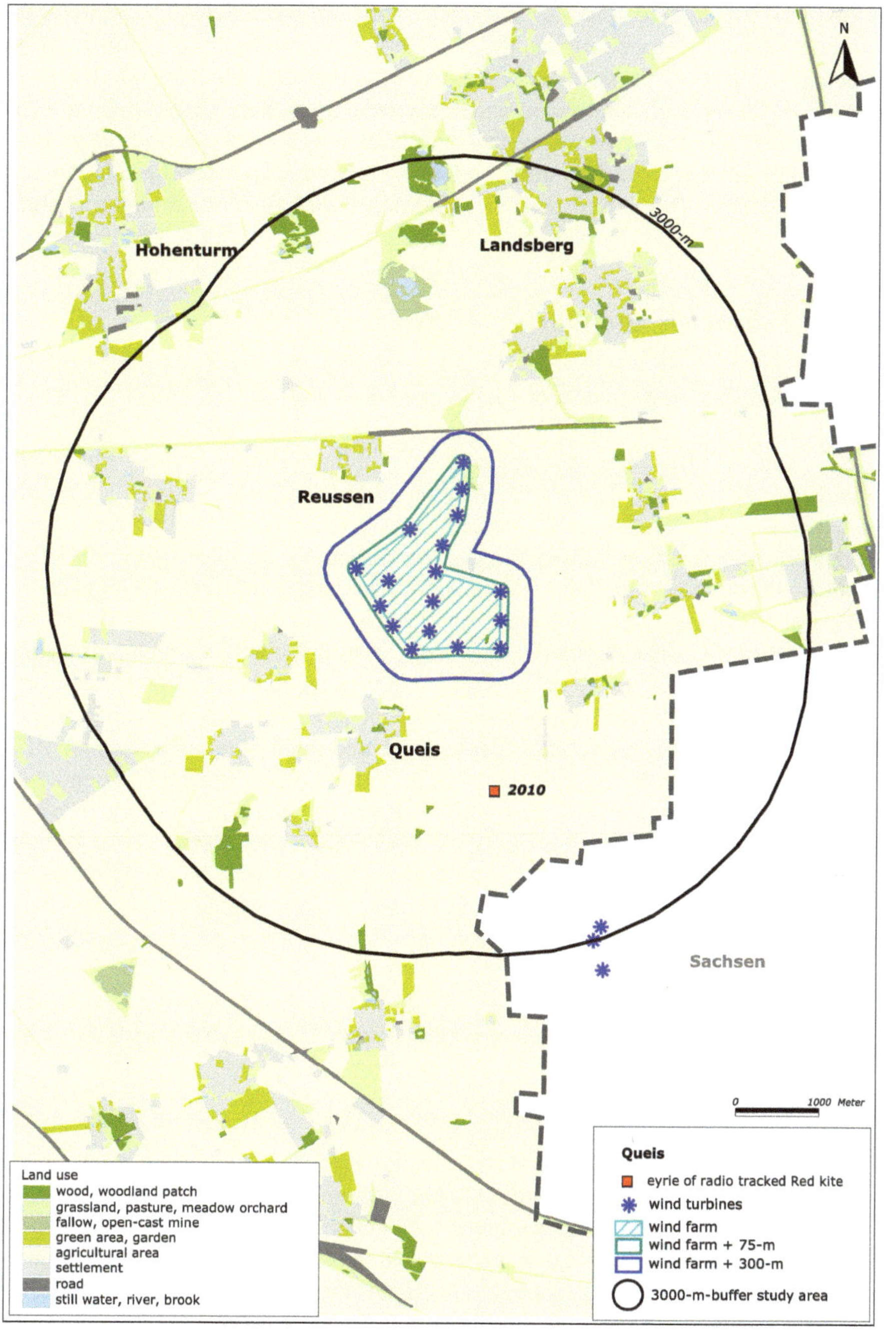

Fig. 3.11 Location and boundaries of the Queis study area including wind turbines with the surrounding area of 3 km (study area) as well as the wind farm polygon +75 and +300 m buffer (wind farm area and wind farm impact area)

End of May to mid-June, all occupied nests were revisited to determine the breeding success. Depending on the availability of experienced climbers (from other bird of prey surveys), nests with successful broods were climbed. The exact number of young was determined and the young birds were ringed. Where nests could not be climbed or provided that no climber was available, the branchlings were counted between end of June and early July to document the reproduction success. As single young birds may be overlooked, these data are minimum values.

3.2.3 Telemetric Monitoring of Red Kite

3.2.3.1 Number and Origin of the Tagged Red Kites

In 2007 and 2008, telemetric monitoring was limited to the two main study areas "Querfurter Platte" and "Druiberg" and in 2009 to the study areas "Querfurter Platte" and "Speckberg". In 2010, the less personnel intensive satellite-tags became available for the first time, which allowed to additionally tag birds outside the so far existing study areas. This was all the more necessary due to the extremely low success to catch red kites to be tagged in 2010. Therefore, an adult female was caught and tagged near the Queis wind farm east of Halle. This area had already been voluntarily surveyed on a large scale with regard to birds of prey for several years. A total of seven adult red kites was VHF tagged between 2007 and 2010. In 2010 a further three adult red kites were fitted with Argos/GPS Solar-PTTs (Microwave Telemetry Inc., weight 22 g) (Table 3.1). All transmitters were attached as backpack harness using Teflon ribbons which crossed through a loop over the sternum. Additionally, the birds were ringed with a ring of the ornithological station Hiddensee and fitted with white wing tags with black digits (Stubbe et al. 2006). All of the tagged birds bred in the vicinity of the investigated wind farms.

3.2.3.2 Long-Term Monitoring via VHF and ARGOS GPS Tags

Data Acquisition

VHF transmitters sent beacon intervals in the 150 MHz band and were tracked using a hand-held receiver and a directional antenna (YAESU VR 500 and H type hand-hold antenna or Sika-receiver and 3-element YAGI antenna). The VHF tagged birds were usually radio-located until visual contact. On days with increased mobility of the birds, cross bearing without visual contact was used. Telemetric monitoring started at the time of tagging and was carried out until the birds migrated towards wintering grounds. Location recording followed no fixed time schedule but represented the whole daily activity period (Table 3.2). A larger number of recorded locations per day as well as breaks lasting for several days

Table 3.1 Overview of red kites tagged for the present study between 2007 and 2010

Red Kite individual	Study area	Start of telemetric monitoring	Tag type and frequency	Ring number, wing tag number	Telemetric monitoring seasons
Barbarossa	Druiberg	01.06.2007	VHF 150.592 Hz	EA 136086 Wing tag: 274	2007 2008
Arthur	Querfurt	02.06.2007	VHF 150.573 Hz	EA 136088 Wing tag: 276	2007
Ramona	Querfurt	03.06.2007	VHF 150.553 Hz	EA 136090 Wing tag: 278	2007 2008
Karl	Druiberg	11.06.2007	VHF 150.634 Hz	EA 136094 Wing tag: 282	2007
Barbara	Druiberg	15.06.2007	VHF 150.614 Hz	EA 136098 Wing tag: 286	2007
Gishild	Querfurt	05.06.2008	VHF 150.695 Hz	EA 143891 Wing tag: 288	2008 2009
Alte Dame	Speckberg	10.06.2009	VHF 150.063 Hz	EA 012136 Wing tag: 437	2009 2010
Lui	Speckberg	03.06.2010	GPS-satellite-tag	EA 176801 Wing tag: 501	2010 ff.
Erik	Querfurt	06.06.2010	GPS-satellite-tag	EA 176802 Wing tag: 502	2010 ff.
Svenja	Queis	17.06.2010	GPS-satellite-tag	EA 176859 Wing tag: 503	2010 ff.

occurred. Excess accumulation of recorded locations for single days was avoided. On days with continuous tracking of the tagged bird (continuous monitoring, see Sect. 3.3.2.2), five locations were randomly selected and included in the long-term tracking data pool. In case of visual contact, behaviour of the bird, habitat quality, events (mowing, ploughing) and the presence of other red kites were noted. In case of no visual contact it was at least determined whether the bird moved or not based on the signal. A minimum interval of 15 or most of the time 30 min was observed between recorded locations. As successively recorded bird locations are not independent of each other, the choice of a sufficiently large interval enables the bird to theoretically traverse its entire home range and thus freely choose a location between records (Kenward 2001). Therefore, underestimation of home ranges and kernel home ranges due to autocorrelated recorded locations is to a large extent avoided. After the red kites left the breeding areas for winter, home ranges used in summer and autumn as well as known roosting sites were extensively searched for tagged birds. From February the following year, weekly telemetric monitoring was carried out until the red kites returned to the breeding area (or definitely did not return) before subsequently resuming regular monitoring. The transmitters had a battery life span of 12–24 months. All adult tagged birds returned from wintering grounds (a total of four) could be monitored in the year of tagging and in the subsequent season until autumn. Three of the tagged birds (all of year 2007) did not return to the breeding area after winter. The reason could only be determined in one case (death in the wintering area). For the remaining two birds a similar reason, e.g. death on migration or in the wintering area, is assumed.

The ARGOS GPS transmitters (ARGOS PTT) were equipped with solar cells and therefore independent of the life span of a battery. Satellite signals with standard location classes of between LC1 (100 m) and LCB (unknown accuracy) were transmitted even under poor lighting conditions. In good weather conditions, additional GPS signals with ±15 m accuracy were transmitted four times a day. During the summer season, locations were recorded at 09:00, 11:00, 14:00 and 17:00 o'clock and during the winter season at 10:00, 13:00, 15:00 and 18:00 o'clock. The locations recorded at 17:00 and 18:00 o'clock were roosting sites of the red kite in most cases. Depending on lighting conditions, i.e. low charge of the battery it was not always possible to record locations for all times (Table 3.2).

Evaluation of Telemetric Data

Calculation of the home ranges

Calculations of the habitat use were carried out using the ArcView 3.2 extension AnimalMovement 2.04beta (Hooge et al. 2000). Home ranges were calculated separately for breeding (locations recorded between tagging or return to breeding area up to 31 July) and post-breeding season (locations recorded between 1 August and return to wintering areas). The breeding season comprised mating, courtship, nest occupation, brood and fledging young including the branchling stage for birds present at the respective time. Further breaking down the stages of the breeding season (e. g, see Nachtigall 1999) was not possible due to small sample sizes resulting from the varying starts of monitoring. Birds were captured at different times (birds were tagged between 1 and 17 June) and in the subsequent year, monitoring of four of the birds started between March and May.

Calculation of the home ranges was based on a 95% minimum convex polygon (95% MCP, Harris et al. 1990; Kenward 2001). In a first step, 5% of the recorded locations were excluded from the data set. The outermost locations, which can be regarded as one-off and short-time excursions from a more intensively used home range, were not included in the calculation of the home range (Kenward 1992). The outermost of the remaining recorded locations are connected to form a convex polygon outlining the minimum area. The resulting area is the home range of the tagged bird. Large areas, not used by the tagged animal, are often included in the MCP, which still quite well reflect the total home range or habitat use.

Additionally, the home ranges with a 95% probability of use were calculated using the fixed kernel method (see Kenward 2001). A probability distribution is calculated, filtered and weighted by number and distance between recorded locations. Areas, which are used less frequently or return only few locations are less probable. As a rule, kernel home ranges are much smaller than MCP home ranges and identify the centre of activity. To determine the core areas (preferred areas)

Table 3.2 Amount of data collected for long-term monitoring via VHF and ARGOS GPS tags in 2007–2010

							Number of locations									
Individual	Gender	Year	Method	Season	Days with telemetry data	Total	Minimum per day	Maximum per day	<8:00 o'clock	8:00–10:00 o'clock	10:00–12:00 o'clock	12:00–14:00 o'clock	14:00–16:00 o'clock	16:00–18:00 o'clock	18:00–20:00 o'clock	>20:00 o'clock
Barbarossa	m	2007	VHF	bs	27	199	2	15	8	12	29	27	35	23	35	30
Barbarossa	m	2007	VHF	post	38	186	1	14	5	17	35	22	25	31	31	19
Barbarossa	m	2008	VHF	bs	45	197	1	11	8	34	24	32	26	34	27	12
Barbarossa	m	2008	VHF	post	22	89	1	11	4	11	11	18	10	20	11	3
Arthur	m	2007	VHF	bs	14	56	1	8	0	5	11	5	11	12	7	5
Arthur	m	2007	VHF	post	31	138	1	28	0	9	13	25	35	44	10	0
Ramona	f	2007	VHF	bs	12	64	1	26	0	0	11	12	14	16	9	2
Ramona	f	2007	VHF	post	21	110	1	23	1	15	10	5	21	37	20	1
Ramona	f	2008	VHF	bs	9	42	1	10	0	3	3	2	13	19	2	0
Ramona	f	2008	VHF	post	3	5	1	2	0	0	0	2	0	0	3	0
Karl	m	2007	VHF	bs	23	153	1	23	1	13	24	31	35	21	14	14
Karl	m	2007	VHF	post	31	144	1	13	2	11	22	29	15	24	23	18
Barbara	f	2007	VHF	bs	19	169	2	19	0	9	28	23	38	35	21	15
Barbara	f	2007	VHF	post	45	229	1	14	5	24	44	49	36	21	33	17
Gishild	f	2008	VHF	bs	18	104	1	14	2	10	14	21	17	17	18	5
Gishild	f	2008	VHF	post	15	73	1	9	1	1	11	11	20	16	10	3
Gishild	f	2009	VHF	bs	14	114	1	18	8	13	8	20	27	33	5	0
Gishild	f	2009	VHF	post	1	1	1	1	0	0	0	0	0	0	1	0
Alte Dame	f	2009	VHF	bs	4	25	5	8	1	1	2	1	7	11	2	0
Alte Dame	f	2009	VHF	post	5	28	2	10	0	0	0	3	23	2	0	0
Alte Dame	f	2009	VHF	wi	8	28	1	10	0	0	0	2	11	14	1	0
Alte Dame	f	2010	VHF	bs	32	151	1	14	11	19	18	35	31	22	12	3
Alte Dame	f	2010	VHF	post	12	42	2	8	0	0	8	14	3	11	6	0

(continued)

Table 3.2 (continued)

							Number of locations									
Lui	m	2010	GPS	bs	53	292	2	4	0	0	8	14	3	11	6	0
Lui	m	2010	GPS	post	75	120	1	4								
Erik	m	2010	GPS	bs	46	177	3	4								
Erik	m	2010	GPS	post	65	211	2	4								
Svenja	f	2010	GPS	bs	34	117	2	4								
Svenja	f	2010	GPS	post	56	192	2	4								

Breeding season: bs, post-breeding season: post, timing of GPS locations in summer: one each at 8:00–10:00, 10:00–12:00, 12:00–14:00, 16:00–18:00, in winter: one each at 8:00–10:00, 12:00–14:00, 14:00–16:00, 16:00–18:00

used by red kites, kernels with 50, 60 and 95% probability of use were calculated. As the MCP and the different kernel home ranges are intercorrelated, mainly the 95% kernel home ranges were used in statistical analysis.

Distance of recorded locations from the nest

The distance from the nest was calculated for each recorded location. Distances <50 m from the nest were defined as close-up area around the nest and were disregarded. For each of the tagged birds, the distances between recorded locations and nest were added up proportionally in 100 m distance classes.

Approach of wind turbines and stay in the wind farm

The area of the wind farm was defined by a polygon connecting the centres of the outermost wind turbines. In a first step, a minimum convex polygon was calculated. The partially very large boundary areas without wind turbines (between unaligned outermost wind turbines) were then disregarded. Two buffer zones of the size of a single rotor diameter (75 m) and four times the rotor diameter (300 m), respectively, were added around the wind farm polygon. The rotor diameter used corresponds to the most common wind turbine models in the study areas (Enercon E-66/E-70). The rotor diameter of the additionally present wind turbines of other manufacturers (Vestas, Fuhrländer, Repower) usually is by approximately 10 m larger. Distances were defined following the NRW Wind Energy Decree (2000, 2005, 2011, draft). Up to a distance four times the rotor diameter, air turbulence caused by the moving rotor will affect neighbouring wind turbines (±30° around the axis up to a distance of eight times the rotor diameter). Effects on comparably tiny objects like birds of prey gliding on air currents will be likely in this area. Experts assume that approaches to wind turbines (distance < single rotor diameter) or stays inside wind farms cause potential risks. In the following, the wind farm polygon plus an additional 75 m buffer zone will be referred to as wind farm area, and the wind farm polygon plus an additional 300 m buffer zone as wind farm impact area.

For each location it was determined, whether it was recorded within the wind farm area or wind farm impact area. Additionally, the distance from the next wind turbine was calculated for all recorded locations. Furthermore, each of the tagged birds was checked for spatial overlap of the kernel home range with wind farm areas or wind farm impact areas, determining the size of the spatial overlap. Subsequently, all wind turbines within a radius of 1,000 and 1,250 m around the respective nest were masked out, to test for the resulting reduction in size of the spatial overlap with the kernel home range.

3.2.3.3 Continuous Monitoring

Data Acquisition

Continuous monitoring aimed to gain specific and as detailed as possible information on the daily habitat use, complementing the telemetric data on seasonal home ranges. These data are important for further evaluation (see Chap. 14). Focus was on the following questions: What is the total distance red kites travel per day? How much time do they spend in flight or sitting? How large are individual differences?

In continuous monitoring, the tagged bird is watched as uninterruptedly as possible by two to four observers between sunrise and sunset. Continuous monitoring was carried out on days with no rainfall to be expected. The previous evening, the roosting site of the respective red kites was determined as a starting point for monitoring the following morning. By permanent contact between observers (via mobile phone or handheld transceivers), the birds could be continuously located for most of the day. Care was taken not to flush birds by observers getting too close. Locations, behaviour as well as activity and resting phases of the red kite were noted.

On total of 20 days, continuous monitoring was carried out for six red kites, three male (Karl and Barbarossa in the Druiberg study area and Arthur in the Querfurt study area) and three female birds (Ramona and Gishild in the Querfurt study area and Alte Dame in the Speckberg study area).

Evaluation

The data were entered in a table and all directly observed time units were processed. Not every flight circle of a red kite can exactly be recorded in the field and times with no visual contact to the bird were unavoidable in case of quick changes in location. Therefore, an average speed of 25 km/h was assumed for "circling" and "flight without visual contact" in the calculation of the actual flight distance. This is an expert estimate, as no literature is available on the subject. The flight time for these time units was then added to the total flight distance per day. Providing the visual contact or contact to the transmitter was lost, but the bird retrieved later in a different location, the direct distance (straight line) between the last contact and the new location was used. This calculated distance will be a minimum distance. From the real distances found in continuous monitoring, daily home ranges were calculated for each of the monitoring days. As the focus in continuous monitoring is on distances rather than points, a distance-based calculation was used, based on the area surrounded by the line of flight determined for the particular day. For areas traversed several times, the outermost line was used as a boundary so that areas used several times were only included once. Flight circles outside the boundaries were included in the daily home range.

3.2.4 *Individual-Related Behavioural Observations*

For individual-related behavioural observations of red kites within wind farms, the Noldus Observer software was used, which is especially designed to collect and analyse behavioural data. Observer XT and Observer 5 (configuration of data acquisition, transmission to Psion handheld computer, data downloads, data analysis) with Observer Mobile (mobile data acquisition with Psion handheld computer) were used. The coding scheme was set up in Observer XT and loaded onto the handheld for data acquisition in the field. The Observer system allows for parallel records of several different parameters accurate to the second. Variables can be set up in codes (e.g. sit, fly, plume) or classes (x1–x2, x2–x3) and recorded as either event (without duration) or duration (including start and end time). Within this partial study on red kite, the following behaviours were recorded:

- Flight height, categories: ground, 1–10, 11–25, 26–50, 51–100, 101–150, >151 m
- Behaviour, categories: traversing, foraging, thermal soaring, hunting stoop, landing, sitting, miscellaneous
- Interaction/aggression, categories: bird of prey/same species, bird of prey/other species, crow, other species
- Approach of wind turbine, categories: rotor sphere, rotor disc*, crossing rotor disc, turbulence (*Rotor disc: rotor plane ±12.5 m to all sides, total disk thickness: 30 m)

Number and duration of parameters was calculated using Observer XT. In addition, general information was noted for each observation unit (potentially covering several flight observations each).

Behavioural observations of individual birds were carried out in 2007 and 2008 only and primarily within the main study areas of that time, Druiberg and Querfurt. In 2007, flight activity of red kite within the wind farm was considerably lower in the Druiberg study area as compared to Querfurt study area. In 2008 flight activity decreased in both areas, and additional observations could be carried out in Speckberg study area to increase sample size (Table 3.3). The wind farms were subdivided into observation areas to comprehensively record the behaviour of each red kite entering the wind farm in observation units of 45 min. As estimating the distances was of vital importance (flight height, distance from rotor), the observation areas were smaller than in area-related behavioural observations (see Sect. 3.2.5). At greater distances between bird and observer (>800 m) distances cannot be estimated reliably and it will sometimes even be difficult to determine, if a bird is located just in front or behind a wind turbine or crosses the rotor swept area.

Table 3.3 Sample size of individual-based behavioural observations of red kite in 2007/2008

	Druiberg	Querfurt	Speckberg	Σ
Recorded individuals (= flight sequences)	161	29	14	204
Behavioural record				
– in seconds – in hours	24,710 6 h 52 min	4,700 1 h 18 min	1,031 17 min	30,442 8 h 27 min

3.2.5 Area-Related Behavioural Observations

3.2.5.1 Data Acquisition

The area-related behavioural observations were carried out in observation areas of between 34 and 190 ha (Figs. 3.12, 3.13, 3.14 and 3.15). In the study areas Druiberg, Querfurt, Dessau and Speckberg, areas within the wind farms and in the study areas Druiberg and Querfurt additional reference areas in the surrounding area outside the wind farms were observed. Each of the separate areas included different crops and, where appropriate, edge structures as agricultural roads, hedgerows, etc., clearly visible from a vantage point.

In 2007 and 2008 all areas were watched several times per month for 45 min of an hour (overview in Table 3.4). In 2007, watches were carried out between May and November and in 2008 between March and September. A total of 966 area-related behavioural observations were performed in 2007 and 534 in 2008. Most of the observations took place in Druiberg.

During 45 min watches, all birds of prey observed in the respective area were recorded in maps, and selected behaviours including duration in seconds and exact location were noted. Distinction was made between the following behaviours: Traversing flight (the bird crossed the area), foraging flight, thermal soaring and sitting. In some cases, e.g. several foraging birds of prey were present on a freshly mown meadow, the exact behaviour could not be determined for each of the birds and the duration of the presence was then noted as "undefined presence". Apart from behaviours, the current state of the area used by red kites (crop, stubbles, ploughed, freshly sown, etc.) and current farming activities were noted. For areas within the wind farm, dangerous situations, as flights near rotor blades or turbulences caused by the moving rotor affecting the bird, were documented exact to the second.

Fig. 3.12 Overview of the observation areas (*purple lines* with labelling) of the area-related behavioural observations in the Querfurt study area

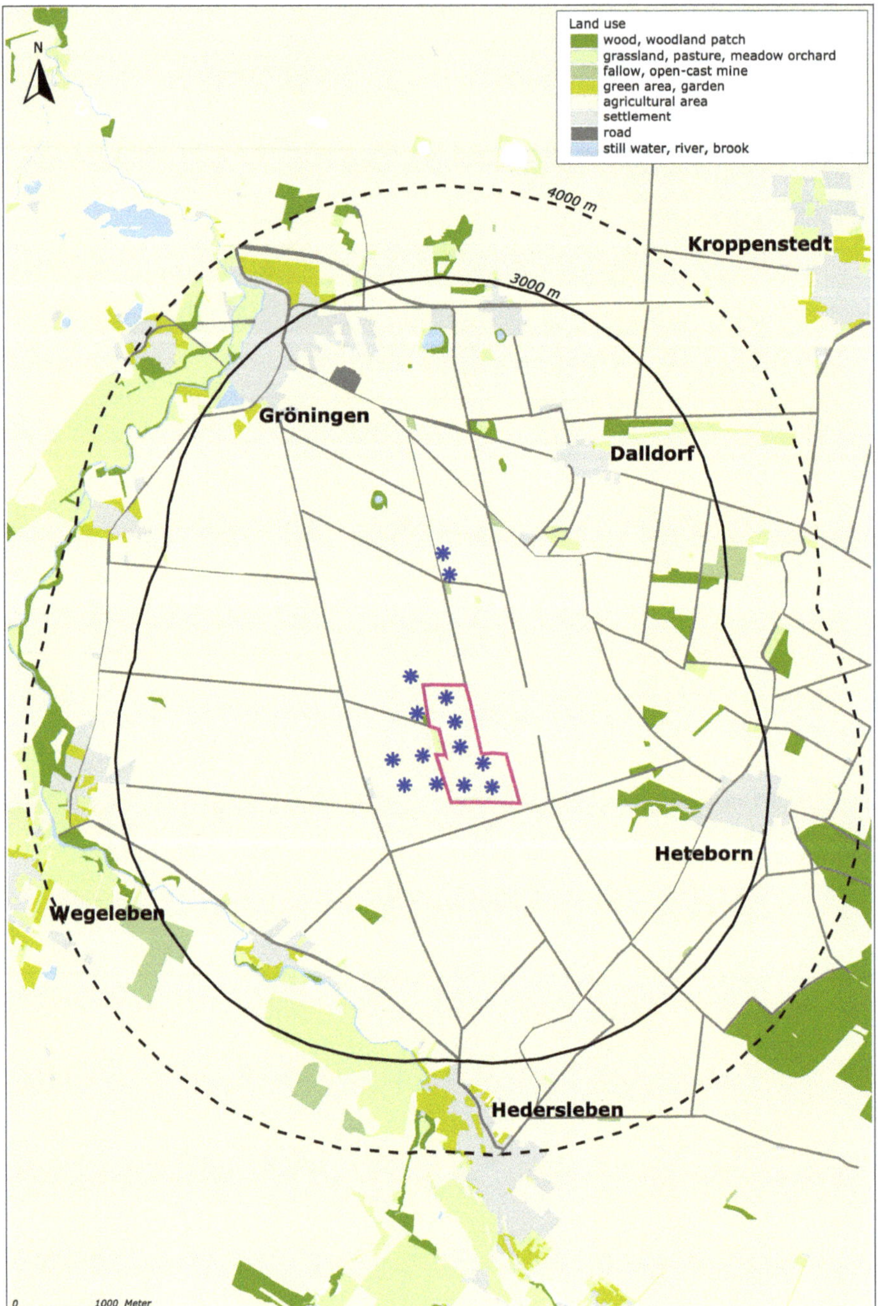

Fig. 3.13 Overview of the observation areas (*purple lines* with labelling) of the area-related behavioural observations in the Speckberg study area

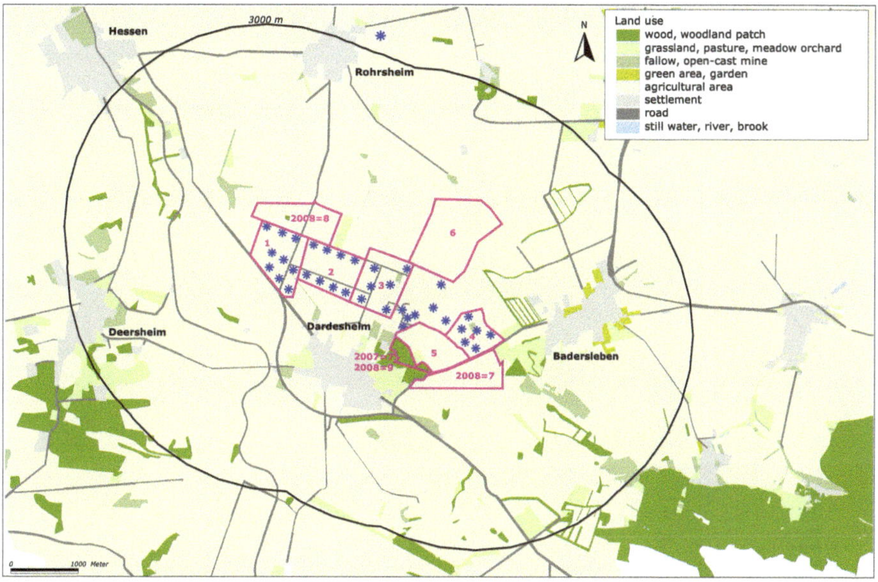

Fig. 3.14 Overview of the observation areas (*purple lines* with labelling) of the area-related behavioural observations in the Druiberg study area

Evaluation

The duration in seconds of behaviours of red kites recorded in the field was evaluated for areas within and outside the wind farm on a monthly basis. To ensure comparability of the results, the behaviours summed up per month were set in relation to the investigated area in s/km^2 and 45 min watch.

To determine preference or avoidance of certain crops for red kite, availability of the particular crop (area) and use of these areas (duration of use in seconds) were set in proportion using the Jacobs index (Jacobs 1974). Positive values of the Jacobs index indicated preference for, negative values avoidance of the respective crop. The Jacobs index was calculated separately per crop and survey month. Apart from crops, edge structures as agricultural roads and hedgerows as well as fallows around the turbine base and access routes to wind turbines were included.

Furthermore, the number of red kite sightings (number of approaches without duration) for areas within and outside the wind farms was calculated per km^2 and 45 min watch and set in relation to the breeding pair density.

Within the wind farm, the proportion of approaches to the 50 m area around wind turbines was calculated.

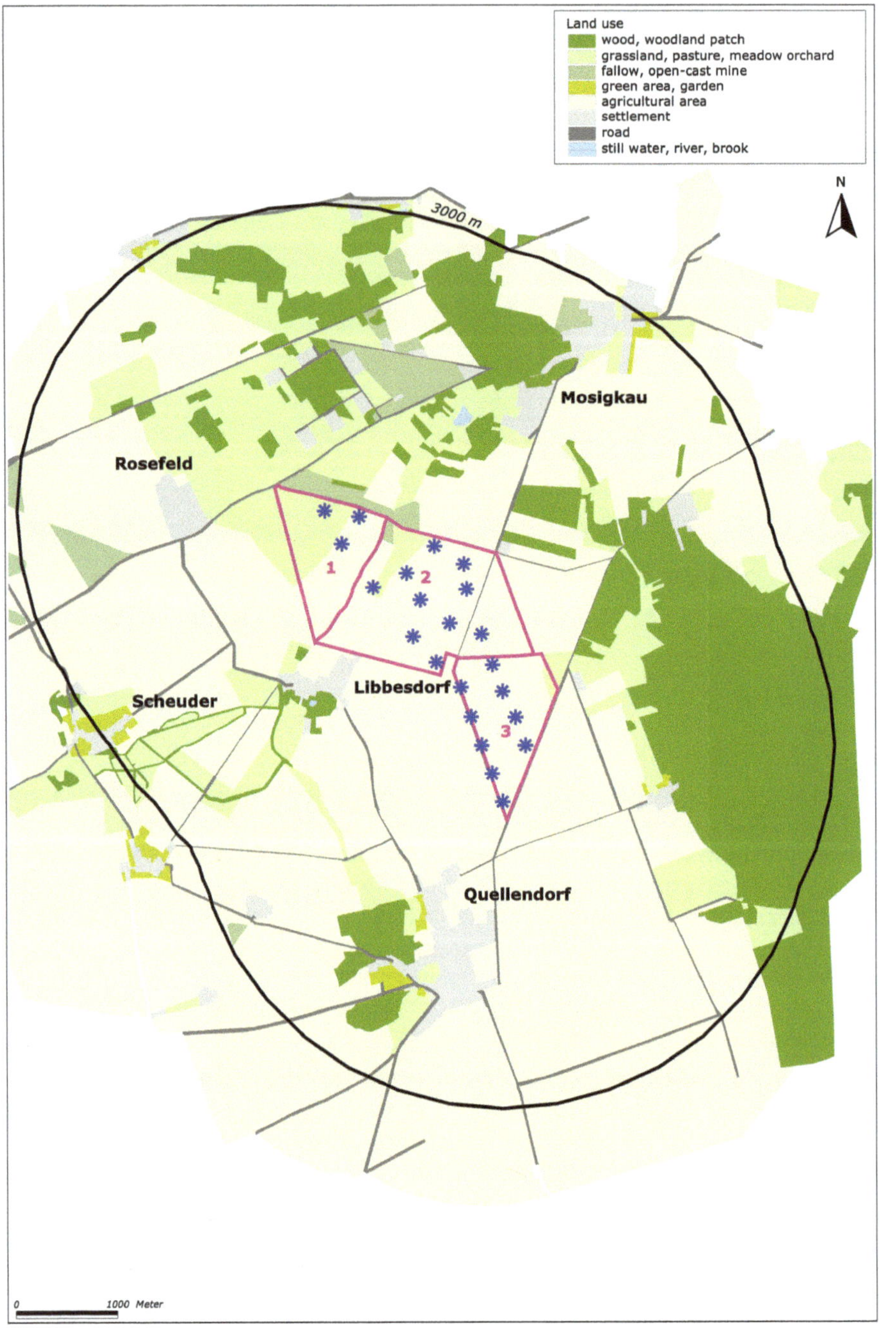

Fig. 3.15 Overview of the observation areas (*purple lines* with labelling) of the area-related behavioural observations in the Dessau study area

Table 3.4 Sample size of area-related behavioural observations of red kites in 2007/2008

	Year	Total observation area (km²)	Start	End	Number of 45 min watches
Querfurt	2007	8.2	10.06.	30.11.	185
	2008	8.2	20.03.	26.09.	180
Speckberg	2007	0.7	28.06.	28.11.	86
	2008	0.7	22.03.	29.09.	112
Druiberg	2007	3.5	31.05.	20.11.	623
	2008	4.4	21.03.	26.09.	154
Dessau	2007	3.7	20.06.	22.11.	72
	2008	3.7	20.04.	28.09.	88

3.2.6 Covering the Fallow Around Wind Turbine Towers with Agricultural Film

In 2008, a field study was carried out in the Druiberg study area to identify possible influences on the flight and hunting behaviour of red kites in the close-up area of wind turbines.

The test was based on the assumption that red kites will be attracted to the turbine base area, if they expect to find food. If the turbine base fallow is as unattractive as possible, red kites will approach the close-up area less often. Gravelled turbine base fallows would have been ideal, as they will not be a suitable habitat for small mammals (prey animals for birds of prey), making the areas unattractive to red kites.

Gravelling the fallows or planting shrubs, which will have the same effect, was not possible in this study, but the owner of several wind turbines within the wind farm Druiberg agreed to cover the turbine base area of three wind turbines with agricultural film for one season. Wind turbines within the observation areas of the 45 min watches were chosen for this purpose (see Fig. 3.16). The turbine areas were covered with agricultural film (manufacturer: Klojer, Weisenheim am Berg) on 07. and 08.05.2008.

Three further wind turbines in the close vicinity of the covered wind turbines were chosen as a reference. Additionally, wind turbine 28, an Enercon E112 turbine with a gravelled forecourt of approximately 0.4 ha (Figs. 3.17, 3.18 and 3.19), was observed separately.

Data was collected following the method described for area-related behavioural observations (see Sect. 3.2.5), but flight movements of birds of prey were recorded for radii of 150, 100 and 50 m around the wind turbines. 12 of the observation units were carried out prior to covering the fallows (mid-March to early May), 29 thereafter (mid-May to end of September) (Tables 3.5 and 3.6).

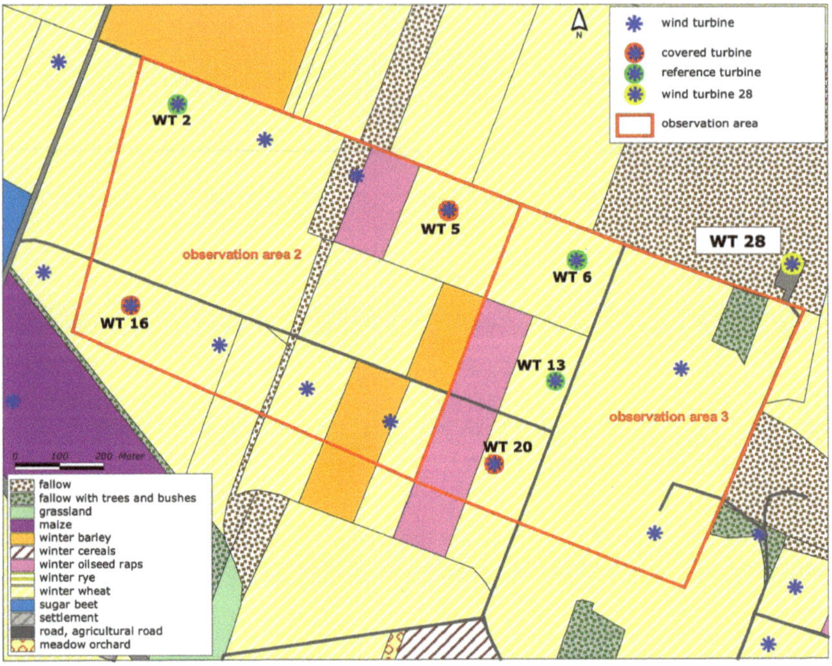

Fig. 3.16 Position of the covered wind turbines, the reference turbines and wind turbine 28 as well as crops in 2008

Fig. 3.17 Wind turbine fallow covered with agricultural film

Fig. 3.18 Uncovered wind turbine fallow with ruderal vegetation

Fig. 3.19 Tower of wind turbine 28 (Enercon E112) in the Druiberg study area with large-scale gravelled forecourt

Table 3.5 Recorded locations for covered and uncovered wind turbine fallows

Observation area	Wind turbine number (covered area)	Wind turbine number (uncovered)
2/3	Wind turbine 5 (381.5 m^2)	Wind turbine 6
2	Wind turbine 16 (521.5 m^2)	Wind turbine 2
3	Wind turbine 20 (721.5 m^2)	Wind turbine 13

Table 3.6 Sample size of area-related behavioural observations for the field test with covered wind turbine fallows 2008

Number of 45 min watches	Area 2	Area 3	Wind turbine 28	Σ
Prior to coverage	6	6	5	17
After coverage	15	14	14	43
Total	21	20	19	60

3.2.7 *Use of Specific Food Resources in the Speckberg Study Area*

In foraging, red kites particularly often visit alfalfa crops. Systematic creation of such areas may be used to dislocate red kites from wind farm sites to minimise the collision risk. Therefore, the use of alfalfa crop areas was thoroughly investigated in the present study.

3.2.7.1 Vantage Point Counts

In vantage point counts the use of alfalfa crops was recorded in relation to the use of other crops. In 2010, clearly visible, distinct landscape sections were watched from 29 vantage points for five minutes each along a defined route within the Speckberg study area. During these five minutes, all birds of prey present in the respective area were recorded. Species, behaviour, flight height as well as information about crop, coverage and farming state of the areas were noted. Vantage points were chosen to include characteristic landscape features of the study area.

The first visual contact for each red kite was included in the evaluation, avoiding double counts for the single points but not for the entire study area. The first visual contact for each red kite was included in the evaluation. Vantage point counts were carried out between 20.05.2010 and 30.09.2010 once per pentade. From nine of the points, alfalfa crops covering a total of 86 ha on nine fields were watched.

To determine a preference for or avoidance of certain areas the number of observed birds was set in relation to the size of the crop area using the index of Jacobs (Jacobs 1974).

3.2.7.2 Area-Related Behavioural Observations

Five alfalfa areas in 2010

In 2010, the use of alfalfa crops by red kites was investigated in the Speckberg study area during and after the breeding season. In this year, alfalfa was cultivated on a total of 175 ha distributed over 18 fields within the study area. 48.8 ha (five fields) thereof were covered in standardised 45 min watches on a regular basis (Table 3.7). Observations were made from a car or a raised hide to avoid additional disturbance. Between 20.05.2010 and 30.09.2010 a total of 157 observation units were carried out on 53 days. Areas 10 and 11 could always be watched simultaneously, areas 3 and 13 on days with no excess flight activity. From mowing, the areas were watched for at least eight days in a row.

Flight activity in the alfalfa areas was entered into day field maps exact to the second. The intensity of use of the alfalfa areas by red kite was evaluated.

Composting plant 2010

The composting plant is located 300 m from the southernmost wind turbine within the Speckberg study area. In 2010, use of this plant by red kites was investigated in standardised 45 min watches. Between 23.05. and 30.09.2010, the composting plant was watched in 48 observation units on 38 days. Behaviour, duration of stay, direction of approach and departure as well as larger anthropogenic activities on the site of the composting plant (e.g. restacking) or neighbouring fields (e.g. mowing) were noted. Red kites clearly leaving the site of the plant were noted as "departing". They were watched as continuously as possible to avoid double counts.

3.2.8 *Statistical Analysis*

Statistical analysis was carried out using software package R (R version 2.11.1 (2010) The R Foundation for Statistical Computing 2010).

Table 3.7 Sample size of area-related behavioural observations for alfalfa crops

Area number	Area size (ha)	Number of 45 min watches
3	7.8	49
8	8.8	44
10	14.5	45
11	9.1	45
13	8.6	47

3.3 Results

3.3.1 Breeding Population and Reproduction

The bird of prey populations developed differently within the respective study areas (Table 3.8). In the Querfurt core area, a considerable decline in the breeding population of all species (red kite, black kite (*Milvus migrans*), common buzzard) could be observed in 2007 in comparison to the following years. In Speckberg, red and black kite occurred in quite different numbers but showed a slight positive trend. In Druiberg, red kite, black kite and common buzzard showed no notable difference in population between the two years of the study. Two (2007, one occupied by a tagged bird) or three (2008) further red kite nests were located just outside the study area. In Dessau, the populations of all three bird of prey species increased from 2007 to 2008.

As the exact number of fledged young could not be determined for all study areas in all years, the proportion of breeding success (proportion of successfully

Table 3.8 Breeding population of red kite (breeding pairs/100 km^2) for all study areas between 2007 and 2010

Species	Study site	Area (km^2)	2007	2008	2009	2010	Change in population size (1st to last study year) (%)
Red kite	Querfurt core area	61.3	13.1	8.2	6.5	6.5	−50
	Querfurt extension	27.3		22	14.7	22	0
	Speckberg	47.8	14.6	14.6	16.7	20.9	43
	Druiberg	54.2	9.2	11.1			21
	Dessau	51.5	5.8	9.7			67
Black kite	Querfurt core area	61.3	24.5	21.2	16.3	14.7	−40
	Querfurt extension	27.3		0.0	0.0	0.0	
	Speckberg	47.8	2.1	6.3	4.2	4.2	100
	Druiberg	54.2	3.7	3.7			0
	Dessau	51.5	1.9	5.8			205
Common buzzard	Querfurt core area	61.3	44.0	6.5	9.8	14.7	−67
	Querfurt extension	27.3		18.3	18.3	18.3	0
	Speckberg	47.8	29.3	41.8	14.6	27.2	−7
	Druiberg	54.2	33.2	27.7			−17
	Dessau	51.5	27.3	58.2			113

Table 3.9 Breeding success of common buzzard, red kite and black kite in all study areas between 2007 and 2010

	2007	2008	2009	2010	
Red kite (proportion of success in %)	75.0	63.3	50.0	30.0	Querfurt, total
	75.0	20.0	25.0	25.0	Querfurt, core area
		100.0	75.0	33.3	Querfurt, extension
	71.4	57.1	37.5	40.0	Speckberg
	40.0	66.6			Druiberg
	100.0	80.0			Dessau
Black kite (proportion of success in %)	73.3	30.8	40.0	44.4	Querfurt
	100.0	33.3	50.0	50.0	Speckberg
	50.0	50.0			Druiberg
	100.0	100.0			Dessau
Common buzzard (proportion of success in %)	81.5	11.1	45.5	50.0	Querfurt
	50.0	50.0	42.9	46.2	Speckberg
	50.0	86.7			Druiberg
	57.1	76.7			Dessau

breeding pairs of the total number of breeding pairs) was used to determine the reproductive success (Table 3.9).

In all but one (Druiberg) study area (Querfurt, Speckberg, Dessau), the reproductive success of red kite was decreasing with no direct influence on density and abundance. Only within the core area of the Querfurt study area (3 km around the area with wind turbines since 2001) the decreasing reproductive success coincided temporally with a decrease in the breeding population. Within the area of the 2007 extension of the wind farm, the reproductive success was likewise decreasing without decrease, but displacement of the breeding population. Groves in the vicinity of the wind farm were barely used for breeding since.

The reproduction rate of black kite varied quite strongly between the study areas, but the sample size was rather small in some of the study areas. The reproductive success of the common buzzard (as well as the density of breeding pairs) varied strongly between years and was found to be different in all study areas. A very low reproduction rate with 89% of the broods not being successful was found for the Querfurt study area in 2008.

3.3.2 *Results—Telemetric Monitoring of Red Kite*

3.3.2.1 Seasonal Habitat Use

Home Ranges

Detailed information on the tagged red kites is given in Sect. 3.2.3 of an annex to this report (https://bergenhusen.nabu.de/imperia/md/nabu/images/nabu/einrichtungen/bergenhusen/projekte/bmugreif/endbericht_greifvogelprojekt_anhang.pdf). Relevant data on the size of the home ranges is given in Table 3.10. As the number of recorded locations and the duration of observations differed between tagged individuals, the influence of these parameters as well as the influences of the different kinds of telemetric monitoring (VHF versus GPS satellite), the different study areas or years on the results were investigated in a first step.

As the sample size (the number of telemetrically monitored red kite) was relatively low to be used in statistical analysis, a multivariate method (multiple regression analysis) as well as univariate non parametric tests were carried out.

None of the variables "number of recorded locations", "type of transmitter", "year", "study area", "region" (the northern study areas or the southern study areas), "pre-breeding season" (start of monitoring prior to the breeding season versus start after tagging during the breeding season) and "gender" had a significant influence on the size of the 95% kernel home ranges (Table 3.11). Figures 3.20 and 3.21 indicate that a lower number of recorded locations may result in a smaller calculated home range, but a multiple regression analysis showed that there is no distinct relation between the number of recorded locations and the size of the home range (Table 3.12).

The statistical tests carried out in the evaluation of the red kite telemetric data are non-parametric methods (Spearman rank correlation, Kruskal-Wallis ANOVA, Mann-Whitney U test). These methods were used, because the available data was not suitable for parametric tests, as dependent variables were not normally distributed and an equality of variances could not be assumed (Levene's test for significance). Conservative methods as the Kruskal-Wallis and the Mann-Whitney U test are robust towards outliers and small sample sizes. Determination of the statistical power is difficult in non-parametric tests and can only be done indirectly. Mumby (2002) gives some guidance. The sample size is a decisive factor. Following the calculations of Castelloe (2000) for pair-wise t tests, it can be assumed that a sample size (n) of 8 should be reached for each category to achieve a statistical power of 85% at a significance level of 0.05. The number of tagged red kite equalled 10. In the calculation of the home ranges, the data was evaluated per season, which resulted in a sample size of up to $n = 26$ for calculation. Therefore, the results of most of the tests can be assumed to be sufficiently reliable. Comparisons of the two telemetric methods (VHF versus Satellite) need to be looked at more critically, as data collected in satellite telemetry is based on three birds only. It cannot be excluded with absolute certainty that there is no difference

Table 3.10 Home ranges of tagged red kites (size in ha)

								Breeding season					Post-breeding season				
Individual	Year	Study area	Status	Distance nest—nearest wind turbine (m)	Number of wind turbines ≤ 1 km from nest	Telemetry period	Tag type	Number of locations	MCP (ha)	95% Kernel (ha)	60% Kernel (ha)	50% Kernel (ha)	Number of fixes	MCP (ha)	95% Kernel (ha)	60% Kernel (ha)	50% Kernel (ha)
Barbarossa	2007	Druiberg	Breeding, successful	740	5	02.06.—11.11.07	VHF	199	283	63	22	18	186	1017	206	26	20
	2008		Breeding, successful	780	5	26.03.—30.09.08	VHF	197	233	49	12	8	89	206	340	108	80
	2009		Seen in study area														
Arthur	2007	Querfurt	Breeding, successful	430	7	02.06.—19.11.07	VHF	56	415	616	112	68	138	357	302	67	37
Ramona	2007	Querfurt	Breeding, successful	730	6	03.06.—06.11.07	VHF	64	174	183	14	11	110	424	264	39	26
	2008		Non-breeder	260		07.03.—25.09.08		47	Homeranges not calculated								
Karl	2007	Druiberg	Breeding, successful	3090	0	11.06.—09.10.07	VHF	153	5494	2572	359	283	144	5527	2002	346	243
Barbara	2007	Druiberg	Breeding, successful	740	5	15.06.—10.12.07	VHF	169	7442	1872	427	339	229	3130	292	80	63
Gishild	2008	Querfurt	Breeding, successful	1070	0	05.06.—18.09.08	VHF	104	1803	728	221	173	73	21335	24836	5367	4115
	2009		Breeding, successful	1070	0	02.04.—17.09.09	VHF	114	1118	368	56	43					
Alte Dame	2009	Speckberg	Breeding, successful	2270	0	10.06.—27.12.09	VHF	25	51	96	18	14	28	222	144	47	37
	2010		Breeding, successful	2276	0	07.01.—29.09.10	VHF	151	72	12	5	4	42	102	116	14	10
Lui	2010	Speckberg	Breeding, no success	1510	0	03.06.—30.10.10	Sat	292	11765	1170	256	195	120	333	408	81	61
Erik	2010	Querfurt	Breeding, successful	820	3	06.06.—10.10.10	Sat	177	5319	1348	351	268	211	364	179	39	20
Svenja	2010	Queis	Breeding, successful?	1430	0	17.06.—26.09.10	Sat	117	718	56	22	17	192	716	242	42	25

Table 3.11 Statistical verification of a relation between the size of the 95% kernel home ranges of red kite and several parameters

Data	Variable	Statistical test			df	*p*
Home range breeding season	Number of locations	Spearman rang correlation	rho	0.11		0.72
	Pre-breeding season	Kruskal-Wallis	Chi^2	3.4571	1	0.063
	Type of tag	Kruskal-Wallis	Chi^2	0.2571	1	0.612
	Year	Kruskal-Wallis	Chi^2	1.4209	3	0.701
	Study site	Kruskal-Wallis	Chi^2	2.2275	3	0.527
	Region	Kruskal-Wallis	Chi^2	0.0816	1	0.775
	Sex	Kruskal-Wallis	Chi^2	0.0816	1	0.775
Home range post-breeding season	Number of locations	Spearman rang correlation	rho	0.042		0.9
	Type of tag	Kruskal-Wallis	Chi^2	0.0769	1	0.7815
	Year	Kruskal-Wallis	Chi^2	5.15	3	0.1611
	Study site	Kruskal-Wallis	Chi^2	1.9103	3	0.5912
	Region	Kruskal-Wallis	Chi^2	0.0593	1	0.8075
	Sex	Kruskal-Wallis	Chi^2	0.5341	1	0.4649
All home ranges	Number of locations	Spearman rang correlation	rho	0.025		0.91
	Type of tag	Kruskal-Wallis	Chi^2	0.0162	1	0.899
	Year	Kruskal-Wallis	Chi^2	2.4518	3	0.484
	Study site	Kruskal-Wallis	Chi^2	4.2874	3	0.232
	Region	Kruskal-Wallis	Chi^2	0.5065	1	0.477
	Sex	Kruskal-Wallis	Chi^2	0.1918	1	0.661

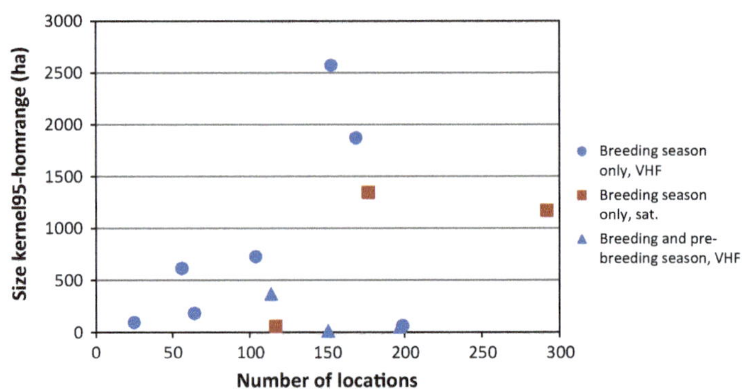

Fig. 3.20 Size of the 95% kernel home ranges of red kite in relation to the number of recorded locations during the breeding season

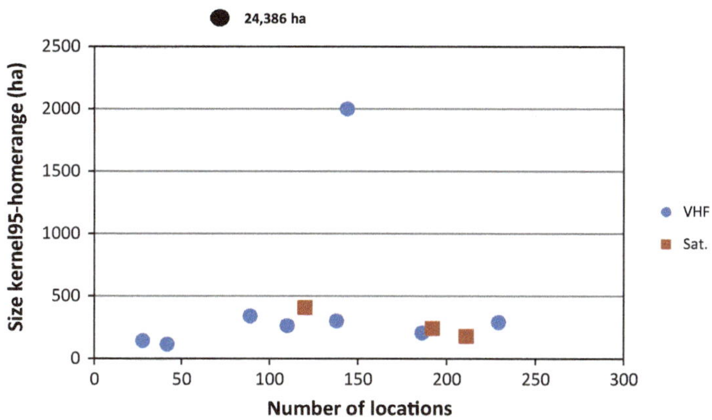

Fig. 3.21 Size of the 95% kernel home ranges of red kite in relation to the number of recorded locations during the post-breeding season. The *black symbol* indicates an outlier with 24,836 ha (Gishild in 2008)

Table 3.12 Results of a multiple regression analysis to determine the size of red kite 95% kernel home ranges

Parameter		B	SE	$t_{(21)}$	*p*
Constant		−182,727	411,091	−0.444	0.66
Sex	m/f	911	2217	0.411	0.69
Type of tag	VHS/Sat.	−588	2634	−0.223	0.83
Pre-breeding season	Yes/no	1497	2011	0.744	0.47
Number of locations	Stetig	−5	18.4	−0.295	0.77

between the two methods, but the use of satellite transmitters did not result in the calculation of larger home ranges as could have been expected. As significance levels were very low (see Table 3.11), the data could be evaluated together without distortion of the results.

All statistical results in Table 3.11 therefore are independent from the type of transmitter, the time of recording (already in the pre-breeding season or during the breeding season) or number of recorded locations. Region or study area as well as gender did not result in significant differences either.

The mean home ranges of tagged red kites were very large throughout the year for both male and female birds (Table 3.13). The home range sizes determined for the years 2007–2010 represented either lowest or highest values of the entire range with only few medium sizes. The standard deviation for almost all calculated home ranges therefore was above the mean. All tagged birds were breeding birds successfully fledging young or in single cases loosing young in the late nestling stage (after ringing). A bias of home range sizes due to non-breeding birds was therefore excluded. During the breeding season male birds had larger home ranges, after the breeding season female birds, but differences were not significant (see Table 3.13).

Table 3.13 Comparison of the home range sizes of male and female red kites for breeding and post-breeding season (telemetric data collected for all tagged birds between 2007 and 2010)

	Breeding season				Post-breeding season			
	95% MCP	Kernel (km²)			95% MCP	Kernel (km²)		
	(km²)	95%	60%	50%	(km²)	95%	60%	50%
♂♂ (n = 6)								
X	39.18	9.70	1.86	1.40	12.98	5.73	1.10	0.77
Min	2.33	0.49	0.12	0.08	2.06	1.79	0.26	0.20
Max	117.65	25.72	3.59	2.83	55.27	20.02	3.46	2.43
SD	45.84	9.53	1.58	1.24	20.92	7.05	1.19	0.85
♀♀ (n = 7)								
X	16.25	4.74	1.09	0.86	43.22	43.17	9.32	7.13
Min	0.51	0.12	0.05	0.04	1.02	1.16	0.14	0.10
Max	74.42	18.72	4.27	3.39	213.35	248.36	53.67	41.15
SD	26.43	6.64	1.59	1.26	84.10	100.52	21.73	16.67

Different kinds of habitat use were even determined between breeding partners. Whereas Barbarossa had a very small home range with clear reference points, Barbara used a spatially separate, very large area and partly ranged over eight to 13 km to reach mown grassland areas.

All three birds, which were monitored over two subsequent years (one male and two female birds) showed similar habitat use between years in terms of landscape sections used as well as home range size. This was equally true for the post-breeding season. Additionally, these red kite were focused on one type of habitat each for foraging and maintained this preference in both years. Barbarossa mainly used the residential area, Gishild the open pit mine and Alte Dame the alfalfa crops in the surrounding area of the nest as well as the edges of residential areas.

After the breeding season red kites with a previously very small home range extended their home range (Barbarossa, Ramona, Alte Dame, Svenja), whereas red kites with a previously large home range reduced their home range (Karl, Barbara, Lui, Erik, as well as Arthur, who had a large 95% kernel home range during the breeding season). The differences in home range size were smaller during the post-breeding season. Only Karl and Gishild still had considerably larger home ranges than the other birds. Furthermore, Gishild was the only red kite using a large home range during the breeding season and an even larger home range in the post-breeding season. This was due to alternate use of two spatially separate core home ranges. The area of the nest was visited repeatedly for several days before migrating towards wintering grounds (see also Sect. 3.3.2.2). Barbara's 95% MCP home range was still large during the post-breeding season, but the 95% kernel fell within the range of the other red kites.

All in all, no significant differences in mean home range size (95% MCP and 50, 60, 95% kernel, $p > 0.05$, pair-wise Wilcoxon test all) between breeding and post-breeding season could be determined for male or female birds.

The location of the nest within the home range varied between relatively central to peripheral and even changed in some cases during the year. During the breeding season Karl's and Barbara's, during the post-breeding season Lui's and over the whole year Gishild's and Alte Dame's nest was located peripheral, but otherwise more central.

All tagged birds showed considerable site fidelity regarding the nest or the close-up area of the breeding site. After the breeding season, the nest was still located within the 50% kernel or at least the 60% kernel home range. The red kites were recorded near the nest up until they finally left the area to migrate to wintering grounds. The only red kite wintering in the breeding area (Alte Dame) stayed in the same area during the winter months and the home range only slightly varied over the two years of telemetric monitoring.

Distance of Recorded Locations from the Nest

Figures 3.22 and 3.23 illustrate the telemetric data of ten tagged birds in relation to the distance between recorded locations and nest. As shown in Section "Home Ranges" (last paragraph), all tagged birds show distinct site fidelity between spring (from the arrival in the breeding area or tagging) and departure to wintering grounds. Therefore, the recorded locations during breeding and post-breeding season were evaluated together. Only locations recorded in the close-up area of the nest (<50 m) were excluded, as they were attributed to presence near the nest and do not represent active movement into the landscape. For three of the red kites

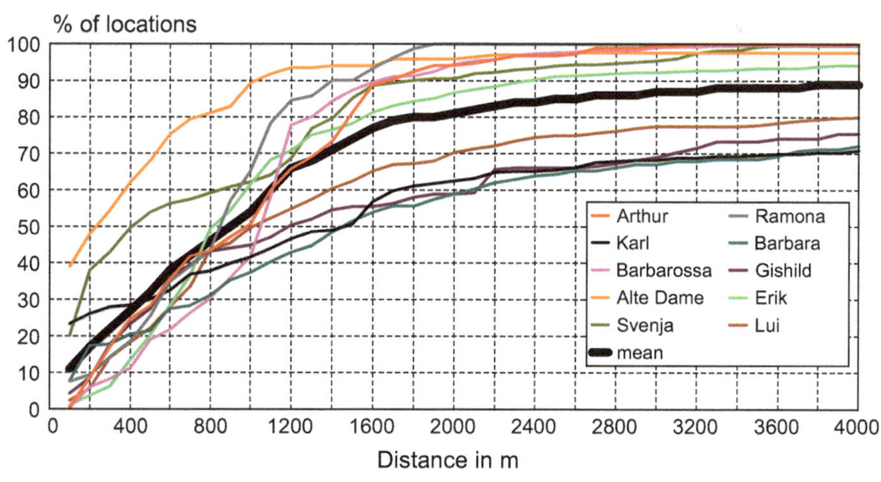

Fig. 3.22 Distance of recorded red kite locations from the nest based on telemetric data collected between 2007 and 2010. Percentages of locations recorded for each tagged bird by increasing distance up to 4,000 m

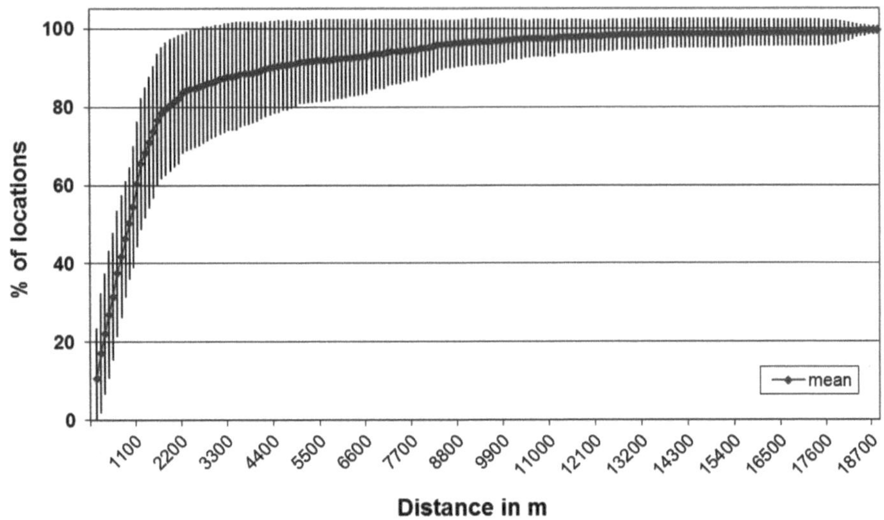

Fig. 3.23 Distance of recorded red kite locations from the nest based on telemetric data collected between 2007 and 2010, mean of all tagged birds with standard deviation by increasing distance up to the maximum distance

(Barbarossa, Gishild and Alte Dame) data collected over two years was included in the calculation of the individual distance graph.

The mean value graph is relatively steeply ascending up to a distance of approximately 1,700 m from the nest and then gradually flattens. Individual variations are considerable and the standard deviation is relatively large. Red kites either show a large proportion of recorded locations in the vicinity of the eyrie (>90% of the locations are recorded up to a distance of 1,700 m, Alte Dame, Svenja) or only few recorded locations in the surrounding area of the eyrie. The latter either return even curves with only 55–67% of the locations recorded up to a distance of 1,700 m, evenly distributed over all distance classes with a slightly flattening curve (Karl, Barbara, Gishild, Lui, i.e. the red kite with the largest home ranges) or curves strongly ascending between 900 and 1,700 m with >80% of the recorded locations (Arthur, Barbarossa, Ramona, Erik).

A mean of 54% of all locations were recorded within a distance of 1,000 m from the nest. 100% of all locations were recorded up to a distance of 18,400 m (Fig. 3.23). 54% of the locations recorded for all tagged birds occurred at a distance of 0–1.0 km, 27% at 1.01–2.0 km, 6% at 2.01–3.0 km, 4% at 3.01–5.0 km and 9% at >5.01 km. Only 3% of all recorded locations occurred at distances >9 km.

Approach of Wind Turbines and Stay in the Wind Farm

Approaches of wind farms or wind turbines up to a distance of below 300 m (wind farm impact area) were verified based on the recorded locations for all of the tagged birds (Table 3.14; Fig. 3.24). Even red kites breeding at larger distances from wind turbines (2–3 km, see Section "Home Ranges") as e.g. Karl and Alte Dame occasionally approached wind farms or turbines up to distances below 300 m or even 75 m. Birds breeding at distances of <1,000 m from wind farms approached wind farms or turbines on a regular basis within the 300 m buffer zone as well as the close-up area of <75 m. Only Barbarossa (small home range in the breeding and post-breeding season oriented away from the wind farm) and Barbara (wide ranging foraging flights in the breeding season away from the wind farm and small home range oriented away from the wind farm later) showed a deviant behaviour. The highest frequencies of use of a wind farm were found for Arthur and Ramona (both Querfurt study areas) breeding at a distance of <500 m from wind turbines. The connection between the distance between nest and wind turbine, the number of recorded locations within the wind farm plus the additional +300 m or +75 m buffer zone and the closest approach of wind turbines was statistically significant in each case (Table 3.15).

Table 3.14 Locations of tagged red kites recorded within the wind farm (wind farm site plus buffer zone) based on the telemetric data collected between 2007 and 2010

	Year	Total number of recorded locations	Locations within wind farm + +300 m buffer zone		Locations within wind farm + +75 m buffer zone		Closest approach of wind turbine (m)
			Number	%	Number	%	
Karl	2007	297	1	0.34	1	0.34	189
Alte Dame	2009	72	0	0	0	0	1,943
	2010	202	1	0.50	0	0	218
Lui	2010	412	20	4.85	1	0.24	39
Svenja	2010	309	13	4.21	6	1.94	78
Gishild	2008	177	8	4.52	1	0.56	196
	2009	115	2	1.74	2	1.74	371
Erik	2010	388	64	16.49	19	4.90	5
Barbara	2007	398	7	1.76	1	0.25	103
Barbarossa	2007	385	15	3.90	0	0	106
	2008	285	9	3.16	1	0.35	114
Arthur	2007	194	100	51.55	35	18.04	30
Ramona	2007	174	77	44.25	21	12.07	9
	2008[a]	47	8	17.02	3	6.38	85

Birds in order of distance between nest and wind turbine

[a]Red kite left study site in the beginning of the breeding season

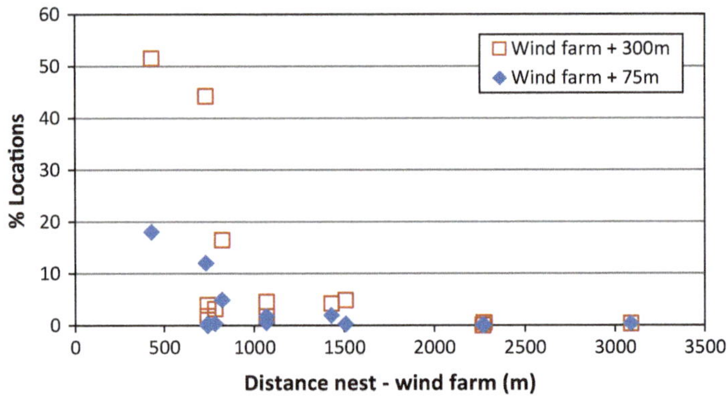

Fig. 3.24 Proportion of red kite locations recorded within wind farms (wind farm site +75 or +300 m buffer zone) in relation to distance between nest and nearest wind turbine within the wind farm based on telemetric data collected between 2007 and 2010

Table 3.15 Results of the Spearman rank correlation test on the correlation between distance between red kite nest and wind farm and overlap of 95% kernel home ranges with wind farm areas or proportion of recorded locations

Parameter	Reference	Season	rho	*p*
Overlap homerange	Wind farm + 75 m	Breeding	−0.62	0.023
Overlap homerange	Wind farm + 300 m	Breeding	−0.63	0.022
Overlap homerange	Wind farm + 75 m	Post-breeding	−0.67	0.012
Overlap homerange	Wind farm + 300 m	Post-breeding	−0.72	0.005
Percentage of locations	Wind farm + 75 m		−0.60	0.023
Percentage of locations	Wind farm + 300 m		−0.71	0.004

It needs to be taken into account that locations may have been recorded within the wind farm + 75 m buffer, although the next wind turbine was located more than 75 m away due to different approaches to calculate the parameters "closest approach of wind turbine" (bilateral distance measurements between recorded location and centre of the wind turbine) and "locations within wind farm + buffer zone" (position within the wind farm area). This will occur, if birds stay in wind farms with a turbine spacing of more than 150 m, a fairly common scenario due to wind yield. A description of the determination of the wind farm + 75 m and wind farm + 300 m reference areas can be found in Section "Evaluation of Telemetric Data" under "Approach of wind turbines and stay in the wind farm".

The relatively large proportion of recorded locations in the wind farm + 75 m close-up area and frequent approaches of the wind turbine base found for red kite Erik in 2010 (Table 3.15) will have been partly due to a manure heap temporarily stored in this area. In 2011, a dunghill was located in the same place at about 30 m from the turbine base, which was likewise intensely frequented by Erik. Erik was

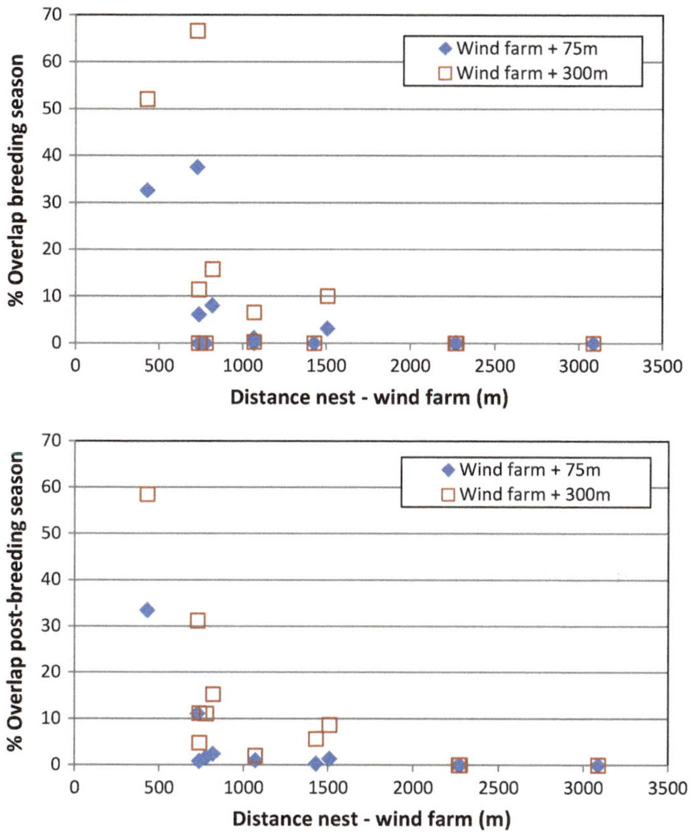

Fig. 3.25 Overlap of red kite 95% kernel home ranges with wind farm areas (wind farm site +75 or +300 m buffer zone) in relation to the distance between nest and nearest wind turbine within the wind farm based on telemetric data collected between 2007 and 2010

last GPS located in mid-May 2011 directly at this turbine and has not been (ARGOS or GPS) located since. On 21 May 2011, a dead red kite was found in a search of the surrounding area of this turbine. As this bird was not Erik, it is likely that at least two red kites were killed at this wind turbine within a short period of time in breeding season 2011.

The 95% kernel home ranges of six birds during the breeding season and eight birds during the post-breeding season overlapped with the wind farm or the wind farm impact area (Fig. 3.25). This was true for all red kites breeding within a distance of up to 1.5 km from wind turbines. Only for Karl and Alte Dame, breeding a distance of >2 km from wind turbines, no overlap was found.

During the breeding season, overlaps with the 60% kernel home ranges as well as the 50% kernel home ranges were found for all birds showing overlaps of the 95% kernel home ranges with the exception of Gishild (60% kernel) and Lui (50% kernel).

For more details see Annex Sect. 1.2, Tables 1.17 and 1.18, https://bergenhusen.nabu.de/imperia/md/nabu/images/nabu/einrichtungen/bergenhusen/projekte/bmugreif/endbericht_greifvogelprojekt_anhang.pdf.

The size of the area overlap depended statistically significantly on the distance between nest and wind turbine (Table 3.15).

After the breeding season, the size of the overlap decreased for all kernel home ranges (95, 60, 50%) with no statistically significance (pair-wise Wilcoxon test; wind farm + +75 m: V = 31, $p = 0.34$; wind farm + +300 m: V = 19, $p = 0.72$). For the 50–60% kernel home ranges, the decrease in overlap was the most significant with only 10% (+75 m) or >30% overlap with the wind farm area (+300 m) in the post-breeding season for the birds breeding closest to wind turbines (Arthur, Ramona).

Apart from the distance between nest and wind turbine and the size of the home range, the spatial orientation of the home range (preferred use of certain directions, spatial distribution of attractive or unattractive foraging grounds) and of the wind farm (number of wind turbines in the close-up area <1,000 m, arrangement of the wind turbines) has an influence on the size of the overlap area. Most significant is the influence of the distance between nest and wind turbine and the number of wind turbines within a distance of 1,000 m around the nest. Extensive overlap with wind farms occurred for all red kites breeding within a distance of <1,000 m from the next wind turbine. This was also true for core areas with a 50% probability of use. After the breeding season, with its increased demand of food to feed the young, these areas are used less frequently. In all cases, several wind turbines were located within a radius of 1,000 m around the nest. Barbarossa was the only exception amongst the red kites breeding in the vicinity of wind turbines with a small home range oriented away from the wind farm (settlement area Badersleben) in both years. This bird only approached the wind farm area after the breeding season and even then only small overlap occurred with the 95% kernel home range.

The importance of size and orientation of the home range for areas overlapping with the wind farm was obvious for all tagged birds. For red kites breeding near wind turbines, even smaller kernel home ranges will result in a large proportionate overlap (e.g. Arthur, Ramona) and all the more for birds using large home ranges (e.g. Barbara). Exceptions only occurred due to individual preferences for food resources in directions away from the wind farm (Barbarossa, settlement area Baderleben).

In red kites breeding at a distance of between 1,000 and 2,000 m from wind turbines, even large home ranges resulted in a considerably smaller overlap (e.g. Gishild, Svenja, Lui). Due to the extensive mobility especially during the breeding season, Lui's home range showed by far the largest overlap.

In red kites breeding >2,000 m away from wind turbines, neither large (Karl) nor small (Alte Dame) home ranges resulted in an overlap with wind farm areas, even though single locations were recorded at a distance of below 300 m from wind turbines.

Possibilities to Reduce Overlap of Home Ranges with Wind Farm Areas by Keeping Areas Free of Wind Turbines

To what extent keeping an area of 1,000 m around nests free from wind turbines would have prevented or reduced overlap of red kite kernel home ranges determined based on telemetric monitoring of habitat use, is presented in Table 3.16 for the area wind farm +75 m and in Table 3.17 for the area wind farm +300 m for a 1,000 m or 1,250 m wind turbine free zone, respectively. Eight red kites with an overlap of home range and wind farm area during the breeding and/or post-breeding season could be included in the evaluation.

For four of the tagged birds with overlap of kernel home range and wind farm area during the breeding season, breeding at a distance of <1 km from wind turbines, a wind turbine free zone of 1,000 m or more would have reduced the overlap. For two red kites, breeding >1,000 m from wind turbines neither a 1,000 m nor 1,250 m zone would have resulted in a reduction of overlap during the breeding season. The remaining two red kites showed no overlap during the breeding season.

In the post-breeding season, the existing overlap of all five red kites breeding at a distance of <1 km from wind turbines could be reduced by a wind turbine free zone of 1,000 m or more. For the three red kites, breeding at a distance of >1 km from wind turbines, a wind turbine free zone would not result in a reduced overlap.

Table 3.16 Reduction in overlap of 95% kernel home ranges of tagged red kites (breeding and post-breeding season) with wind farm areas (wind farm site + +75 m buffer zone) by wind turbine free zones around the nest

	1,000 m wind turbine free zone		1,250 m wind turbine free zone	
95% kernel, reduction of overlap with wind farm + 75 m in %				
n = 6	BS	PBS	BS	PBS
Mean	−24.79	−74.10	−41.30	−92.23
SD	30.67	27.73	37.77	16.27
Min	0	−27.4	0	−59.39
Max	−82.9	−100	−100	−100

Birds with existing overlap were included only
BS breeding season; *PBS* post-breeding season

Table 3.17 Reduction in overlap of 95% kernel home ranges of tagged red kites (breeding and post-breeding season) with wind farm impact areas (wind farm site + +300 m buffer zone) by wind turbine free zones around the nest

	1,000 m wind turbine free zone		1,250 m wind turbine free zone	
95% kernel, reduction of overlap with wind farm + 300 m in %				
n = 6	BS	PBS	BS	PBS
Mean	−27.78	−78.89	−41.28	−91.68
SD	29.89	22.04	37.75	14.85
Min	0	−40.42	0	−63.36
Max	−81.43	−100	−99.54	−100

Birds with existing overlap were included only
BS breeding season; *PBS* post-breeding season

For none of the red kites, a wind turbine free zone of 1,000 m would result in no overlap of the 95% kernel home ranges with wind farm areas at all during the breeding season. A 1,250 m zone resulted only for Ramona in a complete avoidance of overlap. The considerably smaller 50–60% kernel home ranges of only two red kites (Ramona 1,000 m, Barbara 1,250 m) would not overlap during the breeding season. In all other cases, overlap still occurred. Even by observing wind turbine free zones, wind turbines will still remain within red kite home ranges, imposing an additional risk for red kites breeding in the surrounding area, but wind turbine free zones may considerably reduce if not avoid overlap between home ranges and wind farm areas.

3.3.2.2 Continuous Monitoring

Daily Home Ranges of Individual Birds

On total of 20 days, continuous monitoring was carried out for six red kites, three male (Karl and Barbarossa in the Druiberg study area and Arthur in the Querfurt study area) and three female birds (Ramona and Gishild in the Querfurt study area and Alte Dame in the Speckberg study area). The proportion of flight movement within the 500 m distance classes around the nest are presented in Figs. 3.26 and 3.27 for each continuous tracking.

Daily Home Range and Activity

Table 3.18 gives an overview of the 20 continuous tracking sessions carried out between 2007 and 2010.

On average, the birds spent one quarter of the day (mean: 24.5%, SD: 21.2%, n = 20 days) between sunrise and sunset in flight and three quarters of the day sitting. The values showed large variance with a daily flight times of between 11 min and 12 h 20.

The three male birds spent more time in flight (mean: 31.2%, SD: 22.6%, n = 8) than the three female birds (mean: 20.0%, SD: 20.0%, n = 12). Breeding females were hardly flying at all (e.g. Alte Dame in 2010), whereas male birds spent a large proportion of time in flight especially during incubation and raising the young (see Barbarossa in 2008) with decreasing flight activity after fledging the young. While raising young in June, female birds tended to fly considerably more than in the following months. Strong individual differences occurred.

On average visual contact with the birds was lost for 2 h 12 min (n = 20) per day. Assuming that birds behaved in the same way during that time, this would result in a further 33 flight minutes and a total mean flight activity of 3 h 47.

Summarizing all flight distances of all days of continuous monitoring, the red kites were watched over a total distance of 1,618 km. On 04.09.2008 Gishild was an exception, as the nest was not in the centre of activity for that day. 36.0% of the

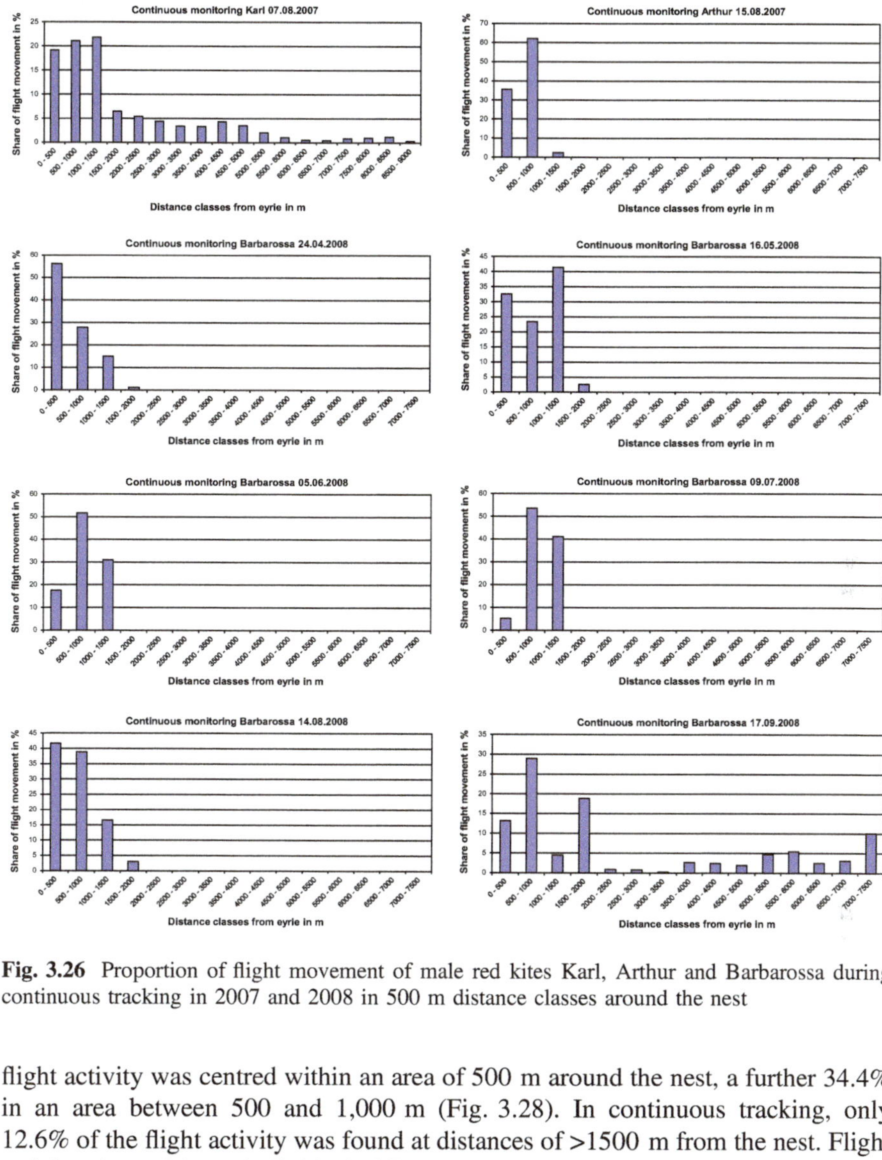

Fig. 3.26 Proportion of flight movement of male red kites Karl, Arthur and Barbarossa during continuous tracking in 2007 and 2008 in 500 m distance classes around the nest

flight activity was centred within an area of 500 m around the nest, a further 34.4% in an area between 500 and 1,000 m (Fig. 3.28). In continuous tracking, only 12.6% of the flight activity was found at distances of >1500 m from the nest. Flight activity showed large individual differences as well. The average daily flight distance was 90.1 km (SD 76.8 km, n = 20), varying between 6.7 and 320.1 km.

Daily home ranges, i.e. the area used per day, covered on average 298 ha (SD 306 ha, n = 20), varying between 1 ha and 1,135 ha. On average, home ranges of the three male birds were smaller (mean: 236 ha, SD: 196 ha, n = 8) than those of the three female birds (mean: 340 ha, SD: 365 ha, n = 12). Comparable to the time spent in flight, the daily home ranges varied considerably.

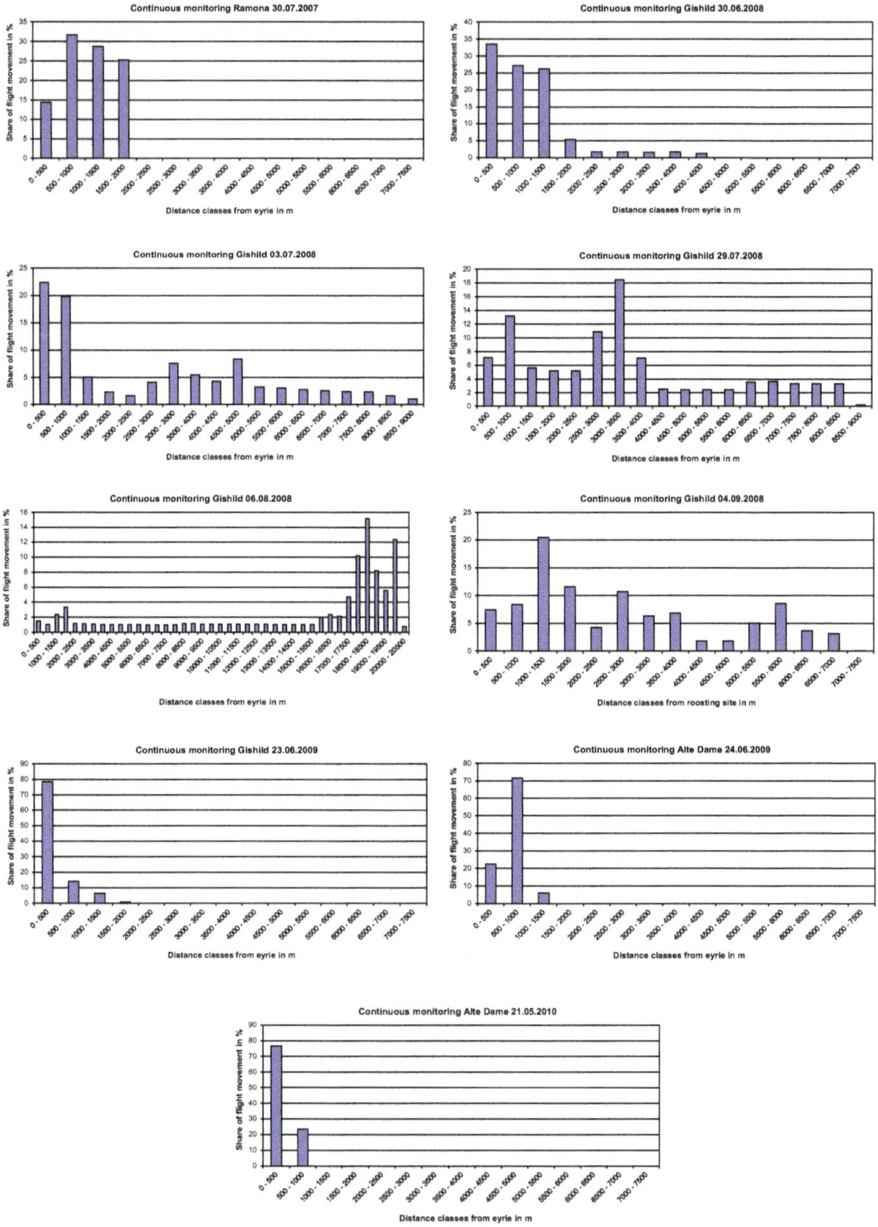

Fig. 3.27 Proportion of flight movement of female red kites Ramona, Gishild and Alte Dame during continuous tracking between 2007 and 2010 in 500 m classes around the nest or the roosting site. Two days of continuous tracking of Alte Dame are not included, as the bird was only foraging over the alfalfa field next to the nest with only minimal flight movement involved

Table 3.18 Daily activities of red kites during continuous tracking

Bird	Date	Max distance from the nest (m)	Daily home range (km^2)	Length of day	Duration sitting		Duration flight		No visual contact	Covered distance (km)
					h	%	h	%	h	
Karl	07.08.2007	8540	4.48	15:08	09:45	72.0	03:47	28.0	01:36	118.9
Arthur	15.08.2007	1330	0.75	14:39	06:43	64.7	03:40	35.3	04:16	93.0
Barbarossa	24.04.2008	1620	2.01	14:24	08:42	64.9	04:42	35.1	01:00	115.9
Barbarossa	16.05.2008	1870	1.11	15:37	10:50	69.4	04:47	30.6	00:00	119.6
Barbarossa	05.06.2008	1540	1.88	16:25	03:04	19.9	12:20	80.1	01:01	302.1
Barbarossa	09.07.2008	1410	0.84	16:24	13:49	90.4	01:28	9.6	01:07	55.0
Barbarossa	14.08.2008	1750	1.56	14:44	13:01	94.3	00:47	5.7	00:56	24.0
Barbarossa	17.09.2008	7500	6.22	12:34	06:00	74.8	02:01	25.2	04:33	67.8
Ramona	30.07.2007	1850	3.27	15:33	07:50	86.1	01:16	13.9	06:27	34.1
Gishild	30.06.2008	4300	5.39	16:35	09:46	60.7	06:19	39.3	00:30	159.8
Gishild	03.07.2008	8880	11.35	16:33	03:49	65.6	02:00	34.4	10:44	99.4
Gishild	29.07.2008	8600	7.63	15:37	12:41	91.7	01:09	8.3	01:47	41.3
Gishild	06.08.2008	20050	4.84	15:12	10:49	96.0	00:27	4.0	03:56	50.6
Gishild	04.09.2008	7020[a]	5.69	13:25	07:55	80.8	01:53	19.2	03:37	64.0
Gishild	23.06.2009	1880	1.29	16:39	04:41	32.9	09:32	67.1	02:26	240.7
Alte Dame	24.06.2009	1240	0.90	16:39	10:59	67.2	05:21	32.8	00:19	134.3
Alte Dame	29.03.2010	90	0.01	12:40	12:24	97.9	00:16	2.1	00:00	6.7
Alte Dame	22.04.2010	270	0.02	14:15	14:04	98.7	00:11	1.3	00:00	7.9
Alte Dame	21.05.2010	990	0.45	15:53	13:15	83.4	02:38	16.6	00:00	60.6
Alte Dame	19.06.2010	200	0.01	16:39	16:22	98.3	00:17	1.7	00:00	7.1
Arithmetic mean		3890	2.98	15:16	09:49	75.5	03:14	24.5	02:12	90.1

[a]from roosting site

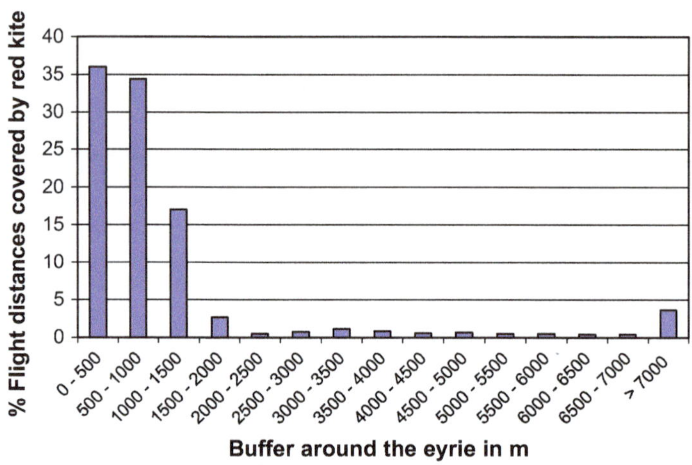

Fig. 3.28 Summarized proportions of distances covered by red kites within the 500 m buffer around the nest (six birds, 20 continuous tracking days)

In 19 of 20 continuous tracking sessions, attachment to the nest was even apparent in the post-breeding season. Only Gishild had no contact to the nest during continuous tracking in September 2009, but again in the following session. With just over 20 km, Gishild reached the largest distance from the nest on 06.08.2008 in the post-breeding season. The mean maximum distance from the nest reached in 19 continuous tracking sessions was 3,890 m (SD 4,940 m). During the breeding season the mean maximum distance reached 2,480 m (SD 2,840 m, n = 14) and during the post-breeding season 7,830 m (SD 3,770 m, n = 5).

3.3.2.3 Flight Altitudes

From the rotor diameter (depending on manufacturer and model year between 66 and 90 m within the investigated wind farms) and hub height (between 65 and 105 m within the investigated wind farms), a rotor swept height band of between 30 and 150 m resulted for the wind turbines constructed after 2001 (Table 3.19).

In only 2.8% of the flight time recorded within wind farms, red kite flew higher than 150 m. A total of 46.6% of this flight time occurred within the height band between 26 and 150 m, bearing a potential risk (Fig. 3.29). The flight height shows distinct seasonal differences, though (Fig. 3.29). Between March and June, the time of establishment of crops forming a dense cover of vegetation and at the same time of maximum food requirement to raise young, red kite spent a considerably larger proportion of time in heights between 50 and 150 m bearing a potential risk of collision with all types of wind turbine than in late summer and autumn months, which are characterised by harvested and ploughed fields. This seasonal shift was also obvious in the adjacent height classes 26–50 and >150 m.

Table 3.19 Rotor swept height band within the investigated wind farms

Study area	Wind turbine model	Hub height (m)	Rotor diameter (m)	Rotor swept height band (m)	
				Wind turbine model	Total study area
Druiberg	Lagerwey	40	27	27–54	19–171 (mostly 79–150)
	Micon	40	43	19–62	
	E-66/E-70	114	70/71	79–150	
	E112	114	114	57–171	
Querfurt	E-66	85	70	50–120	50–150
	E-66	98	70	63–133	
	V90	105	90	60–150	
	FL2500	100	90*	55–145	
Speckberg	RE MM82	100	82	59–141	57–141
	E-70	98	71	63–133	
	E-82	98	82	57–139	
Dessau	E-66	98	66	65–131	65–131
Queis	E-66	65	70	30–100	30–100

*Minimum size according to manufacturer

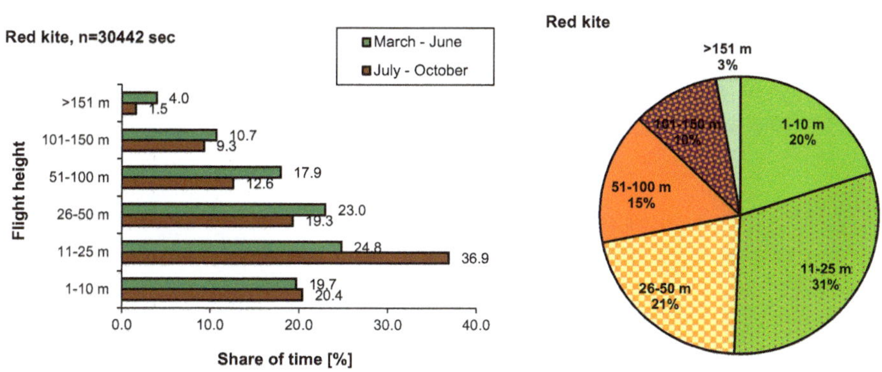

Fig. 3.29 *Left* Seasonal differences in the proportion of time spent in the single height classes. *Right* Proportion of time of flight activity for the single height classes (total time in flight)

3.3.2.4 Dangerous Situations

In 2.46% of the observation sequences, red kites were recorded flying into or crossing the rotor swept area. None of the birds got into an instable flight situation due to the rotor movement.

3.3.3 *Foraging and Land Use*

Use of different crops and other landscape structures by red kites was investigated in the behavioural observations in the study areas in 2007 and 2008 as well as in specific data collected in different locations in 2010.

3.3.3.1 Use of Crops in Behavioural Observations in 2007 and 2008

Figure 3.30 shows the crops within the study areas in 2007 and 2008. Figure 3.31 gives the percentages of use by flying red kites for the same areas and years. Figure 3.32 shows the Jacobs indices, indicating the crop availability in relation to the use for each crop. Positive values of the Jacobs index indicate preference for, negative values avoidance of the respective crop.

Comparison of Figs. 3.30 and 3.31 clearly shows that red kite do not use crops in proportion to availability. Already in March and April, when bare fallows made up a proportion of 30% were preferred with a usage of 50%, whereas winter wheat with a proportion of 45% was avoided with only 30% of use, the disproportionate use was apparent. This effect was even more obvious in April and May, when winter wheat, winter oilseed rape and winter barley made up 65% of the total crop area, but were used by only 15%. Winter cereal crops were clearly avoided during spring, especially in April and May. Already in April and May, these crops were fully established and were not available as a foraging ground for red kite until harvest. In April, bare fallows were clearly preferred (25% availability and 75% of use), as they were suitable foraging grounds due to good visibility. In May, most

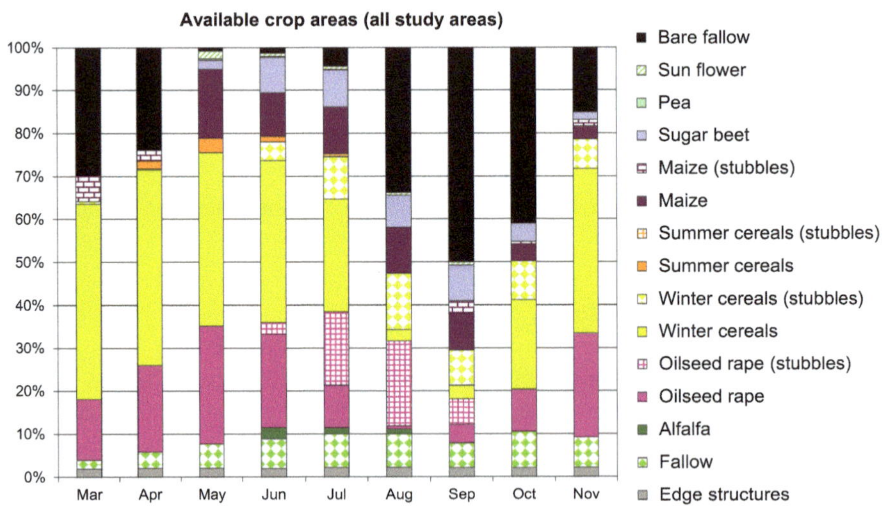

Fig. 3.30 Available crops for all study areas in both 2007 and 2008

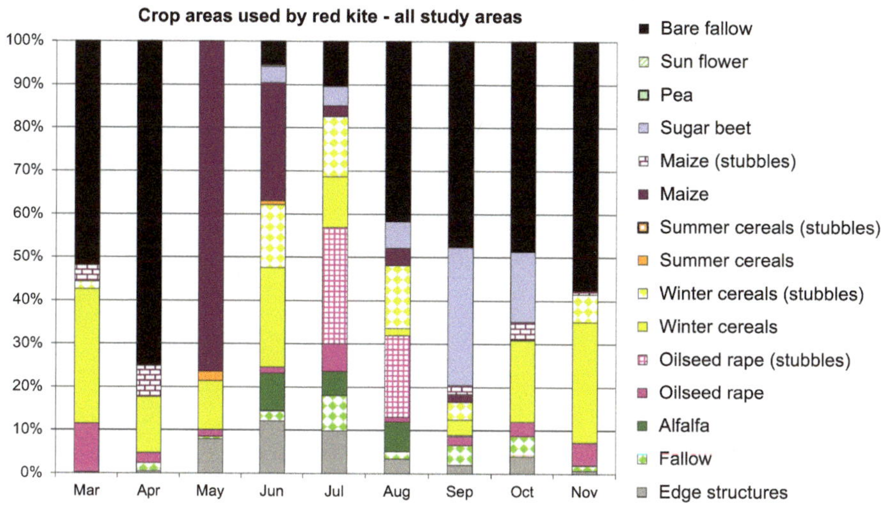

Fig. 3.31 Habitat use of red kite for all study areas in both 2007 and 2008

bare fallows were planted with maize so that only a small proportion of these fallows was still available (Fig. 3.30). As no red kites were recorded in these areas due to the small sample size, the Jacobs index indicated avoidance. The same was true for sugar beet, which in May were only available at a proportion of 2.3%. In June and July the avoidance of winter cereals and the preference for bare fallows or maize crops was continuously decreasing, as harvested oilseed rape and cereal crops were suitable foraging grounds for the red kite. The maize crops on the other hand, which were fully established at that time, were less preferred (June) or avoided (July) (Fig. 3.32). Alfalfa crops were clearly preferred between June and August. Clear preference for edge structures as agricultural roads and hedgerows as well as fallows around turbine bases and access routes to wind turbines was found additionally. In May to July (maximum: June), these edge structures were used three times as often as their percentage of availability. Between May and July, alfalfa crops and edge structures were important foraging grounds for red kites with only few suitable areas available.

Between August and October all crops except alfalfa and sugar beet were used by red kites proportionately to availability, as all crops with the exception of the ripening maize were suitable for foraging. Still, the Jacobs indices for this time show avoidance or preference for different crops. This is mainly due to the vast variety of crops and the resulting low sample size for single crops. The strong preference for sugar beet in September (10% availability, more than 30% of use) was based on seven 45 min watches carried out during beet harvest, thus the harvested areas were preferred instead of the crops themselves. From October, bare fallows were preferred over growing winter oilseed rape and winter wheat.

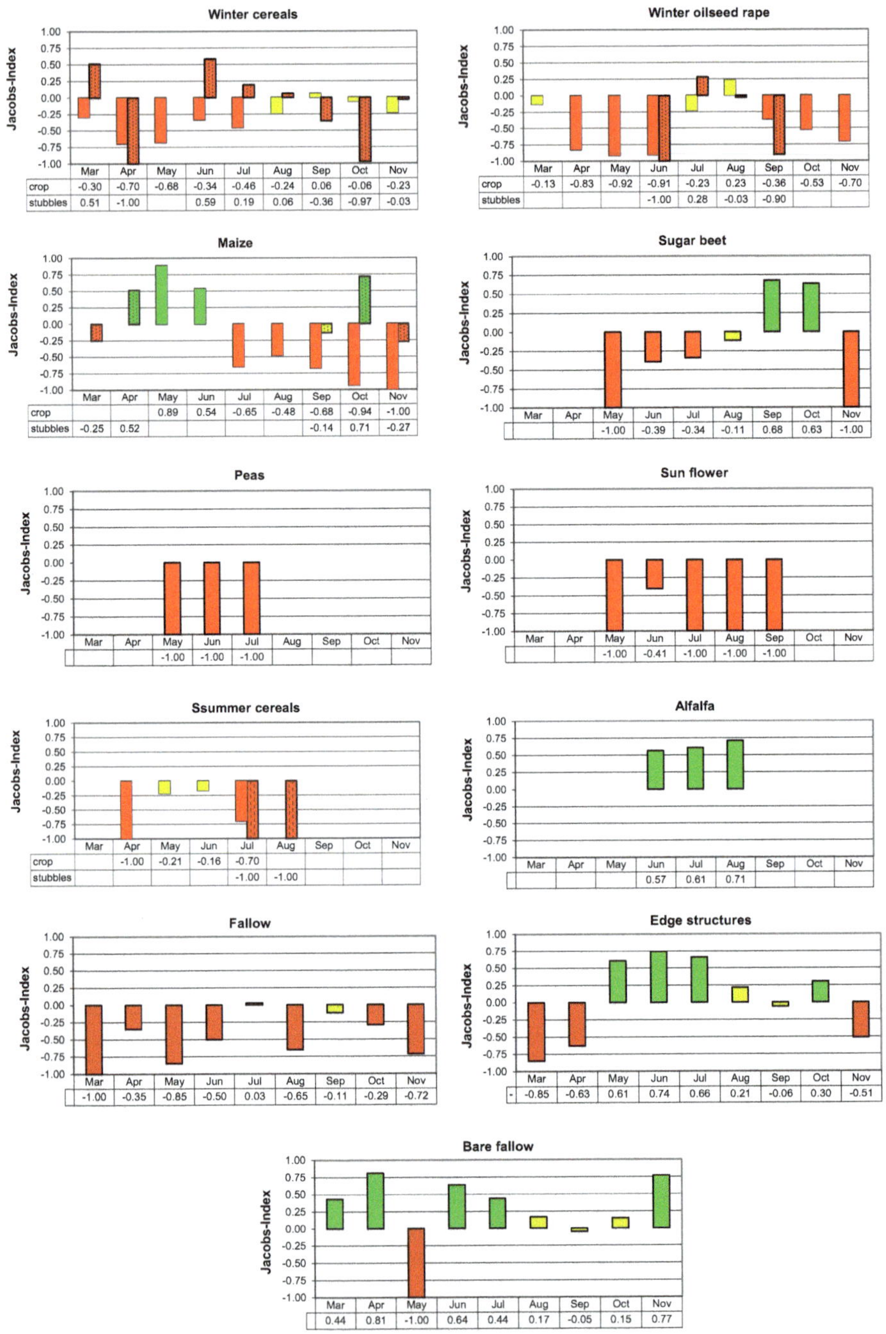

Fig. 3.32 Jacobs indices of the different crops used by red kites for all study areas and both 2007 and 2008; *green* preference; *red* avoidance; *yellow* deviation <0.25, i.e. use equals availability; hatched = harvested fields

3.3.3.2 Use of Alfalfa Crops

The large-scale, area-related behavioural observations in 2007/2008 across all study areas (see Sect. 3.3.4.1) revealed a strong preference of red kite for alfalfa crops despite the fact that these made up a small proportion of all investigated areas.

To further investigate the suitability of alfalfa, 959 ha of agricultural land were investigated within the Speckberg study area in 2010 using a different approach (point counts from a total of 28 vantage points). The availability of crops around the vantage points is presented in Fig. 3.33. With a proportion of over 60%, winter wheat is dominating the crops followed by oilseed rape and alfalfa.

244 sitting or foraging red kites were recorded. Again, the disproportionate use of alfalfa was noticeable especially in May to August (Fig. 3.34), whereas winter cereals were used below average over the whole year. Use of the single crops over the year (Fig. 3.35) can be described as follows.

The proportion of bare fallows continuously increased between May and September. Whereas in May no bare fallow was available and in June only 2 ha, fallows were clearly dominating in September with 537 ha. Bare fallows were only preferred in July. In August and September they were used proportionate to availability. Herbs (thyme and oregano) were present on 4–10 ha between May and September. A preference for these areas in May was derived from use by five birds on 20.05.2010 with no high degree of coverage of the respective areas. In areas with peas (6–8 ha) and potatoes (54 ha between May and September) no red kites were recorded in point counts. Beet crops (26 ha) were not used at all in May, July and August but preferred in June, as crops were not dense yet. The cultivation area of maize equalled 29 ha. Maize crops were preferred by red kites in May and June, but (completely) avoided afterwards. 55.8 ha of summer cereals (summer barley, summer wheat) were investigated. In May, crops were still low and were preferred,

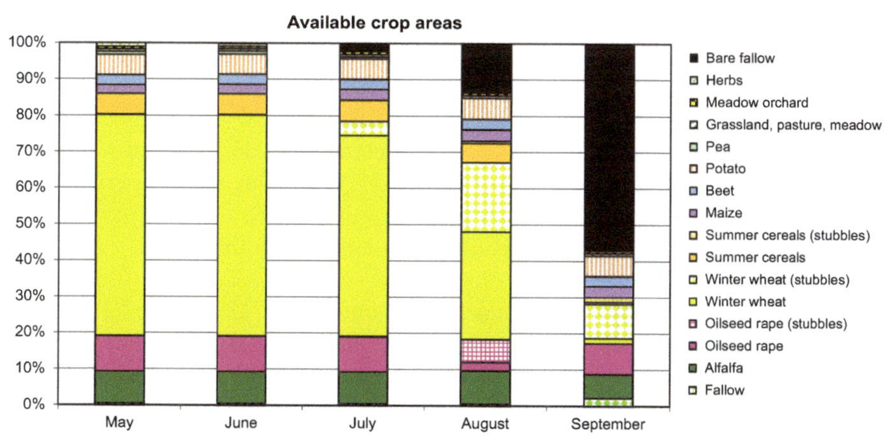

Fig. 3.33 Available areas for 28 vantage points (959 ha all together) within the Speckberg study area between May and September 2010

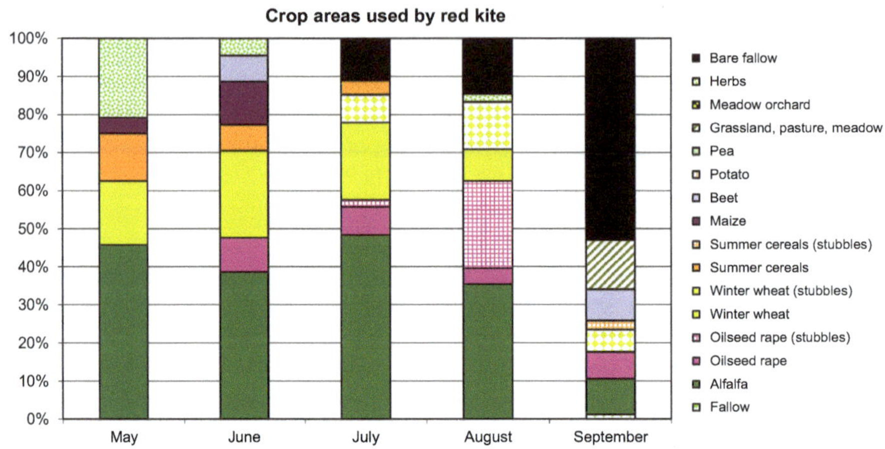

Fig. 3.34 Areas used by red kites (n = 244 observations) for 28 vantage points within the Speckberg study area between May and September 2010

in June and July the use corresponded to the availability and in August and September summer cereals were completely avoided. Harvested summer cereals were avoided in August, which may be due to the small size of the area (5.8 ha) overlapping with other suitable, freely accessible areas. In September harvested summer cereal areas were preferred. Winter cereal crops made up a proportion of over 60% in May and June, of 55% in July, 29% in August and 1.4% in September. These areas were used by red kites to a very limited extent in all months. Cereal stubbles were preferred in July, but were used proportionately in August and September. Oilseed rape was completely avoided in May and proportionately used in June and July. In August, areas with newly drilled rape seed were a good foraging ground and were therefore preferred as well as rape stubbles. Alfalfa was preferred between May and August and used proportionately in September. Between May and August, two small areas of a total of 4.7 ha were fallows, which were not mown and therefore unattractive. In September, a further (bare) fallow was available additionally, which was not used either.

Point counts were carried out in 26 sessions between 20.05.2010 and 30.09.2010 once per pentade. In 21 of the 26 sessions (81%) red kites were observed using alfalfa (Fig. 3.36). The 79 single observations were carried out in 45 areas. With 234 monitoring units of alfalfa fields (26 days, nine fields) this equals a frequency of presence of 19.2%.

More than five red kites per vantage point route (on 27.05., 07.06., 18.08. and 15.09.2010) observed over alfalfa were always associated with cutting, raking or harvesting alfalfa.

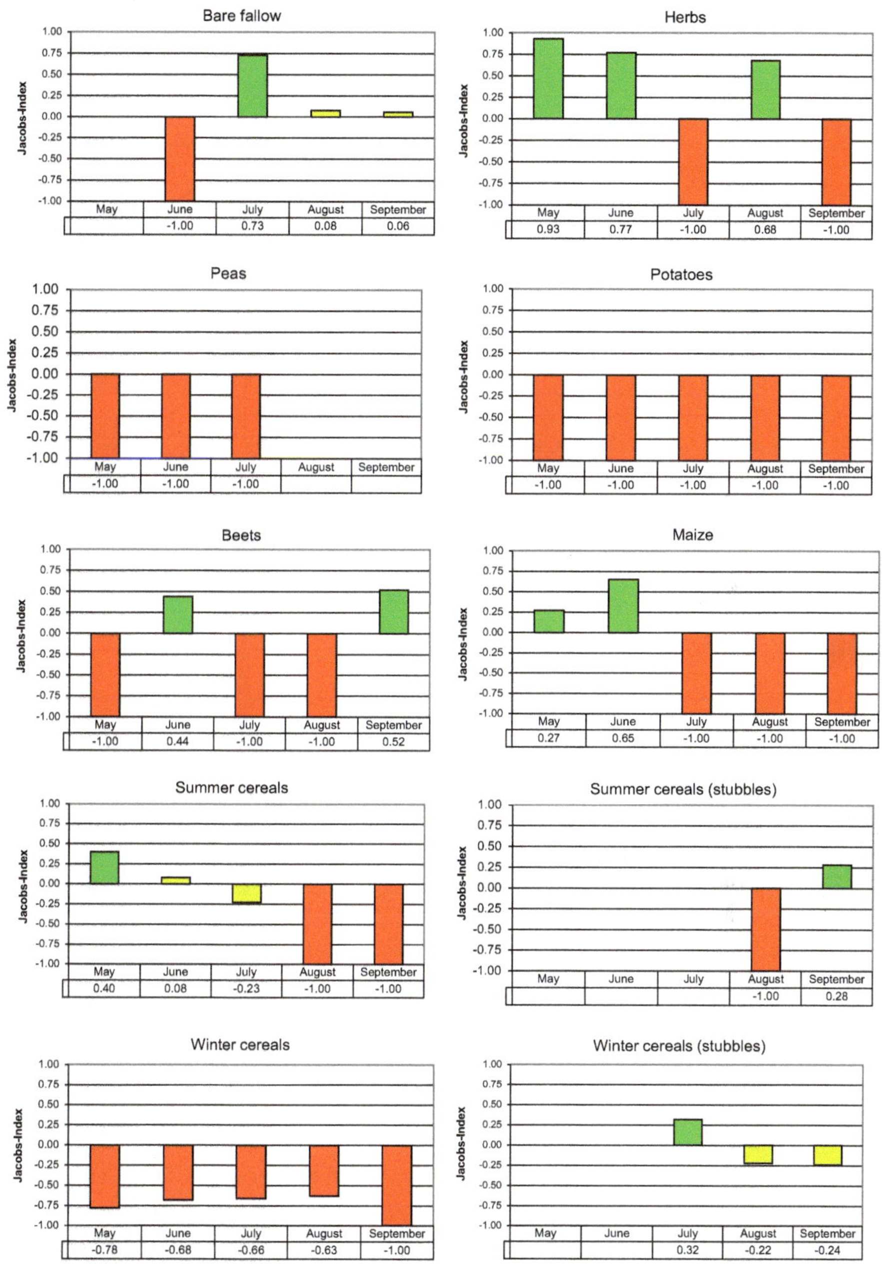

Fig. 3.35 Jacobs indices of the different crops used by red kites for 28 vantage points within the Speckberg study area in 2010; *green* preference; *red* avoidance; *yellow* deviation <0.25, i.e. use equals availability

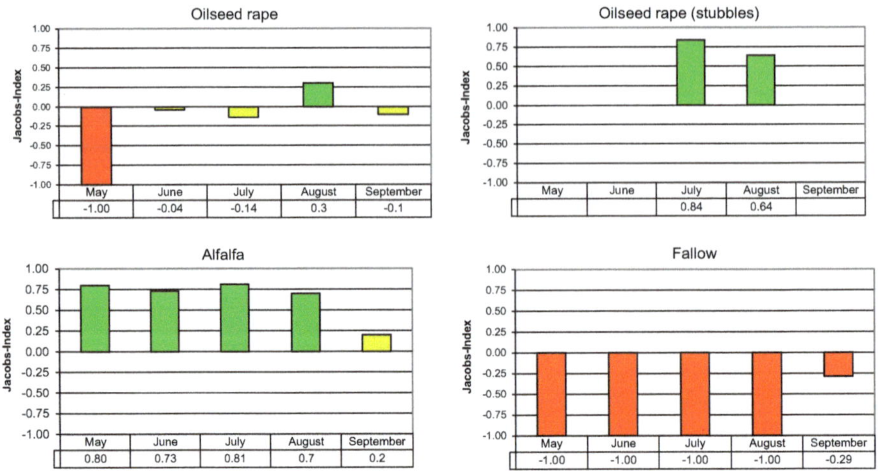

Fig. 3.35 (continued)

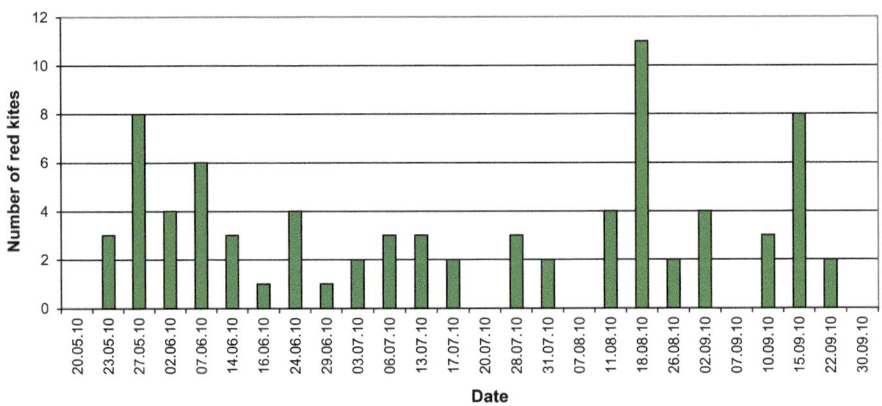

Fig. 3.36 Number of red kites recorded over alfalfa crops during point counts

Use of Selected Alfalfa Crops in 2010

The first cut of all investigated alfalfa crops was performed on 21.05.2010. Whereas the high and densely grown alfalfa was unattractive for red kite, this changed dramatically on the day of cutting. More than 900 s of red kite presence were recorded per hectare and 45 min observation unit (Fig. 3.37). Per crop and 45 min unit, a mean of at least nine different individuals was observed. Already on the day after cutting, the crops were not attractive any more.

The situation reoccurred on the day of the second cut. All five crops (in total 48.8 ha) were cut on 22.06.2010 (Fig. 3.38). No observations were performed on

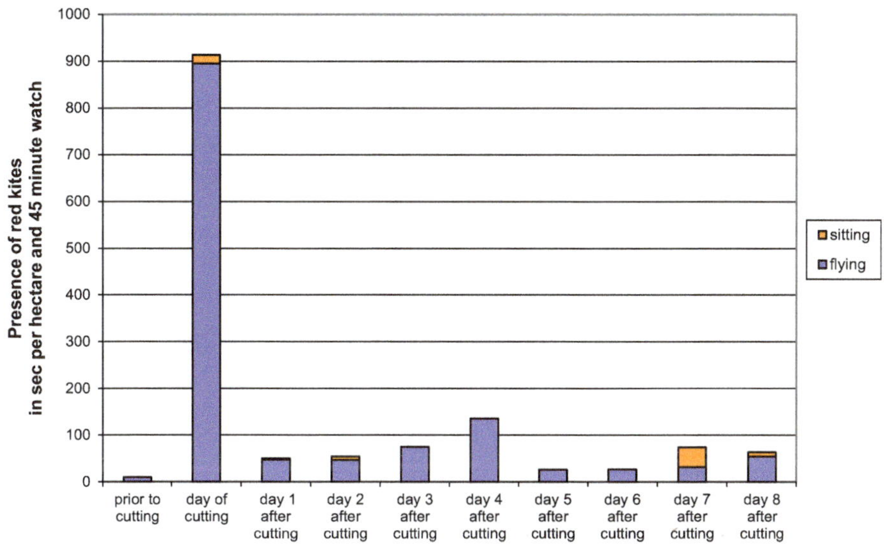

Fig. 3.37 Use of five alfalfa crops (48.8 ha all together) by red kites (in seconds per ha and 45 min watch) prior to cutting, on the day of cutting (21.05.2010) and after cutting

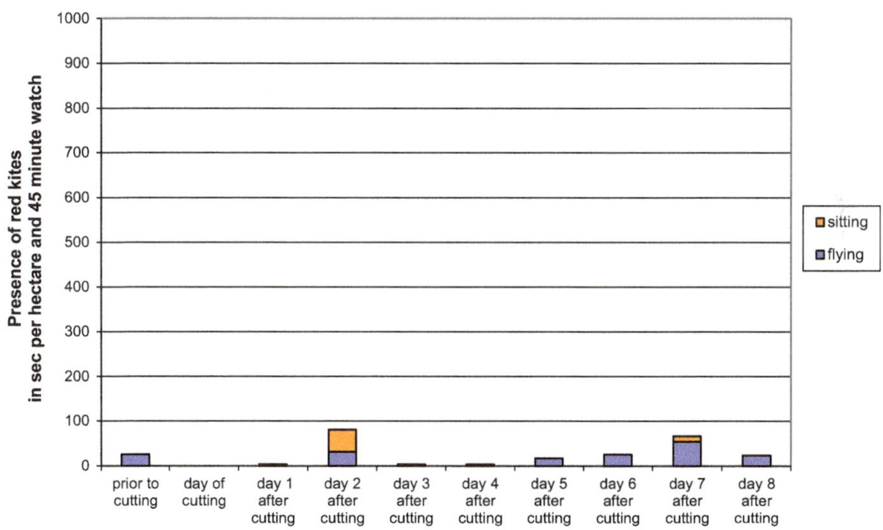

Fig. 3.38 Use of five alfalfa crops (48.8 ha all together) by red kites (in seconds per ha and 45 min watch) prior to and after cutting. No observations were performed on the day of cutting (22.06.2010)

that day, but use of the areas did not differ between prior to and after cutting. It can be assumed that several red kites were foraging over the cut alfalfa crops on the day of the second cut.

GPS locations recorded for male red kite Lui likewise indicated a preferred use of alfalfa crops. Seven locations were recorded directly over alfalfa crops between 04.06. and 18.07.2010. Six of the recorded locations can be associated to cutting and three locations were recorded on the day of cutting.

3.3.3.3 Use of the Composting Plant in 2010

The composting plant was watched in 48 observation units of 45 min each. In 37 of the observation units, the wind farm (the southern wind turbines of the wind farm) was watched simultaneously. In 15 observation units (=31% frequency of use) a total of 33 red kites were observed near the composting plant. Until 16.07.2010, the day large-scale harvesting of cereal fields started, 24 observation units were performed and red kites were observed in the area of the composting plant in nine units (frequency of use: 37.5%). From 17.07.2010 red kites were recorded in six (of 24) observation units, equalling a frequency of use of 25%.

44 targeted arrivals or departures noting the direction were recorded (Fig. 3.39). Most of the flight movement occurred in southwestern and northern direction. The wind farm is located north-east of the composting plant. Only once a bird was recorded moving in north-eastern direction traversing the wind farm directly. On 24.05.2010, a red kite was chased by crows near the composting plant and pursued through the wind farm.

21 red kites were recorded in observation units of the composting plant together with the wind farm. Seven of these birds used the wind farm as well as the composting plant, alternating or successively.

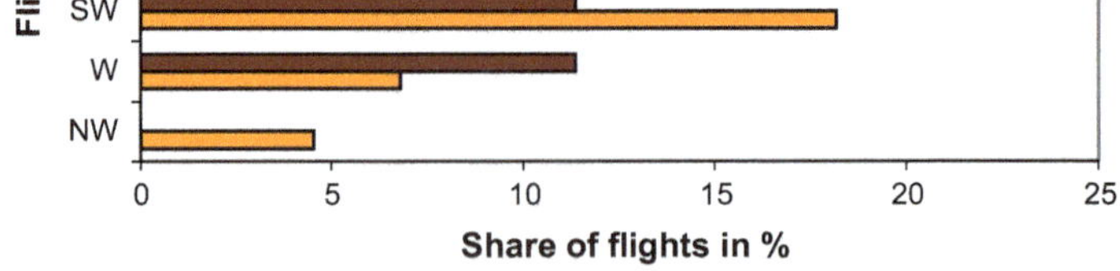

Fig. 3.39 Directions of arrival at or departure from the composting plant

3.3.4 *Use of Wind Farm Sites*

3.3.4.1 Behaviour of Red Kite Within and Outside Wind Farms

Figures 3.40 and 3.41 show the behaviours of red kites recorded in 2007 and 2008 within all investigated wind farms. "Traversing" and "thermal soaring" made up a very small proportion of the recorded behaviours. No differences could be determined between areas within and outside wind farms. The major proportion of behaviours were noted as "undefined presence"—the presence of red kites could be recorded only, due to the high density of birds of prey in the area with no further description of the behaviour—as well as "sitting" and "foraging". "Sitting" (34.2% within or 22.1% outside wind farms) was not relevant for the present issue, as no collision risk results from this behaviour. "Traversing" and "thermal soaring" were secondary due to the low proportion ("traversing" between 1.0 and 1.1%, "thermal soaring" between 3.6 and 3.8%). "Undefined presence" (40.5% within wind farms and 44.2% outside wind farms) included sitting and flying red kites, which could not be distinguished during observations. The distribution of the behaviours "sitting" and "flying" for both study years and all study areas shows that red kites were flying in 52% and sitting in 48% of the observations. As "undefined presence" was only noted for cases where the observer could not differentiate exact to the second between foraging and sitting for all red kite due to the large number of birds of prey being present, whereas traversing flights and thermal soaring could always be recorded exact to the second, it can be assumed that 52% of undefined presence were foraging flights. This assumption is confirmed by the analysis of single undefined behaviours. 97% of the undefined flight behaviour can be attributed to 11

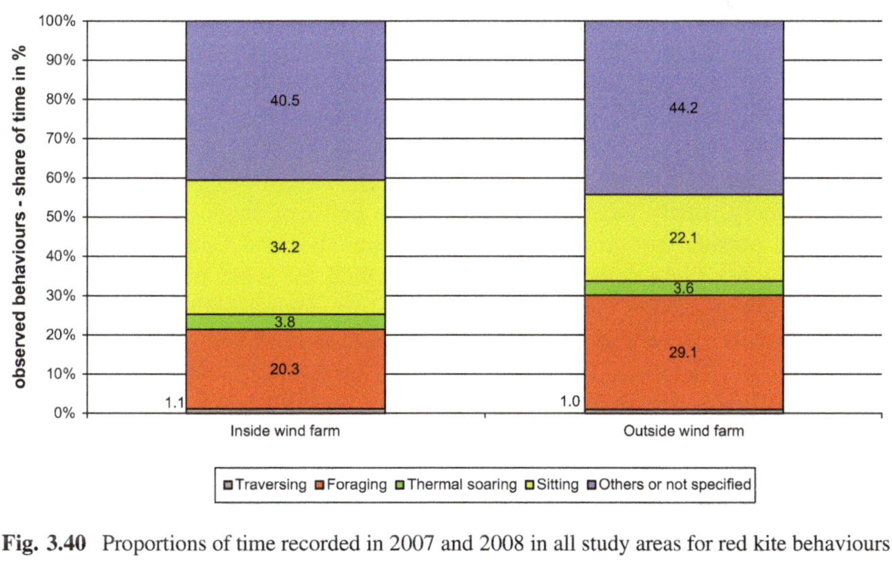

Fig. 3.40 Proportions of time recorded in 2007 and 2008 in all study areas for red kite behaviours

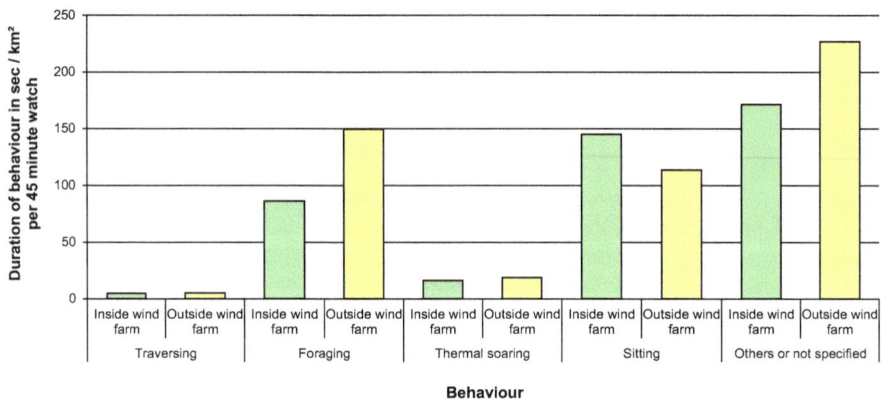

Fig. 3.41 Distribution of recorded behaviours (in s/km^2) of red kites in 2007 and 2008 in all study areas for red kite behaviours

45 min watches of single areas with alfalfa or sugar beet crops being cut or harvested. Due to the considerable activity of birds of prey during these observation units, 45 min of "undefined presence" were noted for each of the birds.

Foraging flights, which made up the largest proportion of flight movement, were of major importance in answering the question, whether wind farm areas were avoided or preferred in comparison to reference areas. Including the proportionate undefined presence at total of 41.4% of all behavioural observations within the wind farm and 52.1% outside wind farms were foraging flights. Comparing the three flight categories "traversing", "thermal soaring" and "foraging" (including the proportion of undefined presence), foraging flights were clearly dominating with 87.9% within or 91.5% outside wind farm areas.

Figure 3.42 shows the duration of foraging flights per km^2 and 45 min watch for the respective months within or outside wind farm sites for the years 2007 and 2008. Large differences in the duration of foraging were found between the study areas. The highest flight activity per km^2 was determined for the study areas Druiberg and Querfurt. Values calculated for these study areas were mainly between 30 and 200 s/km^2.

In Druiberg, partly larger differences occur between single months for areas within and outside the wind farm. In July, September and October 2007 a considerably higher flight activity was recorded within the wind farm. These values (between 137 and 240 s/km^2) can be attributed to single stubble fields or mown fallows (in July) as well as freshly cultivated bare fallows and newly sown crops (September and October). These were preferred by red kites due to the temporary good suitability as foraging grounds. Differences for 2008 can be explained in the same way. In June, July and August differences can be attributed to single areas (e.g. bare fallows, stubble fields and freshly cultivated areas), which were especially attractive foraging grounds for red kites during observation.

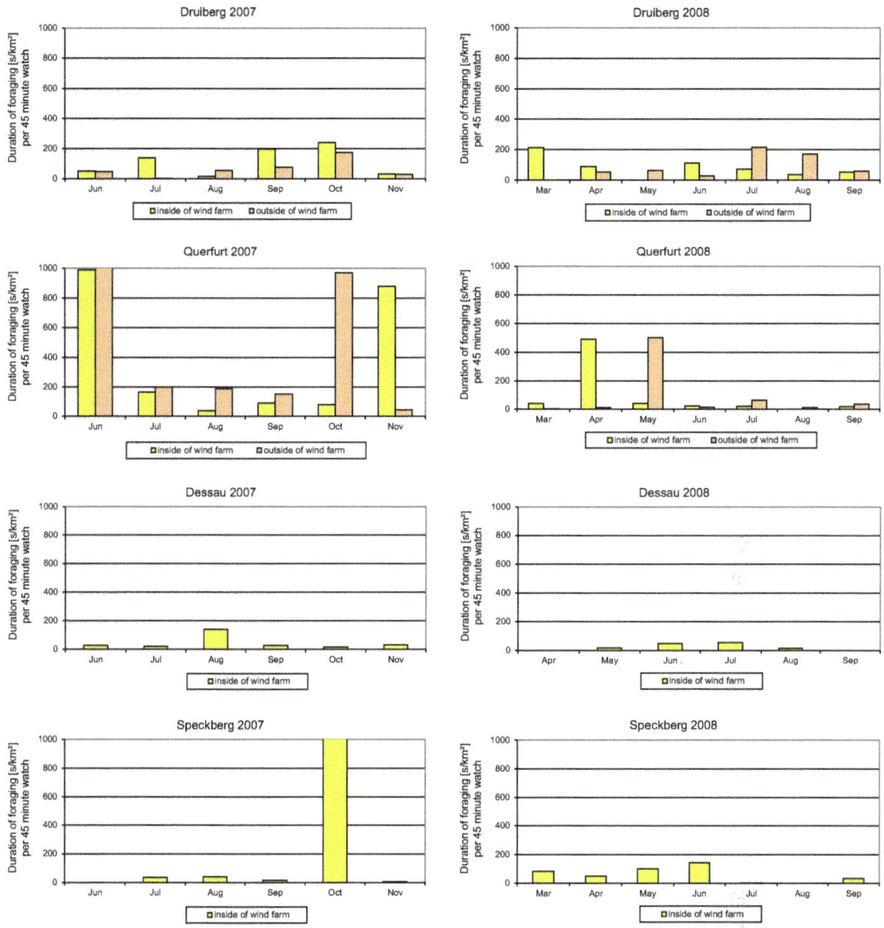

Fig. 3.42 Duration of foraging flights (s/km^2) within the respective study areas in 2007 and 2008

The differences found for March 2008 (more than 200 s within, 0 s outside) cannot be explained by more favourable foraging grounds, as the values are bases on foraging flights over a winter wheat area within the wind farm with comparable foraging areas available outside the wind farm at the same time (61% winter wheat within and 79% outside). These values may be attributed to the low sample size in March 2008 (seven observations within and two outside the wind farm).

In Querfurt, a considerably higher flight activity was found in areas outside the wind farm for almost all months except November in 2007. Especially in June with 2,821 s/km^2 and October with 952 s/km^2, the flight activity within the wind farm exceeded the flight activity outside the wind farm by far. In November a by 20 times higher flight activity was determined with 881 s/km^2 within the wind farm as compared to outside the wind farm (44 s/km^2). These examples underline the fact that the duration of foraging flights depends on the temporary availability of suitable feeding areas and the cultivated crops. The higher flight activity outside the wind farm in June can be attributed to only few 45 min watches of an established maize crop, an early winter wheat stubble field, a freshly cultivated bare fallow and a freshly cut alfalfa field. The likewise high flight activity within the wind farm (987 s/km^2) was due to an edge structure (agricultural road or crop boundary) as well as a winter wheat stubble field. The high flight activity in October (outside) and November (inside) was recorded over freshly harvested crops (sugar beet) and an agricultural road.

In April and May 2008, considerable differences in flight activity were found for areas within and outside the wind farm. In April this was due to a higher flight activity of three red kites over a fresh bare fallow. The high value found in May (502 s/km^2) can almost exclusively be attributed to three red kites foraging over a maize area clearly visible at that time. It is remarkable that a large proportion of the areas within the wind farm was likewise cultivated with maize (over 50%), which were not or only to a small proportion used for foraging (41 s/km^2). The high value found for the area outside the wind farm may be due to the lack of further suitable areas in the vicinity. In areas with many unsuitable foraging grounds, the flight activity of red kites was concentrated on the few suitable areas. In this particular case, 85% of the crops outside the wind farm were winter barley, winter oilseed rape and winter wheat in May 2008, which are no suitable foraging grounds for red kites from May onwards. The establishing maize covering 13% of the area was the only suitable foraging area in the vicinity available at that time. As a result, red kites breeding in the area used these crops disproportionately frequent, which is reflected by the high value.

In Speckberg a comparably low flight activity was recorded in 2007 in comparison to Druiberg and Querfurt. Winter wheat (20%), winter oilseed rape (17%) maize (22%) and beet crops (41%) may have been less attractive for red kites due to more suitable foraging areas in the close vicinity. The very high flight activity of 3,278 s/km^2 in October 2007 is outstanding. This was the highest flight activity found in 45 min watches in both years and can mainly be attributed to five 45 min watches of a freshly harvested sugar beet crop, with between eight and 12 red kites being present at the same time. In 2008, foraging flights in the Speckberg study area likewise occurred over ploughed areas (March and April), clearly visible maize crops (May and June) and stubble fields (September).

For the investigated areas near Dessau, the lowest values for flight activity were found in both years (between 0 and 142 s/km^2). Hardly any differences were found between months as well as between years. In this area, the higher flight activity found for August 2007 was again due to farming activities (freshly cultivated bare fallow and oilseed rape stubble field). The slightly higher flight activity in July 2008 was almost exclusively recorded over harvested oilseed rape or winter barley.

3.3.4.2 Frequency of Red Kites in Wind Farms

Figure 3.43 gives the number of red kites (number of approaches without duration per observations unit divided by observation area) recorded in area-related behavioural observations for all study areas and years. Larger differences in number of approaches were found for the four study areas. The large number of approaches found in 2007 for the Querfurt study area was due to an exceptionally high number of red kites recorded in June (Fig. 3.44) with vast availability of foraging areas. In June 2007, freshly cut alfalfa and not yet established maize areas as well as freshly cultivated bare fallows and mown feed wheat were available. The very high

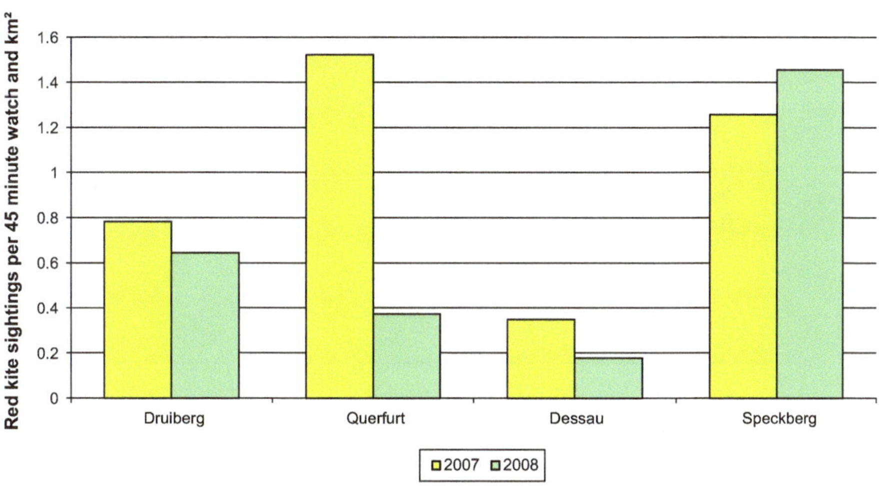

Fig. 3.43 Number of red kites approaching wind farms for the individual study areas in 2007 and 2008

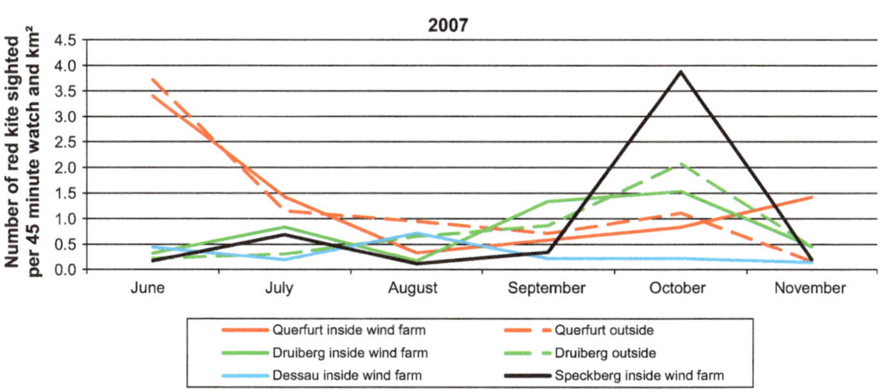

Fig. 3.44 Number of red kites approaching wind farms for the individual study areas per month in 2007

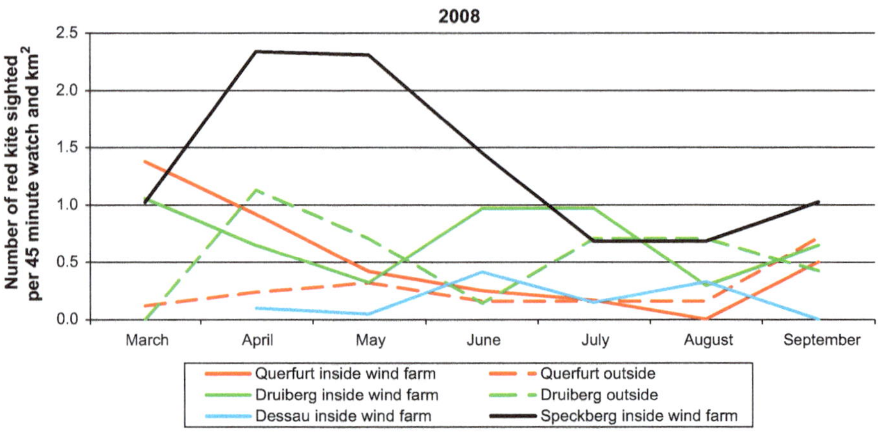

Fig. 3.45 Number of red kites approaching wind farms for the individual study areas per month in 2008

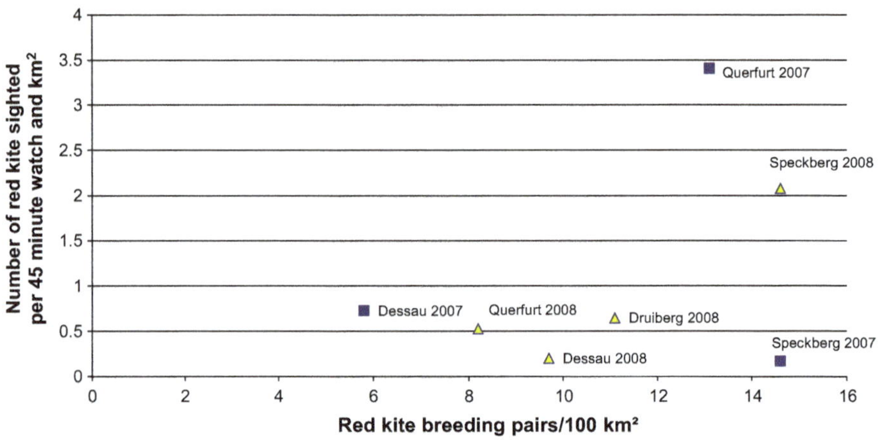

Fig. 3.46 Number of red kites approaching wind farms and breeding population within the individual study areas in 2007 and 2008

numbers recorded in the Speckberg study area in 2007 and 2008 can be attributed to October 2007 and April and May 2008, respectively (see Figs. 3.44 and 3.45). The high number in October 2007 was mainly recorded over a freshly harvested sugar beet crop. The high values in April and May 2008 were almost exclusively found for winter oilseed rape and winter wheat areas.

Apart from depending on the availability of suitable foraging grounds as described above, the number of red kites approaching wind farms showed a slight but statistically insignificant relation to the density of red kite breeding in the investigated areas (Spearman rank correlation test, rho = 0.34, p = 0.42; Fig. 3.46).

3.3.4.3 Flights into the Risk Area of Wind Turbines

Figure 3.47 shows that in 2007 up to 40% of red kite flights recorded in the Druiberg study area lead through the risk area of a wind turbine, and in 2008 it was an even higher share. In the Dessau study area this was true for about 24% of all red kite flights. The numbers recorded in the study areas Querfurt and Speckberg are with 11 and 1% respectively considerably lower. Review of red kite flights within the risk area per month (Fig. 3.47) even shows that in the Druiberg study area the risk area of wind turbines was entered in more than 50% of all flights recorded in June and July 2007. The excess value recorded in July in the Dessau study area can be attributed to the low sample size of only two recorded red kite flights.

3.3.5 Covering the Fallow Around Wind Turbine Towers with Agricultural Film

A total of 23 bird of prey approaches up to a distance of below 150 m from the six wind turbines could be observed (13 red kites, 10 common buzzards; see Table 3.20) equalling 0.56 bird of prey observations (0.31 red kites) per 45 min watch (n = 41). In 30 of the observation units no red kites was recorded. On two days (14.04. and 01.05.2008), red kites were present in the observation area, but did not fly closer than 150 m to the investigated wind turbines.

Three red kite were observed prior to covering the fallow, ten afterwards. Six of the ten approaches after the fallows were covered were aimed to covered wind turbines, four to wind turbines, which were not covered. In five of the six approaches of covered wind turbine fallows, the bird approached the wind turbine tower up to a distance of below 100 m. In all four approaches of turbines without cover the birds stayed at a distance of more than 100 m.

Near wind turbine 28 (gravelled forecourt), eight red kites could be observed within a radius of 150 m in 19 observation units, equalling 0.42 red kite observations per unit.

Figure 3.48 shows the approaches per observation unit and wind turbine before and after covering the turbine fallows. Reference turbines were less approached than covered turbines, prior to (with no difference between both types of turbines) as well as after covering the fallows. Turbines with covered as well as uncovered fallows were more frequently approached from mid-May than up to early May. No difference could be proven between turbines with covered and uncovered fallows for the time with no agricultural film or after (Kruskal-Wallis tests, $p > 0.05$ each).

For wind turbine 28, three to four times more approaches than to the six wind turbines included in the agricultural film test were recorded. Also for this turbine approaches were more frequent from mid-May than before.

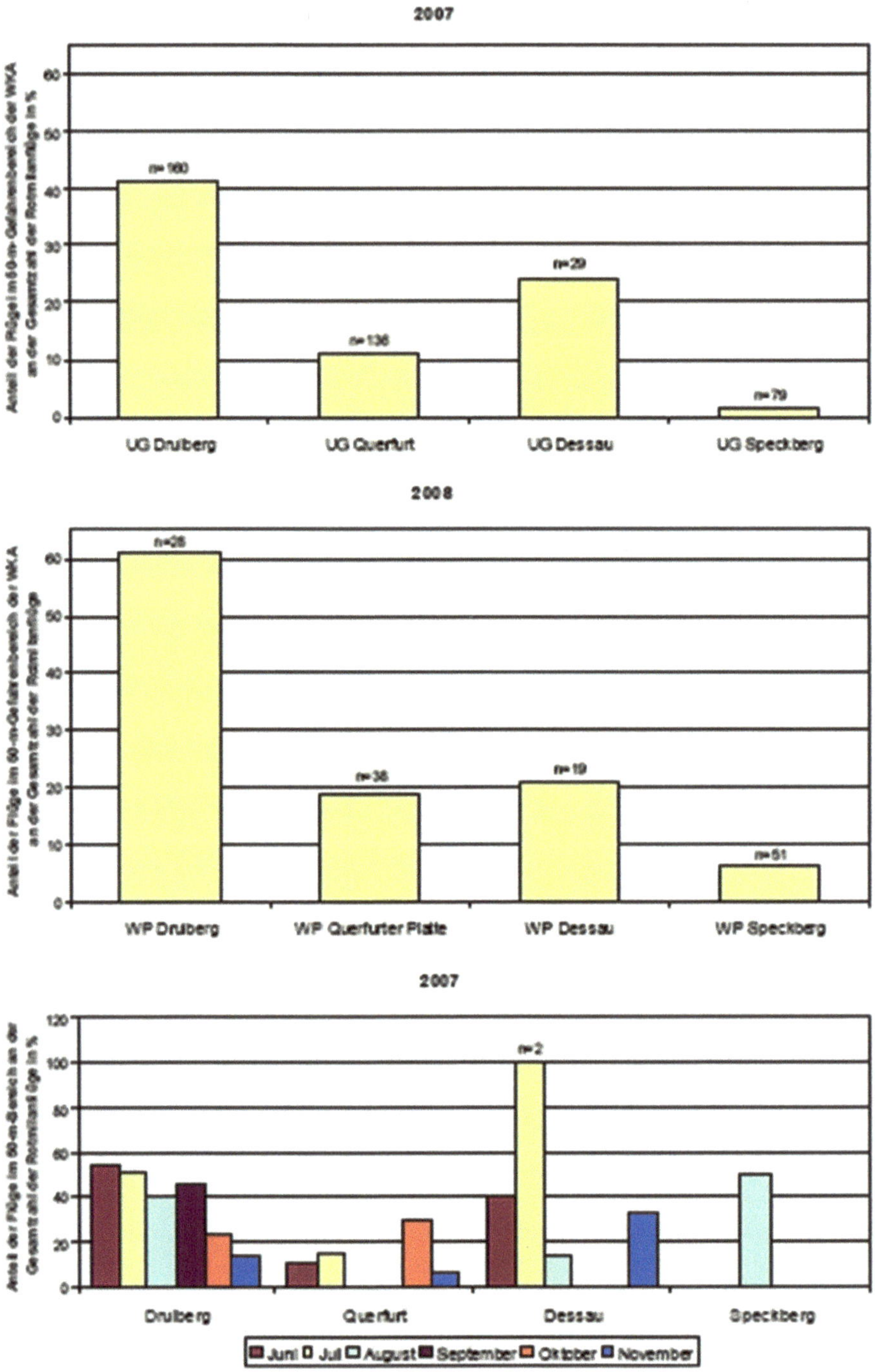

Fig. 3.47 Proportion of red kite flights leading into the risk area of wind turbines (50 m radius) of all flights recorded in 2007 (*above*), 2008 (in the middle) and of the flights recorded per month in 2007 (*below*) in all four study areas

Table 3.20 Birds of prey recorded within the three radii around the investigated wind turbines (observation areas 2 and 3)

Date	Wind turbine no.	Agricultural film	Species	Radius: 150 m	Radius: 100 m	Radius: 50 m
21.03.2008	16 (M)	No	Red kite	1	1	1
21.03.2008	2		Red kite	1	1	
21.03.2008	20 (M)	No	Red kite	1		
30.06.2008	6		Red kite	1		
10.07.2008	20 (M)	Yes	Red kite	1	1	
10.07.2008	20 (M)	Yes	Red kite	1	1	
10.07.2008	20 (M)	Yes	Red kite	1		
17.07.2008	2		Red kite	1		
29.07.2008	20 (M)	Yes	Red kite	1	1	
28.08.2008	5 (M)	Yes	Red kite	1	1	
02.09.2008	13		Red kite	1		
02.09.2008	13		Red kite	1		
18.09.2008	16 (M)	Yes	Red kite	1	1	
14.04.2008	16 (M)	No	Common buzzard	1		
14.04.2008	2		Common buzzard	1	1	
01.05.2008	16 (M)	No	Common buzzard	1		
01.05.2008	2		Common buzzard	1		
29.07.2008	5 (M)	Yes	Common buzzard	1	1	1
30.08.2008	5 (M)	Yes	Common buzzard	1	1	
02.09.2008	13	Yes	Common buzzard	1	1	1
02.09.2008	5 (M)	Yes	Common buzzard	1	1	
02.09.2008	2		Common buzzard	1		
18.09.2008	2		Common buzzard	1		

M "manipulated" turbine base

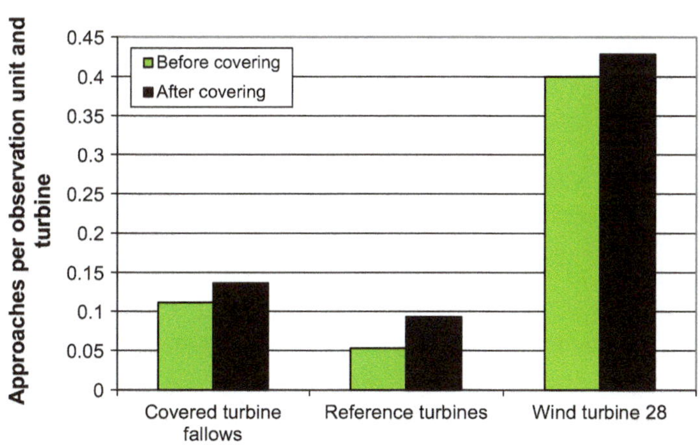

Fig. 3.48 Approaches of red kite per observation unit and wind turbine

3.4 Discussion

3.4.1 Breeding Population and Reproduction

The red kite densities determined for the study areas correspond to those prevailing in Saxony-Anhalt (George and Wadewitz 2001; Mammen and Stubbe 2009a). The study areas can therefore be seen as representative for the main area of distribution of this species.

Whether the considerable decline of the red kite population within the core area of the Querfurt study area temporally coinciding with the construction of the wind farm (Mammen et al. 2009) was also caused by the wind farm, remains unclear. Within this wind farm with 30 wind turbines in operation since 2001 three adult red kite collision victims were found in a systematic search in 2005. The actual number of collided birds was assumed to be five (Mammen et al. 2009).

3.4.2 Home Range Sizes

The mean home ranges of tagged red kites were very large throughout the year for both male and female birds. In some cases, the calculated 95% MCP home ranges substantially exceed the values determined in other study (see Table 3.21), especially the values found for Eastern Saxony (Nachtigall 2008). In all studies within the foothills of the Nordharz (Hagge et al. 2003; Nachtigall 1999), including the study areas Druiberg and Speckberg, home ranges were larger than in the smaller-structured landscape of Eastern Saxony. Likewise, the seasonal maximum MCP and kernel home ranges were considerably larger as well as female birds used smaller home ranges than male birds. No significant differences occurred despite the low variance of single values (Nachtigall 2008).

A male red kite monitored in the open landscapes around the Hakel (Speckberg study area) by Nachtigall (1999) for two subsequent years, used a largely corresponding home range in both years. Due to the high effort involved in land-based telemetry and the limited battery life span of transmitters, only few case studies on habitat use of single individuals covering several years exist so far.

Owing to large variations with at the same time rather small sample sizes, detailed statistical analysis would not have been useful. Given the available data however, location and size of home ranges used by breeding birds during the breeding season seem to strongly depend on the single individual. Reasons may be found in species-specific habitat use (maximum use of small areas or efficient use of widely distributed temporary resources) as well as age and experience of the particular bird. Red kites are excellent gliders so that searching larger areas will not result in energetic problems. Lack of territoriality outside the immediate surrounding area of the nest allows red kites to commonly use far off abundant feeding areas even during the breeding season. An important role plays the availability of

Table 3.21 So far determined home ranges sizes for red kites in Germany

Data source	Hagge et al. (2003)	Nachtigall (1999)	Nachtigall (2008) and Nachtigall and Trapp (2006)
Study area	Nordharz foothills, 12 ♂♀	Nordharz foothills, 2 ♂, 3 ♀	Eastern Saxony, 6 ♂, 2 ♀
Breeding season			
MCP	95% 1.9–62.1 km^2 (♂♀) Mean 12.1 km^2	95% 8.7–38.4 km^2 (♂) 5.5–91.6 km^2 (♀)	100% 3.3–11.1 km^2 (♂) 1.9–12.1 km^2 (♀)
Kernel		60% 0.7–7.2 km^2 (♂) 0.1–11.5 km^2 (♀)	
Post-breeding season			
MCP		95% 5.9–8.0 km^2 (♂) 7.1 km^2 (♀)	100% 2.0–19.0 km^2 (♂) 0.3–0.8 km^2 (♀)
Kernel		60% 1.3–2.0 km^2 (♂) 1.5–3.1 km^2 (♀)	
Breeding and post-breeding season			
MCP			100% 4.7–21.2 km^2 (♂) 2.9–12.5 km^2 (♀)
Kernel			95% 1.0–4.9 km^2 (♂) 4.6–5.6 km^2 (♀) 60% 0.4–1.2 km^2 (♂) 0.9 km^2 (♀)

food during the breeding season (density, distribution and accessibility of prey animals). From an energetic point of view, a smaller home range will be sufficient, if enough food is available. If prey is in short supply, unevenly distributed or not permanently accessible, the bird is forced to search a larger area to cover nutritional needs. It could be proven that feeding conditions in the intensively used agricultural landscape of Central Germany with its large proportion of densely growing crops especially during the breeding season are not ideal for red kites dependent on open landscape and short vegetation (Nachtigall 1999; George 2004; see also Sect. 3.3.3). A higher space requirement than in smaller and richer structured regions as for example Eastern Saxony is obvious and underlined by a review of so far available studies in the Nordharz foothills and present results of telemetric monitoring. Although the landscape of the study areas offered no ideal feeding grounds, all tagged red kites succeeded to fledge young making use of individual strategies of habitat use. Red kite home range sizes vary considerably. Dependence on annually alternating food resources (alternating prey densities, spatially varying suitable feeding areas depending on crop rotation) as well as individual factors complicate an estimate of potential habitat use during the breeding season based on external factors as landscape characteristics in the surrounding area of the nest (e.g. the contrasting habitat use of breeding pair Barbara/Barbarossa or of Barbarossa and Karl breeding in the same area in the same year).

3.4.3 Distance of Recorded Locations from the Nest

The red kites monitored for the present study quite often travelled far from the nest and the number of locations recorded in distant areas is relatively large. A distance similar to the maximum distance of >13 km from the nest determined in the present investigations between 2007 and 2010 has so far only been found by Nachtigall (1999) for a female red kite breeding in the Hakel forest, partly visiting foraging grounds or roosting sites at a distance of about 15 km. On the other hand, Nachtigall (2008) presents a frequency distribution of distances from the nest derived from all locations recorded for nine red kite with 60% (0–1.0 km), 20% (1.01–2.0 km), 15% (2.01–3.0 km) and 5% (>3.01 km), indicating a much lower proportion of distant locations, which were recorded only in single cases at distances of more than 5 km from the nest (Table 3.22). Hagge et al. (2003) found a frequency distribution of about 50% (0–1.0 km), 20% (1.01–3.0 km), 13% (3.01–5.0 km), 10% (5.01–7.00 km) and 2% (7.01–9.00 km) for the Nordharz foothills, which comes closer to the values determined for the study areas Druiberg, Speckberg, Querfurt and Queis.

Table 3.22 So far determined maximum distances from the nest for red kites

Data source	Hagge et al. (2003)	Nachtigall (1999)	Nachtigall (2008)
Study area	Nordharz foothills, 12 ♂♀	Nordharz foothills, 2 ♂, 3 ♀	Eastern Saxony, 6 ♂, 2 ♀
Maximum distance of recorded locations from the nest	7–9 km	♂: 5.6 km or 8.5 km ♀: 3.7 km or 15.5 km Open area breeder, forest breeder	♂: 4.5 km (BS), 8.1 km (PBS) ♀: 3.6 km

3.4.4 Continuous Monitoring

Implementing continuous tracking of highly mobile birds, which can only be equipped with transmitters of about 22 g of weight and a correspondingly limited range involves considerable effort (several observers, complete days). This may be a reason, why no such surveys on red kite were carried out so far. Nachtigall (2008) tried to continuously track tagged birds for about four hours. Up until now, estimates of distances covered by red kites during one day are based on connecting discrete recorded locations, clearly underestimating the actual covered distance of mobile individuals, which are not permanently traceable for a single observer, or of birds circling over a small area for a long time. Even in the present investigations, times with no possible record of the activity of the bird could not be completely avoided. Unfavourable spatial conditions complicated continuous tracking (uneven terrain, abandoned open pit mine, slag heaps, hills, lack of roads). Especially in longer periods with no contact, the bird will probably have shown an "unexpected" behaviour, as it would otherwise have been retrieved more quickly. The linear route assumed for times without contact to the tagged bird can therefore only be minimum distances. The actual distance covered during longer periods with no contact may be considerably longer than estimated this way.

The total of 20 days of continuous tracking carried out for six birds, provides a profound inside view of the daily, individual and seasonally varying habitat use of red kites. It becomes apparent that the distance covered per day may well exceed 100 km in the breeding as well as the post-breeding season and that red kites with smaller home ranges (e.g. Barbarossa) may cover the same total distance as birds with a larger home range (e.g. Karl).

3.4.5 Stay and Habitat Use of Red Kite Within Wind Farms

In foraging, red kites usually circle over the hunting ground for a longer period of time (Walz 2005). As they are unaware of natural enemies approaching in flight, they mainly focus on the ground below (Martin 2011). Due to this behaviour the collision risk for red kites is highest in foraging.

Records of flight heights show that red kites frequently fly at heights within the rotor swept area. Today's standard rotors with low hub heights sweep the air space up to low heights above ground. Higher wind turbines will potentially reduce the risk for red kites, if they have larger distances between the lowest blade tip and the ground. Wind turbines with a hub height of 98 m so far installed within the study areas had larger rotors as well so that the rotor swept area reaches further up in the preferred height band of the red kite (<150 m) but did not substantially shift the lower boundary upwards. All in all, wind turbines with low distances between lower blade tip and ground seem to be the most problematic.

Evaluation of red kite flight movements has shown that up to 50% of all red kite flights recorded per month lead into the risk area of wind turbines (50 m radius) in single study areas. A lower proportion of flights leading into the risk area were also recorded in individual-related behavioural observations, but were related to stays in the rotor swept area. Similar results were presented by Strasser (2006). 10% of 11,467 flight seconds were recorded within the rotor sphere (radius: 35 m). 21 flights crossing the rotor area were observed, and 12 red kites were watched getting into instable flight conditions or tumbling down, generally quickly leaving the wind farm thereafter. Thus, several studies quantitatively document that red kites show no avoidance behaviour towards wind turbines, but instead even use the close vicinity of turbines in flight.

To estimate the collision risk it is therefore important to know the decisive factors for the occurrence of foraging red kites within wind farms. Two main factors are in the focus: the density of the red kite breeding population in the vicinity and the availability of prey within the wind farm.

A certain relationship could be proven between the number of breeding pairs in the vicinity and the frequency of use of the wind farm (contact density). A higher frequency of use of wind farms is therefore likely in areas with a high density of breeding pairs.

Apart from the red kite breeding density in the vicinity, the availability of suitable foraging grounds was a major reason for red kites using wind farms. This was especially true between April and June with a general shortage in foraging grounds due to the dominating established winter cereal and winter oilseed rape crops. During that period, maize, alfalfa, grassland, fallows or early harvested cereal crops (mostly for energy use) as well as edge structures were particularly suitable areas. Apart from agricultural roads and hedgerows, edge structures included turbine base fallows which, with the mostly low-growing vegetation, were suitable foraging areas for red kite throughout the year. Especially in times of a shortage of available foraging areas, these fallows had an attracting effect on red kites. In spring and early summer, the shortage in foraging areas results in concentration of foraging red kites on the few available areas. At the time of harvest in high summer, all harvested or further processed areas were suitable for red kites and preferred over other areas. Several studies gave hints on the importance of food availability (Wuttky et al. 1982; Porstendörfer 1994; Hille 1995a, b; Nachtigall 1999; Mammen 2000; Weber and Stubbe 2000; Walz 2001; Albrecht and von Lindeiner 2002). The present results showed that seeming differences in the use of areas within or outside wind farms can almost entirely be explained by the cultivated crop or the stage of cultivation of the particular area.

Next to agricultural crops and their stage of cultivation, further potential food resources were attracting red kites. Especially on days of larger amounts of compost being re-stacked, which occurred often but irregularly (depending on new deliveries), composting plants were very attractive for kite. Use of composting plants was problematic, as many kites did not stay in or above the plant, but explored the surrounding area as well. The results for Speckberg study area showed that wind farms in the vicinity of composting plants were visited more frequently (Figs. 3.49, 3.50 and 3.51).

Additionally, red kites were interested in farmyard manure heaps, which were often stored temporarily on field edges in spring and summer before they were spread out after harvest. If these heaps are located near wind turbines, red kites will increasingly approach the risk area of the rotor. Lately, common practice was to use crane pads of wind turbines for storage. While searching for tagged bird Erik in the Querfurt study area, a large manure heap was found at the wind turbine, where the bird was located last. In the search for Erik, another red kite was found dead. It is therefore highly probable that two red kite collided with this wind turbine within a few weeks. Already in the year before, a manure heap was located in this place with several red kites circling over the heap observed several times.

To minimise risks for red kites, attractive crops (maize, alfalfa, grassland, fallows) should be avoided in the vicinity of wind farms. Agricultural works (ploughing, early harvest, cutting) should be avoided for the remaining crops until harvest of winter oilseed rape, winter wheat and winter barley in July. As edge structures like areas around wind turbine bases may attract red kites, thus increasing the collision risk, these areas should be kept as small as possible. Fallows around wind turbine bases should be developed into higher growing ruderal grass and herb vegetation to further reduce attractiveness for red kites.

Fig. 3.49 View of the composting plant from a wind turbine near Speckberg. The plant was partly covered (*left* in the picture) to keep kite away from animal substances (03.04.2010, U. Mammen)

Fig. 3.50 View from the composting plant on the wind farm Speckberg (03.04.2010, U. Mammen)

A sufficiently large distance should be kept between composting plants and wind turbines. For technical reasons, complete enclosure of the premises will not be possible.

No manure should be stored within wind farms.

3.4.6 Covering the Turbine Base Area

The results of the field test with covered turbine base fallow were inconclusive. The agricultural film test could only be carried out in the Druiberg study area with a considerably lower abundance of birds of prey than in other study areas. It is questionable, whether covering the turbine base fallows with agricultural film achieved to make these areas unattractive for kites. The opposite might as well be possible and kites may have been attracted due to unevenness or soiling of the film (Fig. 3.17).

Large-scale gravel layers were attracting kites as shown by the large number of observations made near wind turbine 28 (Enercon E112), which had a large gravelled forecourt (Fig. 3.19). Possibly, the actual abundance of prey is less important for search-hunters like the red kite than the availability of open areas,

Fig. 3.51 One of two wind turbines on the site of a composting plant near Hohendodeleben (Börde district, Saxony-Anhalt; 30.07.2008, U. Mammen)

where spotted prey can easily be taken. As red kites cover large distances per day, areas with no available prey are searched as well. For gliders, search of areas with only few prey animals will be energetically worthwhile.

The strong preference for edge structures (see above) including turbine base fallows of foraging red kites shows that these structures may attract red kites to wind farms. Even if the tests with agricultural film did not yield conclusive results, the task to design turbine base areas as unattractive as possible for hunting red kites will remain a major issue in the reduction of the collision risk.

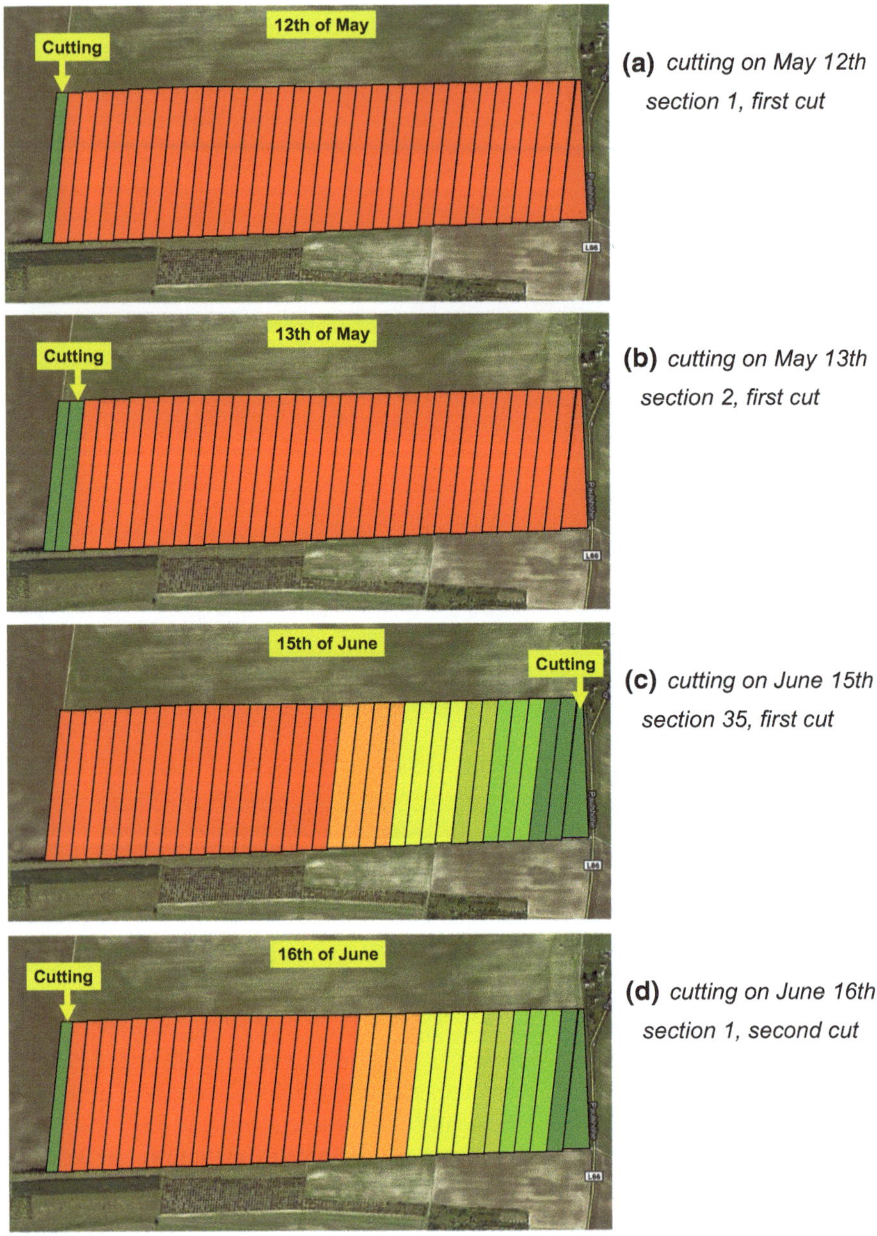

Fig. 3.52 Diagram of a 70 ha alfalfa crop cut section-wise, subdivided into 35 sections of 2 ha each

3.5 Conclusion

The investigations revealed that red kites show no or only limited avoidance behaviour towards wind turbines. They were frequently flying at height of the rotor within the risky height band. Apparently, the risk of collision was mainly determined by the frequency of stays in a wind farm or the number of times the wind farm was crossed in foraging.

How often red kites approached wind farms, significantly depended on the distance between nest and wind farm. This automatically includes a relation between the breeding density in the surrounding area of a wind farm and the occurrence of red kites within the wind farm. This relation could be shown in the present study but could not be further investigated due to the study design (focus on telemetric monitoring). It was possible though, to clearly show the influence of food availability within the wind farm on the occurrence of red kites. Food availability was determined by the kind of crop, attracting red kites each month to a different extent depending on the stage of growth, farming activities (strong preference of freshly mown or ploughed areas) as well as by additional sources of food as manure heaps within the wind farm or composting plants in the vicinity. Food availability in areas outside wind farms may as well have an influence on the number of foraging flights leading into the wind farm, as they may dislocate red kites from the farm depending on attractiveness.

In terms of a possible influence on the collision risk the question arises, which of the identified factors can be influenced by appropriate planning or suitable management. Due to the relative site fidelity, the distance between nest and wind farm may be influenced by the choice of the wind farm site. Likewise, the use of agricultural areas within and outside the wind farm can be changed. Measures to minimise the collision risk are listed in Sect. 3.6.

3.6 Conclusions for Practical Measures

Implications of the conclusions drawn from the results for the practice will not completely resolve the problem of wind turbines and red kites, but will considerably reduce the problem, if implemented consistently.

Distance between red kite nest and wind turbines

The larger the distance between wind turbine and red kite nest, the smaller the overlap between home range and wind farm area (see Sections "Distance of Recorded Locations from the Nest" to "Possibilities to Reduce Overlap of Home Ranges with Wind Farm Areas by Keeping Areas Free of Wind Turbines") and the lower the risk for red kites to collide with a wind turbine rotor. The distance should be at least 1000 m or better still 1250 m.

Avoidance of attractive areas within the wind farm

No field forage or other crops, mown during the breeding season of red kites (until mid-July), should be cultivated within a wind farm.

Harvest within the wind farm should start after harvesting other fields within the region. In Central Germany, this is applies to winter barley and oilseed rape, the first crops to be harvested.

Fallows around wind turbine bases should be as small as possible and designed as unattractive for red kites as possible. They should not be mown or ploughed before end of July.

Manure heaps, which are often stored temporarily on field edges in spring and summer before being spread out on the fields after harvest, are attracting red kites. These heaps should not be stored in the close-up area of wind turbines.

Wind farms should be constructed at a sufficiently large distance from composting plants, as these are visited by red kites on a regular basis due to continuous availability of food. The results for Speckberg study area show that wind farms in the vicinity of composting plants are visited more frequently.

Cultivation of alfalfa

Alfalfa crops are only attracting red kites on the day of cutting. To effectively guide red kites towards such areas (and thus keeping them away from wind farms) a sufficiently large area needs to be cut per day. It was not possible to exactly determine the required size of the area to be cut. If the area is too small, it will not attract red kite from further away. If the area is too large, costs will rise unnecessarily. Based on own assumptions we propose an area of 2 ha to be cut per day. Alfalfa can be cut every five weeks (35 days). This results in a total area of 70 ha (see diagram in Fig. 3.52a–d).

Cutting needs to be performed twice between early May and mid-July. From mid-July to early May of the following year, the area should not be cut to allow vertebrate animals (especially small mammals) to develop sufficiently.

Acknowledgements We would like to thank Nicole Heinrichs, Lukas Kratzsch, Thomas Müller, Leonid Rasran, Annett Schaar, Romy Schiele, Marcel Seyring, Ralf Siano, Eike Steinborn and Konny Suckow for the participation in the collection of data in the field and several stages of the evaluation.

Heinrich Bartelt enabled us to implement the agricultural film test within the wind farm Druiberg. Adrian Aebischer provided his knowledge of GPS satellite telemetry and Gerfried Klammer made the capture of Svenja possible. Thank you for that.

We are grateful to Bernd Nicolai, Egbert Günther, Michael Hellmann, Gunthard Dornbusch, Stefan Fischer and Michael Stubbe for our many inspiring discussions.

References

Aebischer A. (2009) Der Rotmilan – Ein faszinierender Greifvogel. Haupt Verlag, Bern

Albrecht H, von Lindeiner A (2002) Habitatnutzung und Brutbiologie des Rotmilans (*Milvus milvus* Linné, 1758). Projekt im Auftrag des Bezirks Mittelfranken, Abschlussbericht. Landesbund für Vogelschutz in Bayern, Hilpoltstein

Arndt O (2002) Die Entwicklung der Kulturlandschaft auf der Querfurt-Merseburger Platte seit dem Mittelalter. Naturschutz im Land Sachsen-Anhalt 39:3–20

Bergen F (2001) Untersuchungen zum Einfluss der Errichtung und des Betriebs von Windenergieanlagen auf Vögel im Binnenland. Dissertation, Ruhr University of Bochum

Bergen F (2002) Einfluss von Windenergieanlagen auf die Raum-Zeitnutzung von Greifvögeln. In: Oldenburg H (ed) Windenergie und Vögel - Ausmaß und Bewältigung eines Konfliktes; Technische Universität Berlin, p 86–96

Berny P, Gaillet J-R (2008) Acute poisoning of Red Kites (*Milvus milvus*) in France: data from the SAGIR network. J Wildlife Dis 44:417–426

Brauneis W (1999) Der Einfluß von Windkraftanlagen auf die Avifauna am Beispiel der „Solzer Höhe" bei Bebra-Solz im Landkreis Hersfeld-Rotenburg. Untersuchung im Auftrag des Bundes für Umwelt und Naturschutz (BUND) Landesverband Hessen, Ortsverband Alheim-Roteburg-Bebra, pp 91

Castelloe J (2000) Sample Size Computations and Power Analysis with the SAS System. SAS Institute Inc, SUGI Proceedings, paper 265–25

George K (2004) Veränderungen der ostdeutschen Agrarlandschaft und ihrer Vogelwelt insbesondere nach der Wiedervereinigung Deutschlands. Apus 12:1–138

George K, Wadewitz M (2001) Aus ornithologischen Tagebüchern - Bemerkenswerte Beobachtungen 2000 in Sachsen-Anhalt. Apus 11:1–36

Hagge N, Nachtigall W, Herrmann S, Stubbe M (2003) Habitatnutzung und Aktionsraumgrößen telemetrierter Rotmilane (*Milvus milvus*) und Schwarzmilane (*Milvus migrans*) im Nordharzvorland. J Ornithol 145: 44–45

Harris S, Cresswell WJ, Forde PG, Trewhella WJ, Woollard T, Wray S. (1990) Home-range analysis using radio-tracking data - a review of problems and techniques particularly as applied to the study of mammals. Mammal Rev 20:97–123

Hille S (1995a) Untersuchungen zur Ökologie des Rotmilans *Milvus milvus* (1758) in der Rhön. Staatsexamensarbeit, University of Gießen

Hille S (1995b): Nahrungswahl und Jagdstrategie des Rotmilans (*Milvus milvus*) im Biosphärenreservat Rhön/Hessen. Vogel und Umwelt 8, special issue:99–126

Hooge PN, Eichenlaub B, Salomon E (2000) Animal movement extension to ArcView: version 2.04 beta. Alaska Biological Science Center, U.S. Geological Survey, Anchorage

Hötker H, Thomsen K-M, Köster H (2004) Auswirkungen regenerativer Energiegewinnung auf die biologische Vielfalt am Beispiel der Vögel und der Fledermäuse – Fakten, Wissenslücken, Anforderungen an die Forschung, ornithologische Kriterien zum Ausbau von regenerativen Energiegewinnungsformen. Endbericht für das Bundesamt für Naturschutz, Förd. Nr. Z1.3-684 11-5/03. Michael-Otto-Institut im NABU, Bergenhusen, pp 80

Jacobs J (1974) Quantitative measurement of food selection. Oecologia 14:413–417

Kenward RE. (1992) Quantity versus quality: programmed collection and analysis of radio-tacking data. In: Priede IG, Swift SM (eds) Wildlife Telemetry. Remote Monitoring and Tracking of Animals, Ellis Horwood, London, p 231–246

Kenward RE (2001) A manual for wildlife radio tagging. Academic Press, London, pp 311

LAU [Landesamt für Umweltschutz Sachsen-Anhalt] (eds) (2000) Die Landschaftsschutzgebiete Sachsen-Anhalts. Landesamt für Umweltschutz Sachsen-Anhalt, Magdeburg, pp 494

Mammen U (2000) Bestandsabnahme beim Rotmilan *Milvus milvus* von 1994 bis 1997 in Deutschland. Ornithologische Mitteilungen 52:4–13

Mammen U (2009) Quo vadis *Milvus*? Falke 56:56

Mammen U, Stubbe M (2009a) Aktuelle Trends der Bestandsentwicklung der Greifvögel- und Eulenarten Deutschlands. Populationsökologie Greifvogel- und Eulenarten 6:9–25

Mammen U, Stubbe M (2009b) Jahresbericht 2003 und 2004 zum Monitoring Greifvögel und Eulen Europas. Jahresbericht Monitoring Greifvögel Eulen Europas 16/17:1–118

Mammen U, Mammen K, Strasser C, Resetaritz A (2009) Rotmilan und Windkraft – eine Fallstudie in der Querfurter Platte. Populationsökologie Greifvogel- und Eulenarten 6:223–231

Martin GR (2011) Understanding bird collisions with man-made objects: a sensory ecology approach. Ibis 153:239–254
Mebs T, Schmidt D (2006) Die Greifvögel Europas, Nordafrikas und Vorderasiens. Biologie, Kennzeichen und Bestände. Kosmos, Stuttgart, pp 495
Mumby PJ (2002) Statistical power of non-parametric tests: A quick guide for designing sampling strategies. Mar Pollut Bull 44:85–87
Nachtigall W (1999) Aktionsraum und Habitatnutzung des Rotmilans (*Milvus milvus* Linné, 1758) im nordöstlichen Harzvorland. Diploma thesis, University of Halle-Wittenberg
Nachtigall W (2008) Der Rotmilan (*Milvus milvus*, L. 1758) in Sachsen und Südbrandenburg – Untersuchungen zu Verbreitung und Ökologie. Dissertation, University of Halle
Nachtigall W, Stubbe M, Herrmann S (2003) Aktionsraum und Habitatnutzung des Rotmilans (*Milvus milvus*) im Winter - eine telemetrische Studie im Nordharzvorland. J Ornithol 144:284–294
Nachtigall W, Trapp H (2006) Aktionsraum und Habitatnutzung des Rotmilans in Ostsachsen. Poster at the 6. international symposium of population ecology of raptors and owls, Meisdorf
Nicolai B, Mammen U (2009) Dichtezentrum des Rotmilans *Milvus milvus* im Nordharzvorland – Bestandsentwicklung, Ursachen und Aussichten. Informationsdienst Naturschutz Niedersachsen 3/2009:144–150
Nicolai B, Günther E, Hellmann M (2009) Artenschutz beim Rotmilan - Zur aktuellen Situation in seinem Welt-Verbreitungszentrum Deutschland/Sachsen-Anhalt (Grundlagen, Probleme, Aussichten). Naturschutz und Landschaftsplanung 41:69–78
Porstendörfer D (1994) Aktionsraum und Habitatnutzung beim Rotmilan *Milvus milvus* in Süd-Niedersachsen. Vogelwelt 115:293–298
Porstendörfer D (1998) Untersuchungen zum Aktionsraum des Rotmilans (*Milvus milvus*) während der Jungenaufzucht. Vogelkundliche Berichte Niedersachsen 30:15–17
Regionale Planungsgemeinschaft Harz (2007) Umweltbericht zum Regionalen Entwicklungsplan für die Planungsregion Harz. Quedlinburg
Resetaritz A (2006) Ökologie überwinternder Rotmilane *Milvus milvus* (Linné, 1758) im Nordharzvorland. Jahresbericht Monitoring Greifvögel Eulen Europas 4, special issue:1–123
Resetaritz A, Stubbe M, Hagge N, Herrmann S (2006) Aktionsräume im Brutgebiet überwinternder Rotmilane (*Milvus milvus* L.). Populationsökologie Greifvogel- und Eulenarten 5:281–300
Strasser C (2006) Todfundmonitoring und Untersuchung des artspezifischen Verhaltens von Greifvögeln in einem bestehenden Windpark in Sachsen-Anhalt (2005). Diploma thesis, University of Trier
Stubbe M, Stubbe A (2007) Fragmentierung der Landschaft und andere anthropogene Einflüsse auf Wildtierpopulationen und Biodiversität. Beiträge zur Jagd- und Wildforschung 32:31–50
Stubbe M, Stubbe A, Weber M, Mammen U (2006) Farbmarkierte Rotmilane (*Milvus milvus*), Schwarzmilane (*Milvus migrans*) und Mäusebussarde (*Buteo buteo*). Populationsökologie Greifvogel- und Eulenarten 5:324
Stübing S (2001) Untersuchungen zum Einfluss von Windenergieanlagen auf Herbstdurchzügler und Brutvögel am Beispiel des Vogelsberges (Mittelhessen). Diploma thesis, Philipps-University of Marburg
Töpfer S (1995) Beziehungen zwischen Landschaftsstruktur und Vogelbeständen einer Agrarlandschaft im nördlichen Harzvorland. Diploma thesis, University of Halle
Traxler A, Wegleitner S, Jaklitsch H (2004) Vogelschlag, Meideverhalten & Habitatnutzung an bestehenden Windkraftanlagen Prellenkirchen – Obersdorf – Steinberg/Prinzendorf. Biome, Gerasdorf, pp 106
Villafuerte R, Viñuela J, Blanco JC (1998) Extensive predator persecution caused by population crash in a game species: The case of red kites and rabbits in Spain. Biol Conserv 84:181–188
Walz J (2001) Bestand, Ökologie des Nahrungserwerbs und Interaktionen von Rot- und Schwarzmilan 1996-1999 in verschiedenen Landschaften mit unterschiedlicher Milandichte: Obere Gäue, Baar und Bodensee. Ornithologische Jahreshefte Baden-Württemberg 17:1-212

Walz J (2005) Rot- und Schwarzmilan. Flexible Jäger mit Hang zur Geselligkeit. Aula-Verlag, Wiebelsheim, pp 150

Weber M, Stubbe M (2000) Nahrungsangebot und Nahrungswahl von Rotmilan (*Milvus milvus*) und Mäusebussard (*Buteo buteo*) im nordöstlichen Harzvorland nach 1990. Populationsökologie Greifvogel- und Eulenarten 4:203–222

Wuttky K, Stubbe M, Matthes H (1982) Greifvogelbesiedlung des Hakel und Überwinterung des Rotmilans *Milvus milvus* (L., 1758). Hercynia NF 19:121–134

Chapter 4
Telemetric Monitoring of Montagu's Harrier in Schleswig-Holstein

Bodo Grajetzky and Georg Nehls

4.1 Introduction

In addition to the white-tailed sea eagle (*Haliaeetus albicilla*) and the red kite (*Milvus milvus*) the Montagu's harrier (*Circus pygargus*) was investigated within the framework of the joint project "Birds of prey and wind power: Analysis of problems and solution proposals" funded by the German Federal Ministry for the Environment, Nature Conservation and Nuclear Safety.

After a collapse in population numbers in the 19th century and the 1950s the Montagu's harrier today is one of the rarest bird of prey species in Germany. With between 410 and 470 breeding pairs it is listed as endangered (2) in the German Red List (Südbeck et al. 2009). Similar to the rest of Central Europe the distribution is patchy and concentrated on single breeding locations (BirdLife International 2004). The core breeding areas are located in Mainfranken (Pürkhauer et al. 2009), the Hellweg Börde in Westphalia (Joest et al. 2013), in the Oldenburg area in Lower Saxony (Baum and Baum 2011) and on the west coast of Schleswig-Holstein (Hertz-Kleptow and Schmüser 2013).

With the disappearance of natural habitats (e.g. up- and lowland moors) the Montagu's harrier has executed a dramatic habitat shift into agricultural land and is increasingly nesting in cereal fields since the 1970s all across Europe. During the investigation period in 2007 and 2008, 80% of the breeding population in Schleswig-Holstein was nesting in cereal crops mainly winter barley and wheat fields (Hertz-Kleptow and Schmüser 2013).

In Schleswig-Holstein the main period of the habitat shift towards agricultural land coincided with the construction of wind farms in the 90s in nearly the same areas. Today, the Montagu's harrier's main distribution area is the area with the highest density of wind turbines in North Frisia. To date, no reliable data is

B. Grajetzky (✉) · G. Nehls
BioConsult SH, Schobüller Str. 36, 25813 Husum, Germany
e-mail: b.grajetzky@bioconsult-sh.de

H. Hötker et al. (eds.), *Birds of Prey and Wind Farms*,
DOI 10.1007/978-3-319-53402-2_4

available on behaviour and collision risk of Montagu's harriers in the context of wind turbines.

Unlike for the white-tailed sea eagle and the red kite, no collisions of Montagu's harriers have been registered in the central German database of collision victims until 2007 with the beginning of this project (Dürr 2007). Because of the small breeding population and the spatial overlap of breeding areas and wind farms, the species was nonetheless included in the "birds of prey and wind power" project funded by the BMUB.

To investigate the impacts of wind power on the breeding population of the Montagu's harrier, a total of 11 adult Montagu's harriers were fitted with VHF transmitters in two study areas in North Frisia in breeding seasons 2007 and 2008. Flight behaviour and habitat use were documented. The investigations pursued the following aims:

- Investigation of behaviour and habitat use of Montagu's harriers in the close-up area of wind farms
- Identification of impacts of wind farms on activity and habitat use (avoidance, collision risk)
- Derivation of measures to minimize the collision risk and impairments of the habitat.

4.2 Study Areas

The two study areas are located on the North Sea coast of Schleswig-Holstein in the district North Frisia (NF). The landscape is characterized by the salt marshes of the Wadden Sea intertidal zone and polders on the landward side of the dikes, which are mainly agriculturally used.

The salt marshes of the foreshore host high breeding densities of waders (common redshank *Tringa totanus*, Eurasian oystercatcher *Haematopus ostralegus*, northern lapwing *Vanellus vanellus*) and passerines (skylark *Alauda arvensis*, western yellow wagtail *Motacilla flava*, meadow pipit *Anthus pratensis*). These areas therefore are potential hunting habitats for the Montagu's harrier. The nest sites of the telemetrically monitored birds are located behind the dike on agricultural land within the polders.

4.2.1 Southern Study Area: Louisenkoog/Sönke-Nissen-Koog

The southern study area is located in the municipality Reußenköge, North Frisia (Fig. 4.1). The intertidal zone with the salt marsh on the seaward side of the dike is Natura 2000 area, SPA and holds the status of a nature reserve. In addition, the area is part of the Schleswig-Holstein Wadden Sea National Park.

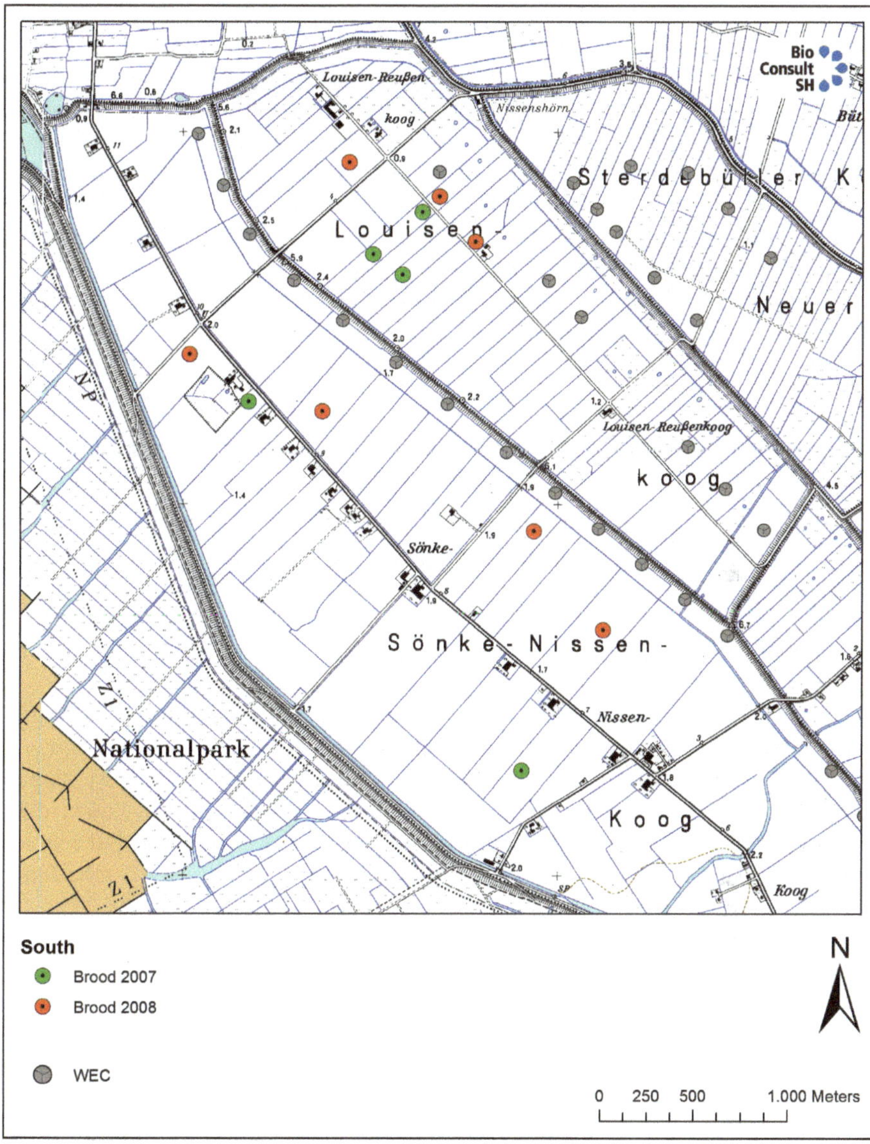

Fig. 4.1 Southern study area, Sönke-Nissen-Koog/Louisenkoog, including breeding sites of Montagu's harriers in the polder in 2007 and 2008 as well as the seaward salt marshes of the intertidal zone

Polders behind the first (Sönke-Nissen-Koog) and second dike line, the sleeper dike (Louisenkoog) are mainly intensively farmed. Winter cereals and oilseed rape clearly dominate with over 90% of the crops. These crops are the breeding habitat of the Montagu's harrier. In this area, the breeding sites of the Montagu's harrier can be found behind the first (in Sönke-Nissen-Koog) as well as behind the second dike line (Louisenkoog).

Wind farms in this study area with a total of 28 wind turbines are mostly linearly arranged along the dike lines. The near-coast row of wind turbines following the line of the second dike consisted of 13 either Vestas V 80 or Repower MM 83 turbines with a total height of 100 m and a rotor diameter of 80 m (Vestas) or 70 m (Repower). For birds breeding in the Louisenkoog the wind farm was located across the flight path between the breeding site and the salt marshes of the foreshore, a potential barrier to be traversed on a regular basis.

Towards the inland, a further 15 Vestas V 66 and Repower MM 70 wind turbines with identical dimensions were present.

4.2.2 Northern Study Area: Friedrich-Wilhelm-Lübke-Koog

The northern study area is located in the municipality Klanxbüll in the north-western part of North Frisia. The polder, bordering with the west side directly on the dike, is mainly agriculturally used. The intertidal zone on the seaward side of the dike is Natura 2000 area, SPA and holds the status of a nature reserve. From 150 m from the dike the intertidal zone is part of the Schleswig-Holstein Wadden Sea National Park.

The Friedrich-Wilhelm-Lübke-Koog is relatively sparsely populated. The polder is mainly intensively farmed. Winter cereals and oilseed rape clearly dominate with over 95% of the crops. These crops are the breeding habitat of the Montagu's harrier. Due to the relatively small fields and a relatively well developed ditch system as well as the adjoining grazed dike and the railway embankment, the area is comparably rich in structures for an agricultural area. Smaller groves can be found near the farmsteads and freshwater sluices and reservoirs alongside the dikes and sleeper dikes.

To the north the polder is bordered by the Hindenburgdamm, a railway embankment. By extensive grazing programmes, the Rickelsbüller Koog following to the north was developed into a wet meadow habitat for resting and breeding birds and is designated as a nature reserve (Fig. 4.2). The presence of this habitat within the home range of the Montagu's harriers breeding in the polder represents an important difference to the southern study area.

With the salt marshes of the foreshore and the Rickelsbüller Koog, two possible important hunting habitats are available to the Montagu's harrier.

Like in the southern study area, the wind farm at a distance of between 700 and 900 m from the breeding sites is linearly aligned along the second dike line stretching over a distance of about 5 km. In contrast to the southern study area, the wind turbines are located east of the breeding sites and therefore not within the flight path towards the feeding habitats in the intertidal zone or the Rückelsbüller Koog (Fig. 4.2). During the assessment period, a total of 19 Enercon E66 wind turbines with a total height of 100 m, a hub height of 65 m and a rotor diameter of

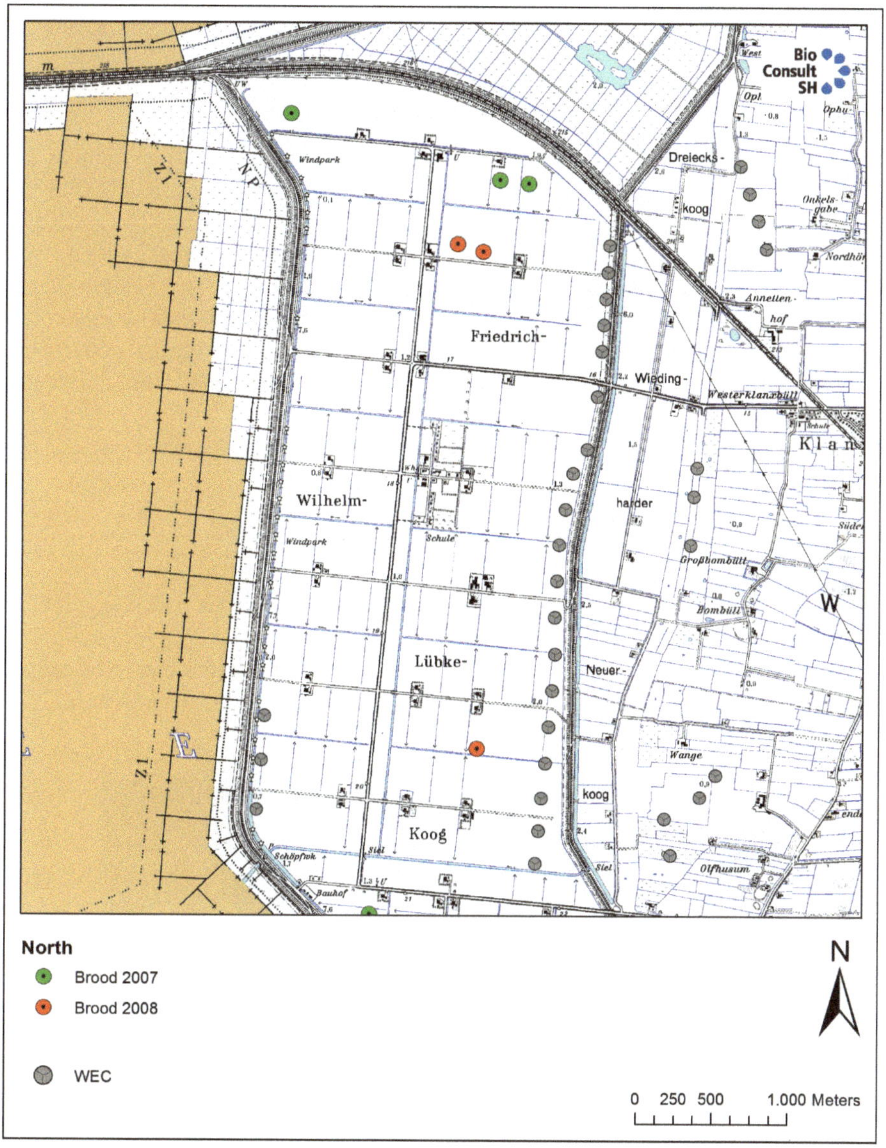

Fig. 4.2 Northern study area, Friedrich-Wilhelm-Lübke-Koog, including the breeding sites of 2007 and 2008 in the polder as well as the seaward salt marshes of the intertidal zone. In the north, the adjoining Rickelsbüller Koog, developed into a wetland habitat and designated a nature reserve

66 m were in operation. The three turbines near the sea dike in the southern part of the study area (see Fig. 4.2) were no longer in operation and were dismantled.

4.3 Materials and Methods

4.3.1 VHF-Tagged Birds

A total of 11 breeding birds were fitted with VHF transmitters, four female and seven male birds. In the breeding season 2007 six birds were tagged, in 2008 five birds. The two investigated areas on the North Sea coast (Sönke-Nissen-Koog/ Louisenkoog, Friedrich-Wilhelm-Lübke-Koog) are breeding areas which were occupied by several breeding pairs for many years (Schmüser and Gahrau 2011). Since 1995, the Schleswig-Holstein wildlife register (WTK, Flintbek) carries out a program to protect birds nesting amongst cereal crops by monitoring populations and specific nest protection measures throughout the federal state (Schmüser and Gahrau 2011).

4.3.2 Catch and Telemetric Monitoring

To catch the birds, prepared wooden posts were installed near the nest site, which in the cleared out agricultural landscape were accepted by the breeding birds after one to three days of familiarization (Fig. 4.3). In some cases the perch had to be moved several times before it was accepted.

Fig. 4.3 Prepared perch in the vicinity of a Montagu's harrier's nest site to catch the breeding birds. The stand is attached to the wooden pole by means of a spring, which is released by the weight of the landing bird triggering a loop closing around the bird's leg

The trap mechanism was triggered by the weight of the landing bird. By the pressure the bird's weight applied to the top section, a spring was released, which closed an attached loop around the bird's leg.

Two VHF transmitter systems by Biotrack (type TW 3 with two different batteries) weighing 6.8 or 8.9 g with a range between 2 and 5 km depending on habitat with a battery life span of 6–8 weeks were used (Fig. 4.4). The weight of the tag was 2–3% of the respective bird's body mass. Therefore, the general recommendation to not exceed 3% of the body mass of birds (e.g. Kenward 2001) was matched.

The chosen type of transmitter had a high signal selectivity allowing to very precisely localize birds hidden in the vegetation in the close-up range of about 100 m and to retrieve lost tags. A general statement on the signal accuracy or localization errors is not possible, as these depend on the distance between sender and receiver as well as the structure of the landscape (relief, presence of vertical structures, etc.). In this study, the signals were mainly used to retrieve hidden individuals and were only used in habitat use analysis in case of exact localisations. About 90% of the spatial localisations were made visually by an observer, who was usually located within a radius of 100 m around the bird in mobile tracking. The spatial inaccuracy depending on distance occurring in visual localisations was corrected or minimized in the analysis of the habitat use by choice of evaluation methods (see Sect. 4.3.5.2).

Depending on moulting stage or sex, the tags were either attached to the two central tail feathers or glued to the back (Fig. 4.5). The tags were fitted following the instructions of the Manual for Wildlife Radio Tagging (Kenward 2001).

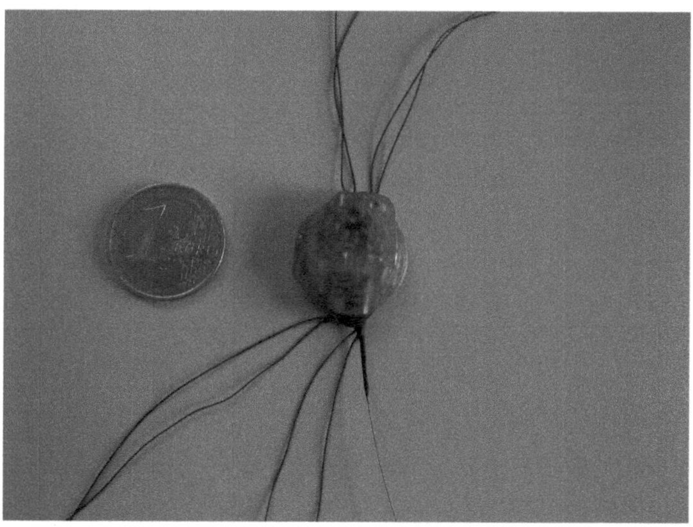

Fig. 4.4 VHF-transmitter used for Montagu's harriers (6.8 g version with attachment strings)

Recording of the telemetric data started one to two days after fitting the tags, to avoid or compensate for recording procedure-related changes in behaviour.

4.3.3 Data Acquisition in the Field

Figure 4.6 shows a field log of the telemetric sessions with recorded parameters and entries. The birds were followed visually whenever possible by bike or by car, if the

Fig. 4.5 Tags fitted to Montagu's harriers. *Left* glued to the back (female). *Right* attached to the base of the two central tail feathers (male, following Kenward 2001)

Telemetriebogen

Individuum: L3w 07 **Senderfrequenz:**150.4839 **Datum:**12.07.07

Erfassungszeitraum: 6:03 – 14:20 Uhr **Bearbeiter:** BG **Blatt** 1 **von** 5

Bemerkung:

Uhrzeit	Karten-punkt	Position (Beschreibung)	Genauigkeit Lokalisierung (m)	Verhalten	Flughöhe (m)	Habitat/ Feldfrucht
6:03	1	W-Feld Schlafdeich	20	na	4	w
04	2	Straße SN-Koog	5	na	3	s
06	3	Fliegt Rtg. Brutplatz, Beute	30	fb	15	-

Fig. 4.6 Telemetric field log for Montagu's harriers, including recorded parameters and sample entries. Codes or abbreviations were used for the parameters to be entered

signals occurred at greater distances. We chose 1 min intervals to record the behaviour especially in the close-up area of wind farms as detailed as possible. In these intervals (snapshots) flight height, behaviour and habitat were recorded and positions entered into a field map at the same time.

4.3.4 *Determination of Flight Heights*

The flight height was visually estimated and logged for each of the 1 min intervals. Structures of known height as dikes and the given dimensions of wind turbines (hub height, height of upper and lower rotor tip) were used as reference points at the site. Absolute heights were recorded to be more flexible in the choice of the size class for evaluation.

Height estimates based on visual observations may result in measurement errors. A general statement on the extent or variance of estimation errors is not possible, as no suitable measurement method is available for individually tracking Montagu's harriers to carry out comparative studies. Tests using a laser rangefinder for height detection in Montagu's harriers proved that focusing this rather small and mobile species in the field was hardly possible.

The extent of the estimation error depends on the absolute flight height and the distance of the observer from the bird. For most of the flights of the Montagu's harriers at a height of up to 20 m, the estimation error is probably not exceeding 5 m, as the numerous wind turbines in the area could be used as a reference point and the spatial distortion was kept rather low due to the short distance from the bird (<50 m). For flight heights >100 m height estimations involve a greater degree of uncertainty. In the representation and evaluation this was compensated for by using broad flight height classes. All flight heights above 100 m were treated in the same way in the evaluation of the collision risk. Flights at high altitudes outside the field of view (>200 up to approximately 500 m) represented a very small proportion of the recordings (<1%) with no relevant influence of the overall results despite a possible underrepresentation of this height class.

Errors by individual observers were minimized by working with small teams of a total of four experienced field ornithologists, who were familiar with the area as well as height measurements of wind turbines. 75% of the work was carried out by only two of them.

4.3.5 *Data Analysis*

4.3.5.1 Determination of Home Ranges

Digitisation and analysis of the positions and home ranges was carried out using ArcView GIS version 3.2. Several extensions were used (e.g. Home Range

Analysis, Animal Movement, Distance). Two different methods were used in the determination of the home range size.

Minimum convex polygon

The minimum convex polygon method (MCP, e.g. Mohr and Stumpf 1966, Schoener 1981) still is the simplest and most widely used method to measure home ranges (White and Garrott 1990; Kenward 2001). The outermost locations are connected to form a convex polygon with exterior angles of >180°. Due to the unweighted inclusion of all locations, the resulting home ranges include areas with only a few or no recorded locations so that the range tends to be overestimated. Low sample sizes, uneven intensity of use or single above-average flight distances (outliers) may lead to a distortion of the results. To be able to compare the determined home ranges to other results, the 95% MCP home ranges are used in addition to kernel home ranges (see below). No further analysis is performed using this method due to the limited informative value.

Kernel contours

The kernel method (kernel contours, e.g. Worton 1987, 1989) includes the intensity of use of different areas of the home range. A probability distribution is calculated, filtered and weighted by number and distance between the recorded locations. Areas, which are used less frequently or return only few locations are less probable and tend to be omitted. The size of the resulting home range usually remains considerably below the values calculated using the MCP method and identifies the main areas of activity. The resulting size and shape of the home range usually describes the actual behaviour of the animals more accurately.

The smoothing parameter to adjust the probability space around single locations was estimated using Least Squares Cross Validation (LSCV) to determine the parameter returning the smallest possible error. This method can be applied to not normally distributed data sets and is widely used in telemetric studies (e.g. Höntsch 2004; Trierweiler et al. 2010).

Several authors repeatedly investigated and assessed the suitability of the kernel method to represent home ranges. According to Kernohan et al. (1998) it is the most suitable of the established methods to realistically describe home ranges, as the method is robust towards low sample sizes, autocorrelation and treatment of outliers. Additionally, core areas of activity can be described (e.g. areas with a 50% probability of use).

Due to the above mentioned advantages as compared to the MCP method, the kernel contours method was used in the present study to determine the individual home ranges of the tagged Montagu's harriers. The 95% home ranges and the core areas of flight activity as 50% kernel were determined for the birds during hatching young. Following the recommendations in Kenward (2001), a data set of at least 10 days with a total of 100 recorded locations was defined as the necessary base of data.

Treatment of autocorrelated data

With short time intervals of only one minute between two recorded locations, successive data points are not independent of each other but highly autocorrelated. Many mathematical testing methods are based on independent single data sets and cannot handle autocorrelated data resulting in errors in the calculation of home ranges and habitat use (White and Garrott 1990; Worton 1989).

In addition, applying the kernel method for autocorrelated data usually leads to underestimation of home ranges. Mathematical independence of telemetric records is achieved by choosing sufficiently large time intervals between two recorded locations. The sampling interval between successive locations needed to attain independence of the observations, can be determined using a function developed by Swihart and Slade (1985, 1997). The TTI (time to independence) interval is the ratio of the mean squared distance between successive observations and the mean squared distance from the centre of activity.

In radio-telemetric studies, statistically independent data may only be achieved by using very large time intervals resulting in a loss of important biological information and therefore unrealistic estimates of the home range (McNay et al. 1994). In tests, the kernel method proved to be the least susceptible to autocorrelation of the commonly used methods (Kernohan et al. 1998). In more and more studies a certain degree of autocorrelation of the data is permitted intentionally to retain important biological information and to calculate more realistic home ranges (Millspaugh and Marzluff 2001).

In the present study, the effects of the chosen time lag on the size of the 95% kernel home ranges was determined by drawing random samples for different sampling intervals. Tests were made with intervals between 2 min. (every second recorded location) and up to 60 min (every 60th recorded location). Due to the correlation of sampling interval and home range size, the smallest possible time lag was determined resulting in a home range covering at least 90% of the maximum size. The threshold was reached as soon as a larger time lag would have resulted in a considerably smaller home range size caused by the reduction of data and the subsequent loss of relevant biological information (Fig. 4.7). In addition, shape and location of the resulting home range was inspected visually. The procedure was performed individually for each of the nine birds with sufficient data available. The home ranges described in Sect. 4.6.2 are based on time lags of between 25 and 40 min (Fig. 4.7).

4.3.5.2 Habitat Analysis

Habitat analysis (availability, use and selection) of the telemetric data was based on the Euclidian distance of the recorded locations by habitat type (Conner and Plowman 2001). Using this method, the spatial distributions of recorded locations are compared to random distributions based on relative distances. Distances from a type of habitat, which are smaller than expected, indicate an association with (a preference for) the habitat. Comparison and analysis of the distances and the resulting ranking of habitat use is performed using a pair-wise t-test or similar non-parametric tests (Mann-Whitney-U test).

In comparison to commonly used methods (e.g. compositional analysis, Neu et al. 1974) this method was relatively rarely applied so far but offers major advantages in the analysis of the present data (Conner et al. 2003).

Use of individual birds as sample unit

In choosing an individual as a sample unit, birds with a different amount of data available (intensity of observation) may be compared without considerable influence on the overall result of the analysis.

Independent analysis for each habitat type

Any number of habitat types can be included in the analysis as long as sufficient data is available. As each habitat type is analysed separately by calculating the Euclidian distances, choice and number of habitats have no influence on the individual results. In composition analysis however, the habitat types included in the analysis are interrelated by their relative ratio (the total number of habitats equals 1, unit-sum constraint). Errors in allocating a recorded location to a habitat will have an influence on the other habitat types. Using Euclidian distances, all habitat types are tested independently and linear habitat types used by the Montagu's harrier can be included in the analysis.

Robustness towards localisation errors

Especially in richly structured habitats with a large number of boundaries, telemetric errors (fuzziness) may lead to errors in the allocation of a recorded location to a habitat. Analysis methods where each recorded location is allocated to a habitat require partly complex correction of telemetric errors before using the data in habitat analysis. Error modelling is not necessary with Euclidian distances, as relative distance errors are minor in relation to the total number of actual distances and have no relevant influence on the overall result (Conner et al. 2003).

Implementation of the analysis

1. Simulation of locations of a uniform random distribution within the home range of each bird (ArcView 3.2, extension "Movement").
2. Calculation of the distance from each random point to each habitat type and calculation of the average distance. The resulting vector (random = ri) represents the expected distances from the habitats for each individual (ArcView 3.2, extensions "Distance" and "Spider Distance Analysis").
3. Repetition of steps 1–3 for the recorded locations. The resulting vector (utilisation = ui) represents the average actual distance of each individual from the habitats.

Creation of a vector of ratios (di) of the expected and actual distances per bird and habitat (ui/ri). If habitat use occurs at random $\sum$di = 1.0. In case of disproportionate use (selection) the distances are smaller than expected, $\sum di < 1.0$. The mean vector ρ is calculated as mean of the di values.

4. Each element of $\sum di(\rho)$ is t-tested for significant differences. The difference between ρ and 1.0 is a measure for disproportionate use (selection, preference).

4.3.5.3 Assessment Areas

To analyse small-scale effects in the spatial distribution of recorded locations, different radii were used in GIS analysis. Depending on problem and scale, ranges of verification of 100 or 300 m around the wind turbine sites were used. The range of verification is given in the respective section. In the analysis of the habitat use, a radius of 100 m around the turbine tower was defined as close-up range (Sect. 4.6.2.3). In the analysis of the spatial distribution of foraging birds in the vicinity of wind turbines a verification range of 300 m was used (Sect. 4.7.2). These verification ranges need to be distinguished from distances from nest site with respect to avoidance behaviour and collision risk mentioned in this report.

4.4 Breeding Biology—Results

4.4.1 Tagged Breeding Birds

The tagged birds are part of the breeding population of Schleswig-Holstein with a mean of 50 breeding pairs. Between 1995 and 2011, annual fluctuations between 31 and 61 territories occurred, which is typical for this species. Accordingly, Schleswig-Holstein hosts between 12 and 15% of the total German breeding population (Fig. 4.7).

A total of 11 breeding birds were fitted with VHF transmitters, four female and seven male birds. In breeding season 2007 six birds were tagged, in 2008 five birds. The periods of data transmission varied between a few days and up to seven weeks (Fig. 4.8).

In the second year some of the birds could already be tagged during egg-laying, making data available over the entire breeding season. In two cases the tag was lost at an early stage. No calculation of the home range was possible for these two birds.

Table 4.1 lists the date of capture, study area and biometric data of the tagged birds.

For two of the tagged birds, a paired male and female originating from different broods in the Netherlands, the place of birth could be traced back because they were ringed as young birds in the nest, 228 and 218 km from their common breeding site in the northern study area (Friedrich-Wilhelm-Lübke-Koog).

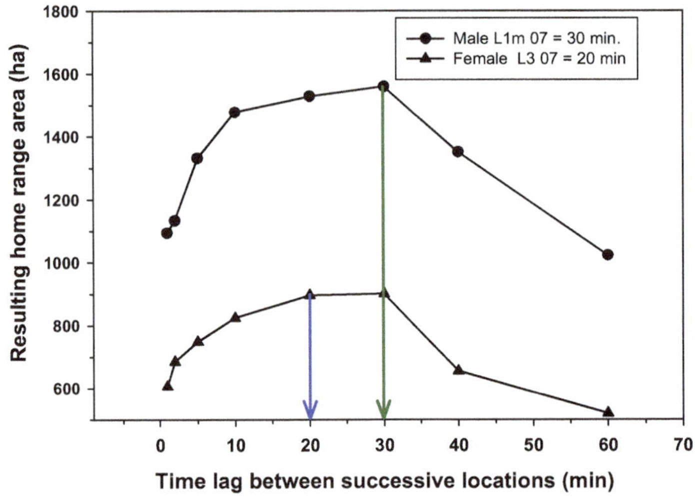

Fig. 4.7 Determination of time lags between successive recorded locations to reduce autocorrelated data in the calculation of the home range (95% kernel) of two birds

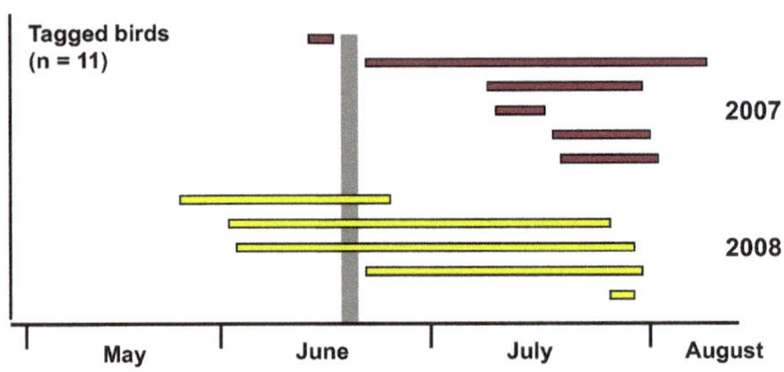

Fig. 4.8 Transmission times of the tagged birds in breeding season 2007 (*red*) and 2008 (*yellow*). The *grey bar* indicates the average (±2 days) time of hatching

Males with a body mass of between 258 and 299 g were considerably lighter than females with a weight between 295 and 362 g (Table 4.1).

Comparative data on the body mass of Montagu's harriers is available from recent published telemetric monitoring studies in the Netherlands (Trierweiler et al. 2010). The 12 male birds monitored in the provinces of Flevoland and Groningen weighed between 250 and 333 g in the nestling period, 9 birds weighed under 300 g (Trierweiler et al. 2010). With the exception of the two males of a weight of considerably below 300 g, the values fell within the range of the present study. Mebs and Schmidt (2006) indicated a range of between 319 and 445 g for female birds with no information on stage or season for the monitored birds. Further comparable data for the breeding season were not published so far.

The present data confirm the considerably higher body mass of females in comparison to male birds. The sample size is rather small, but the trend is obvious for both of the breeding pairs tagged for the present study. The two male weighed 258 and 278 g and the corresponding female birds 352 and 344 g.

Table 4.1 Date of capture, study areas, gender, age, body mass and transmission time of the VHF-tagged birds

Name	Date of tagging	Study area	Gender	Age	Body mass (g)	Transmission time (d)
FWL1m 07	07.06.13	Northern study area	m	ad (≥y 3)	292	2
FWL2m 07	07.07.19	Northern study area	m	ad (y 12)	278	17
FWL1m 08	08.06.03	Northern study area	m	ad (≥y 3)	280	42
FWL2m 08	08.06.02	Northern study area	m	ad (≥y 3)	299	51
FWL3w 08	08.07.22	Northern study area	f	ad (>y 4)	362	4
L1m 07	07.06.19	Southern study area	m	ad (≥y 3)	258	52
L1m 08	08.06.25	Southern study area	m	ad (≥y 3)	260	44
L3m 08	08.05.22	Southern study area	m	ad (≥y 3)	n/a	29
FWL2w 07	07.07.18	Southern study area	f	ad (y 4)	344	23
L1w 07	07.07.13	Southern study area	f	ad (y 3)	352	12
L3w 07	07.07.03	Southern study area	f	ad (y 3)	295	19

Age: y = year n/a = no data available, faulty scale

No conclusions can be drawn on the fitness of the birds or the quality of the breeding habitat due to the small sample size, high variability and the multitude of relevant parameters.

4.4.2 *Breeding and Reproduction*

During telemetric monitoring, a total of 14 broods were tracked within both study areas (five broods in 2007, nine in 2008).

In 2008, only broods were already inspected during incubation and clutch sizes of three and four eggs were found. In two broods single eggs did not hatch. In several broods a reduction in the number of young by death of single nestlings was observed during the breeding season, which may be attributed to a shortage of food. This was particularly significant for two of the broods in the southern study area, where the number of nestlings went successively down from four and three young to only one after the death of the associated polygynous male. The carcasses of the nestlings found at several nest sites served as food for the remaining young or as reserve prey for the adult birds.

The 14 investigated broods (Table 4.2) yielded a numerical reproduction rate of 1.5 fledged young per breeding pair. This value is comparably low, but within the annual fluctuations of the breeding population in Schleswig-Holstein (Gahrau and Schmüser 2012).

Table 4.2 Data on breeding biology for a total of 14 breeding pairs in 2007 and 2008 in both study areas

Name		Transmitter	Breeding season	Nesting habitat	Clutch size	Nestlings	Fledged young
FWL1m 07	m	x	2007	Winter wheat		≥2	2
FWL2m 07	m	x	2007	Winter wheat		≥2	2
FWL2w 07	f	x	2007				
FWL1m 08	m	x	2008	Winter wheat	3	3	2
FWL2m 08	m	x	2008	Winter wheat	4	3	3
FWL3w 08	f	x	2008	Winter barley	3	2	2
L1m 07	m	x	2007	Winter barley		3	2
L1w 07	f	x	2007				
L2m 07	m	–	2007	Winter barley		2	1
L3w 07	f	x	2007	Winter barley		3	3
L1m 08	m	x	2008	Winter barley	4	3	2
L2m 08	m	–	2008	Winter barley	–	–	–
L3m 08	m	x	2008	Winter barley	–	–	–
L3m 08	m	x	2008	Winter wheat	3	3	1
L3m 08	m	x	2008	Winter wheat	4	4	1
L4m 08	m	–	2008	Oilseed rape	–	–	–
Total						30	21

Values in brackets: minimum number, earlier reductions cannot be excluded

4.5 Breeding Biology—Discussion

The breeding biology results for the investigated breeding population and the individual birds widely correspond to known data for Schleswig-Holstein. Due to the special geographical situation in Schleswig-Holstein, the following differences occur in comparison to the southern range of distribution:

In the northernmost German area of the range, the breeding phenology is delayed by one week in comparison to areas in central or southern Germany (see Mebs and Schmidt 2006). Territories are established mainly early to mid-May and eggs are laid end of May to early June. The young usually fledge end of July.

The clutch size is similar to other areas and varies between 3 and 5 eggs. Inspecting the 14 nests of the tagged breeding birds, we found clutch sizes of 3 and 4 eggs. The calculated reproduction success for these broods resulted in 1.5 fledged young per breeding pair, which is within the range of variation for Schleswig-Holstein (Gahrau and Schmüser 2012). The investigated broods represent selective samples taken from two spatially separate breeding areas. The reproduction success of the population of the federal state has an annual average of 2.0 fledged young per breeding pair with considerable annual fluctuations between 1.4 and 2.6. On average, the threshold of 1.8 young required to maintain the population is slightly exceeded (Mebs and Schmidt 2006). Fluctuations in breeding success are in the majority of cases the result of varying clutch losses due to agricultural activity or predators (Schmüser and Gahrau 2011). Seasonal food availability plays an equally important role caused by population changes of small mammals (e.g. Looft and Busche 1980) or other changes over the breeding season. In both years of the study we found broods with a number of young decreasing over the nestling period, indicating a seasonal shortage in food availability.

Reproduction rates are similar to those in other areas within the distribution range, but remain below the values of the Mainfranken population which has since 1994 risen from 4 to now 160 breeding pairs as a result of effective protection measures (Pürkhauer et al. 2009, pers. comm.) with a total breeding success of 2.2–2.3 young per breeding pair, the highest values in entire Germany (Belting and Krüger 2002).

4.6 Results—Displacement and Barrier Effects of Wind Turbines

4.6.1 Spatial Relation Between Breeding Sites and Wind Turbines

In breeding seasons 2007 and 2008, a total of 24 nest sites and their spatial relation to wind turbines were investigated within the study areas. Distances between nests and wind turbines ranged between 76 and 990 m (Fig. 4.9). Six of the 24 nests were located at distances of up to 300 m, a further eleven within a radius of 500 m. The distribution of nests around wind turbines gives no indication of minimum distances or avoidance of turbines. No general statistical analysis is possible though, as the study design only selectively includes certain parts of the study areas rather describing the local situation in the investigated breeding areas. Still, the distribution of breeding sites in both years of the study shows that broods in the close-up area of wind turbines are relatively rare but occur on a regular basis.

According to present results, site selection mainly depends on the presence of farmland with suitable structures. The preference of breeding birds for winter barley was especially apparent in breeding season 2007 (Fig. 4.10). In that year, six of the eight pairs in the southern study area bred in winter barley, which had a proportion of only 4.55% of the total arable crop, indicating a distinct preference for this crop.

In 2008, the second year of the study, only two of the eight pairs bred in winter barley with almost the same proportion of crops (4.8%). Similar to broods on the west coast of Denmark and other parts of Germany (Mainfranken), five of the nests were built in winter wheat in that year, which was probably due to a differently structured growth (delayed growth, coverage) at the time of selecting a nest site.

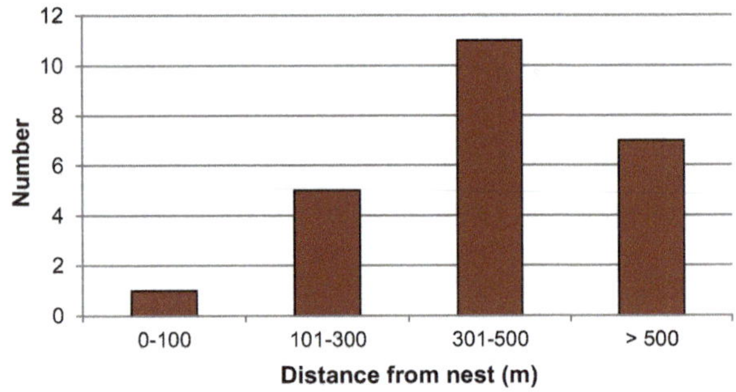

Fig. 4.9 Distance ranges of 24 mapped nest sites of Montagu's harriers from wind turbines in the study areas

4.6.2 Home Range and Centre of Activity in Relation to Wind Farms

4.6.2.1 Home Range Sizes

In the calculation of home range sizes during the breeding season, only birds with data available for at least ten days and a total of 100 independent recorded locations were included (see Sect. 4.3.5.1). This was the case for nine of the 11 tagged birds.

Using the kernel method, the mean 95% home ranges of these nine birds (three female, six male) equalled 976 ± 354 ha in the breeding season (Table 4.3). Variations between 341 ha (female) and 1516 ha (polygynous male) occurred. On average, the three female birds had with a mean of 626 ha considerably smaller territories than the male birds (1151 ± 246 ha).

Calculation of the minimum convex polygons resulted in larger areas and ranges (MCP, Table 4.3). MCP home ranges covered on average 1964 ± 246 ha and the values varied between 522 and 4169 ha. The larger sizes in comparison to the kernel method are the result of unweighted inclusion of the outermost locations so that unused or unusable areas are included.

The maximum flight distance between nest site and foraging ground ranged between 0.75 km (female, hunting in the vicinity of the nest) and an excess value of 15.55 km (male with above-average flight distances). A mean maximum distance of 6.0 ± 3.8 km was measured. The individual flight distances clearly correspond to the MCP home range sizes (including all locations in the calculation of the home range) but not to the kernel areas (Table 4.3).

4.6.2.2 Centre of Activity (Core Area)

The sizes of the centres of flight activity within the home ranges, which is described by kernel areas with a 50% probability of use, varied much less in size than the home range sizes (mean 117.9 ± 63.5 ha). These ranged between 41 and 252 ha,

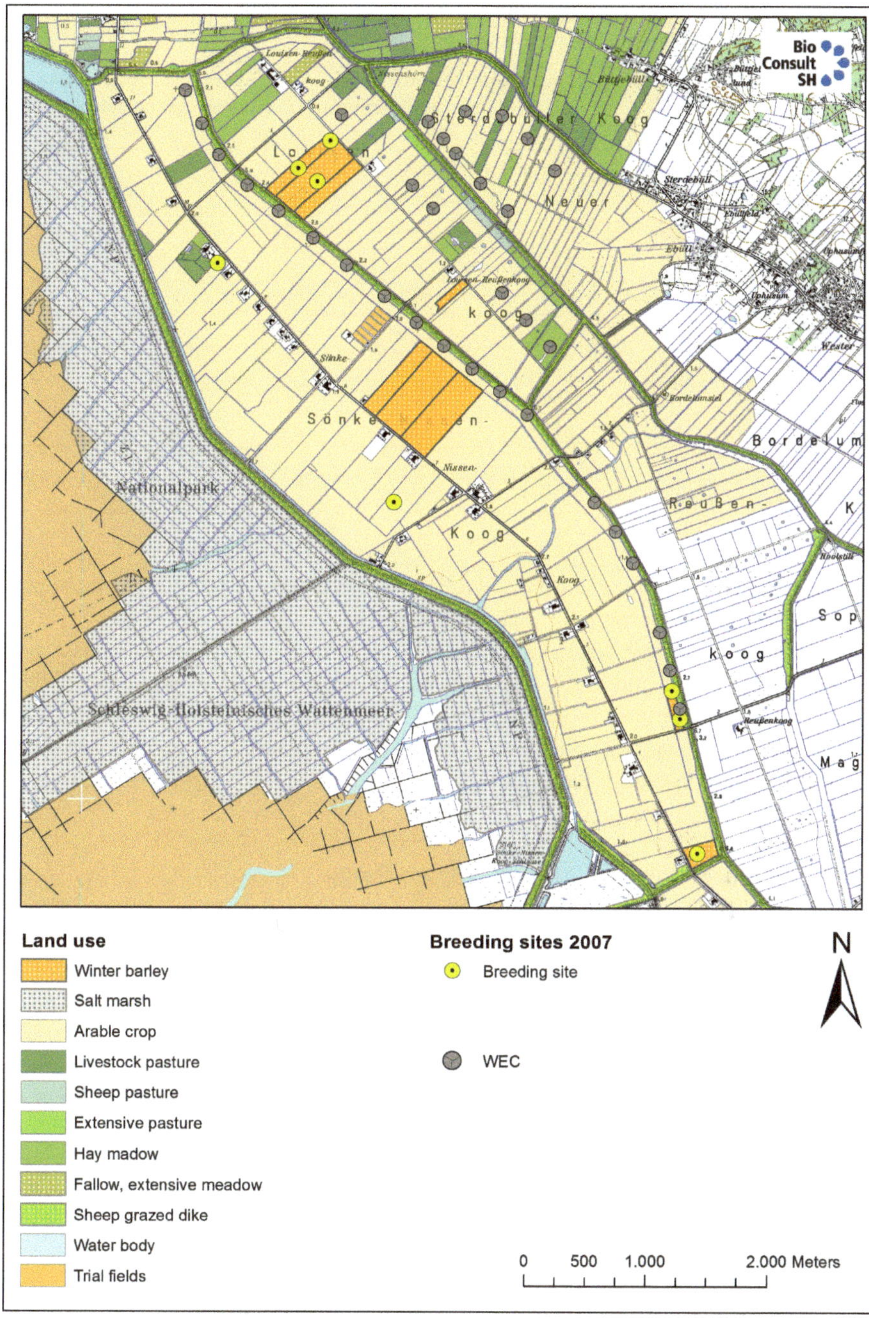

Fig. 4.10 Distribution of breeding sites, wind turbines and winter barley as well as further habitats and uses in the southern study area in 2007. Six of the eight broods were located in winter barley

Table 4.3 Size of the home ranges (MCP, kernel) and centres of activity (core area) as well as maximum flight distances from nest for nine tagged birds

Bird/area	Gender	Home range (MCP) (ha)	Home range (95% kernel) (ha)	Core area (50% kernel) (ha)	Max. flight distance from nest (km)
L1m 2007	m	2909	1516	82	5.50
L3m 2008	m	2083	1297	91	3.34
L1m 2008	m	4169	1180	252	15.55
FWL2m 2007	m	2391	1146	141	9.44
FWL1m 2008	m	1044	914	170	3.71
L3w 2007	f	2445	896	87	5.38
FWL2m 2008	m	726	851	122	2.25
FWL2w 2007	f	1392	642	72	5.34
L1w 2007	f	522	341	41	0.75
Mean		1964.6	976.5	117.9	6.03
Standard deviation		1170.9	354.6	63.5	3.84

whereas the extreme values were calculated for the above mentioned stationary females and the widest ranging male (Table 4.3).

For all nine birds the centre of activity (50% kernel) was around the nest sites (see Figs. 4.11, 4.12 and 4.13). Here male and female birds reached the highest flight activity and presence due to courtship rituals and food passing by male birds and brood care by female birds as well as feeding and nest related territorial activities by both breeding partners.

4.6.2.3 Wind Turbines Within the Home Range

Wind turbines were located in seven of the nine kernel home ranges. Use and approach of wind farm sites differed substantially between territories. Between 1 and 15% of the locations of the individual birds were recorded within the verification radius of 100 m around wind turbines (Table 4.4).

The intensity of use was depending on the distance of the nest site, as the centre of activity and the location of important foraging grounds. The proportion of records within a distance of 100 m around a wind turbine increased with increasing number of wind turbines within the home range of the particular bird. An exception was male L1m 2008, which did not fly close to wind turbines despite the 16 wind turbines in its home range, because the flight corridor between breeding site and foraging ground was free from wind turbines (Table 4.4).

The lines of wind turbines were used by the birds as guidance, foraging ground and transfer space between breeding site and hunting grounds. The different spatial relations and patterns of use are described in detail in the description of individual home ranges (Sect. 4.6.2.4).

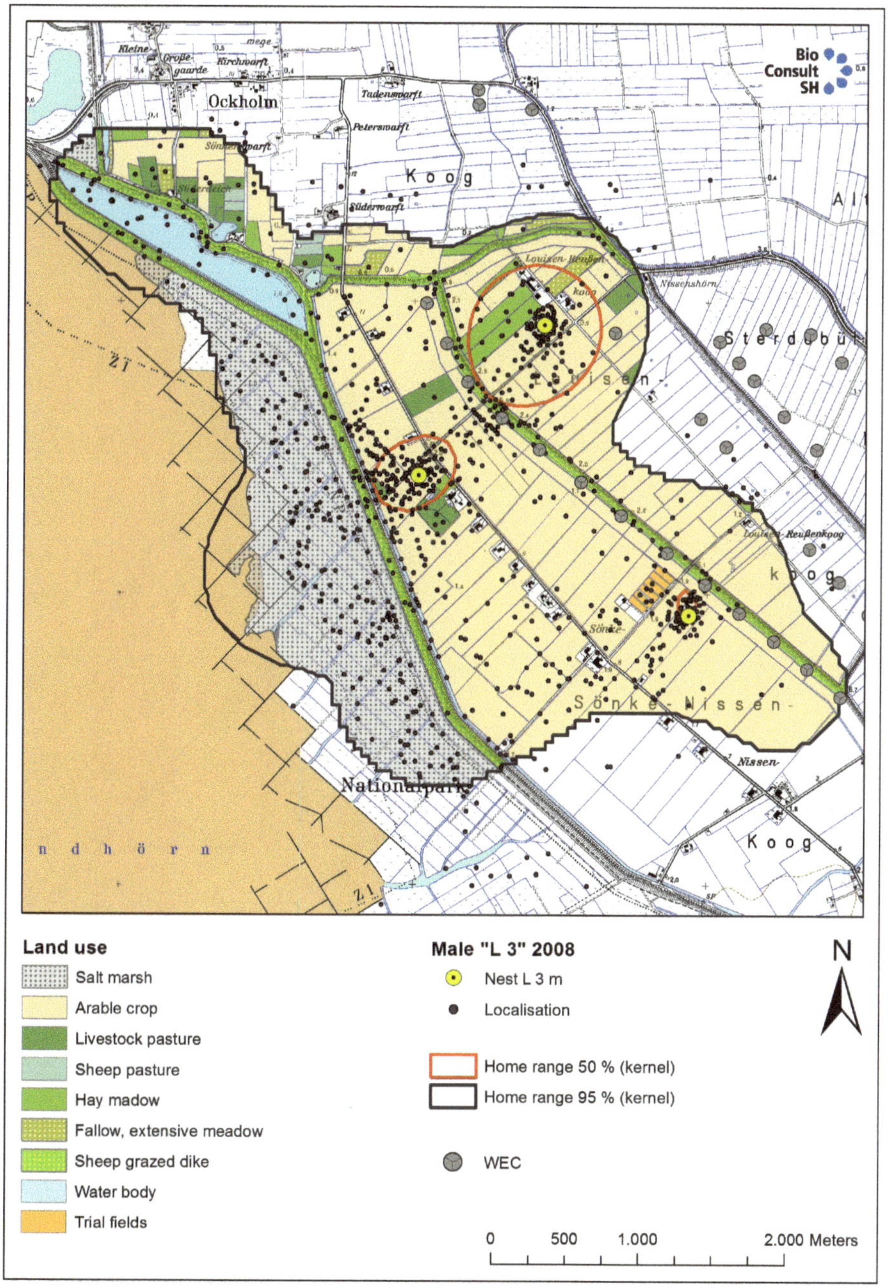

Fig. 4.11 Recorded locations, home range (95% kernel) and centres of activity (50% kernel, *red circles*) of the polygynous male L3m (2008) in the southern study area (Sönke-Nissen-Koog/Louisenkoog)

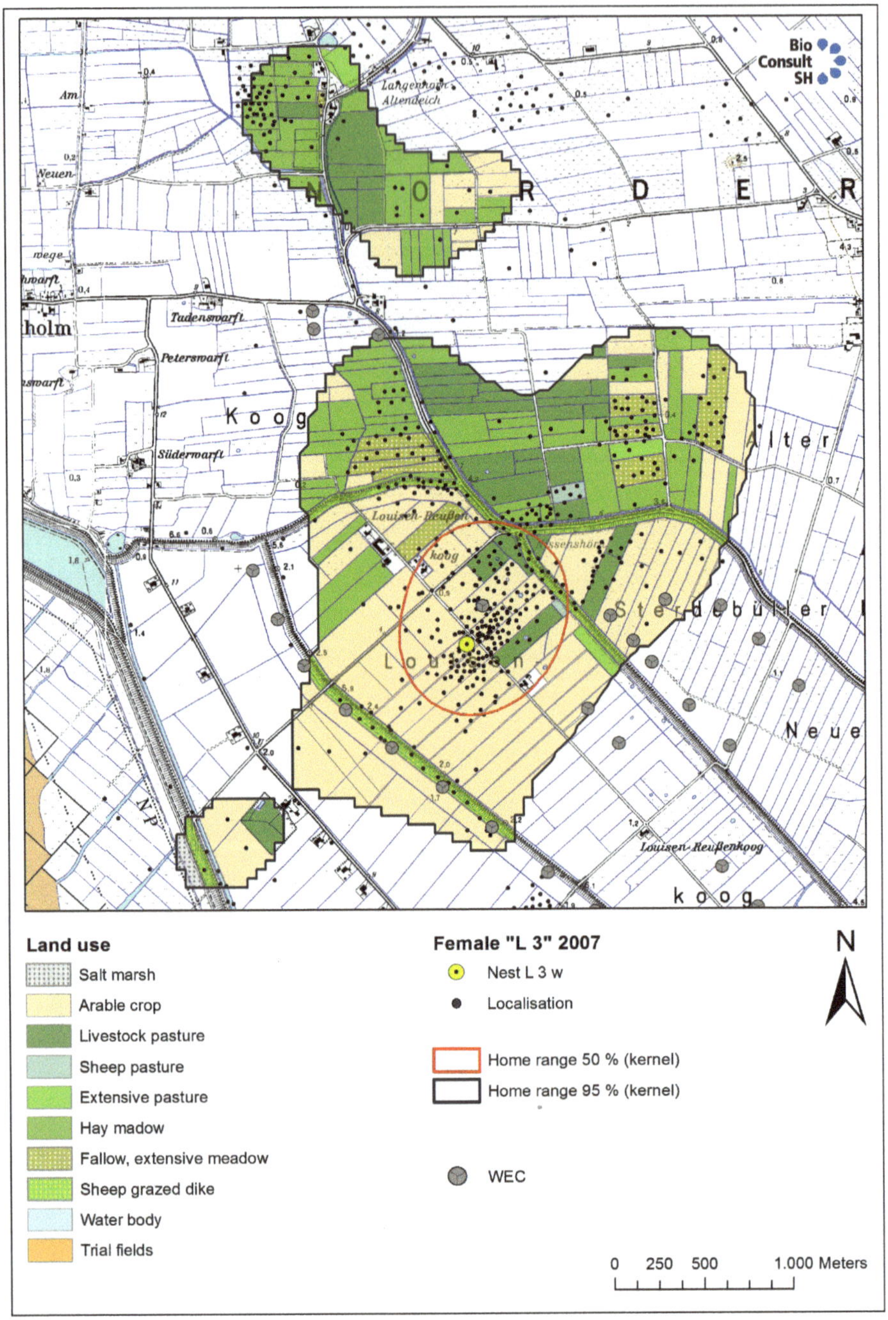

Fig. 4.12 Recorded locations, home range (95% kernel) and centre of activity (50% kernel, *red circle*) of female L3w (2007) in the southern study area (Sönke-Nissen-Koog/Louisenkoog)

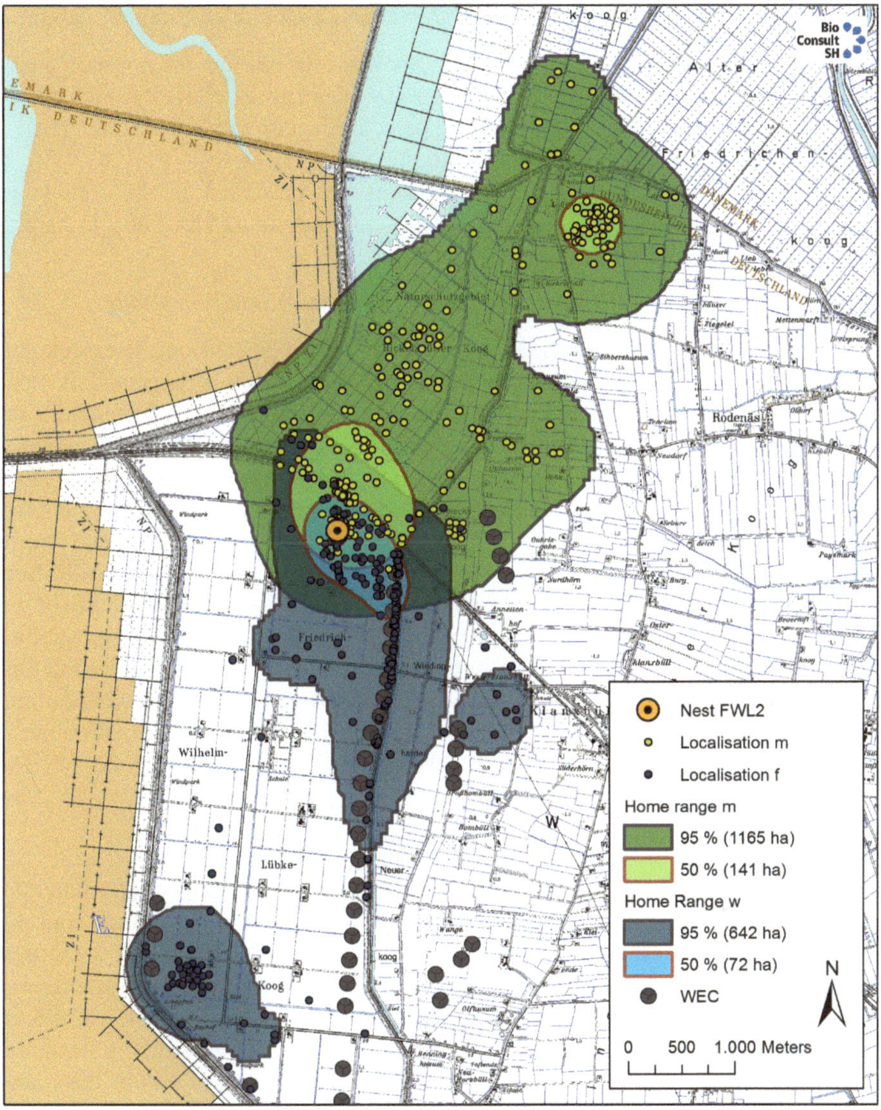

Fig. 4.13 Recorded locations, home range (95% kernel) and centres of activity (50% Kernel) of breeding pair FWL 2 (2007) in the northern study area (Friedrich-Wilhelm-Lübke-Koog)

4.6.2.4 Individual Home Ranges

Between both study areas, two different kinds of home range use occur, which can be distinguished by habitat use and location of areas of activity:

- Home range with salt marsh as the main foraging habitat.

- Home range with extensive pastures or wetland as the main foraging habitat.

Overlap of these two kinds of home ranges was especially in the northern study area with the directly adjacent salt marshes and wetland areas relatively small. Birds from neighbouring territories and breeding pairs used areas with different degrees of intensity resulting in distinct habitat partitioning (see Fig. 4.13).

Three examples of individual home ranges are described in the following, illustrating typical structures, foraging habitats and individual variability in habitat use of Montagu's harriers.

Polygynous male—southern study area (Sönke-Nissen-Koog/Louisenkoog)

The male was paired with three females in Sönke-Nissen-Koog/Louisenkoog. Figure 4.11 illustrates the recorded locations, the resulting home range and the core areas of highest activity in the vicinity of the three nests. The bird spent a significant number of time units in the vicinity of the nests, travelling between the three nest sites, passing prey and elaborating courtship rituals with the three females. This spatial constellation resulted in a high contact rate with the wind farm site and single wind turbines in the close-up area of the nests. 5% of the recorded locations occurred within a radius of 100 m around wind turbines. The wind farm site was used as a thoroughfare and additionally as a hunting ground. This male used the saltmarshes of the intertidal zone as the main foraging ground, where it spent about 60% of its hunting time. Pastures in the northwest of the home range were used as an alternative hunting ground. The bird was finally found dead after a probable collision with a wind turbine at the time of the young hatching (see Sect. 4.9). The location of the southern nest, where the bird was found, was located 300 m from the next wind turbine.

Female—southern study area (Sönke-Nissen-Koog/Louisenkoog)

This female from the same area, caring for the young mainly by herself with only little contribution of the male, had a comparably large home range of 892 ha (95% kernel). The bird used extensive pastures so that the pattern of use differs from the above male despite using the same breeding area (Fig. 4.12). The bird focused towards the inland on an extended, small-scaled grassland area with extensively used meadows, pastures and fallows in north-eastern direction. The salt marshes in the southwest were only used sporadically and are not included in the 95% home ranges (Fig. 4.12). For this bird, the presence of wind turbines within the home range and the centre of activity is obvious too. The single wind turbine located within the core area was regularly passed at a distance of below 20 m on the way to the foraging grounds. A total of 10% of the recorded locations occurred within the 100 m verification radius around wind turbines.

Breeding pair—northern study area (Friedrich-Wilhelm-Lübke-Koog)

The home ranges of a breeding pair in the northern part of the Friedrich-Wilhelm-Lübke-Koog are an example of a far-reaching partitioning of the habitat between partners and neighbouring territories, which is typical for Montagu's harriers depending on habitat conditions (Fig. 4.13). The male partner of

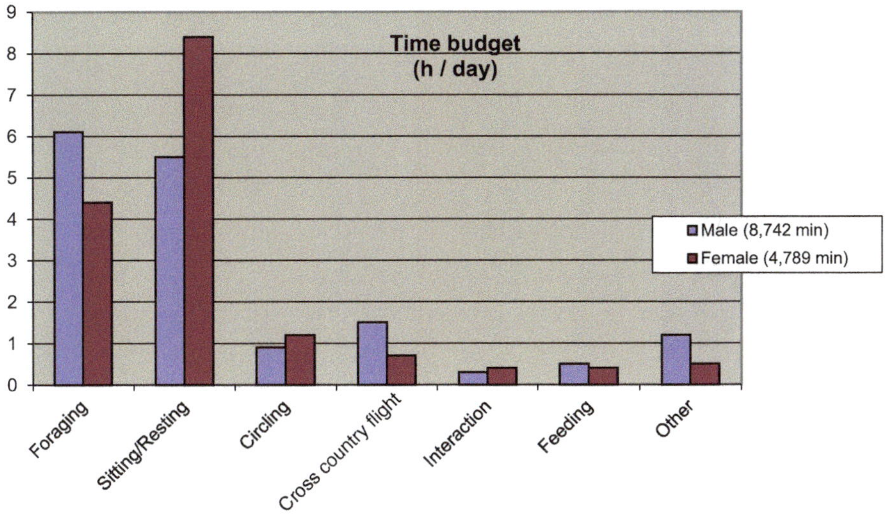

Fig. 4.14 Activity budgets of tagged male and female birds for both study areas. The recorded time intervals were normalised to a 16 h day of activity

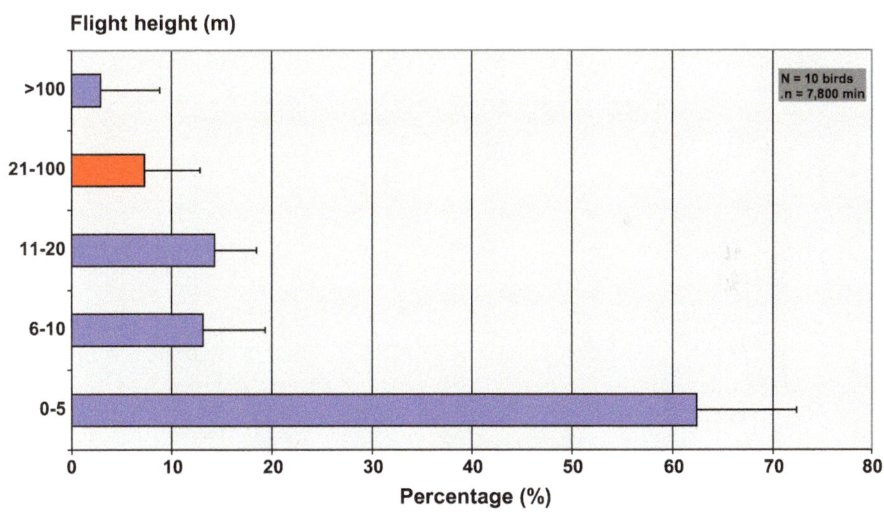

Fig. 4.15 Distribution of heights of Montagu's harriers bases on all flight sequences of years 2007 and 2008 (n = 7800 min, proportion of flight minutes, mean and standard deviation of nine birds). The band between 21 and 100 m marked in *red* indicates the potential risk area of the rotor swept area of many commonly used wind turbine models

this breeding pair mainly used the extensive wetland in the Rickelsbüller Koog at a distance of 400 m north of the nest site. Hunting flights of the female lead into the agricultural area in the south of the polder, regularly using the row of wind turbines aligned in north-south direction as a guidance (Fig. 4.13, see also Fig. 4.22). The

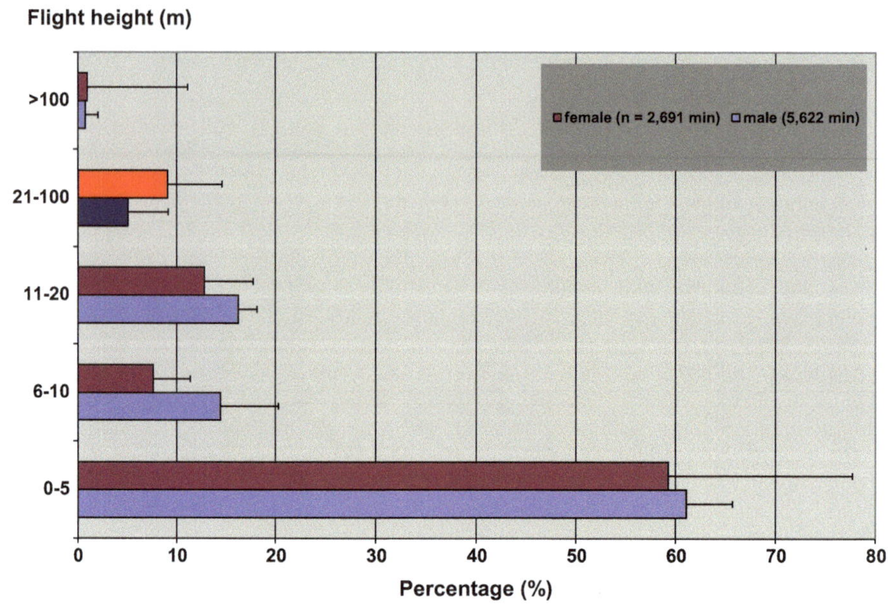

Fig. 4.16 Flight height distribution of male and female birds in 2007 and 2008 (See Fig. 4.15 for more details)

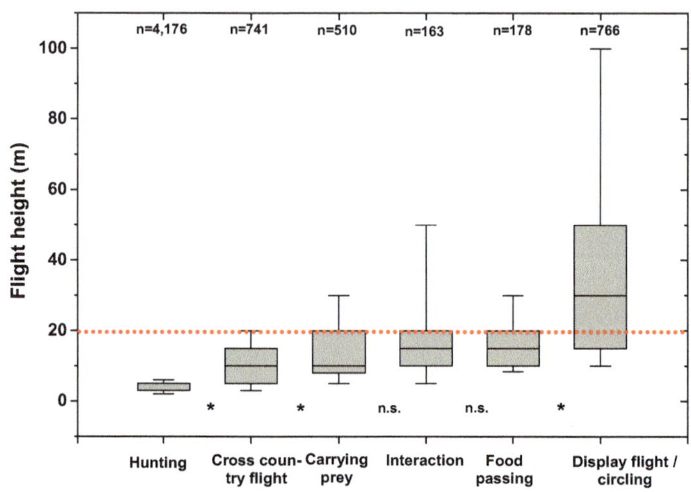

Fig. 4.17 Flight height distribution of different flight behaviours (median, standard deviation as well as 25 and 75% quartiles). Significance levels of pair-wise comparisons of flight heights: n. s. = not significant; * = <0.05, Mann-Whitney-U tests

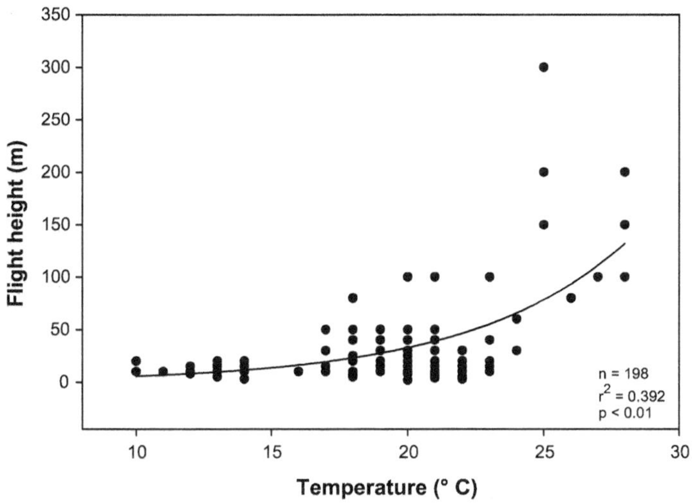

Fig. 4.18 Flight heights of courtship displays and circling depending on temperature

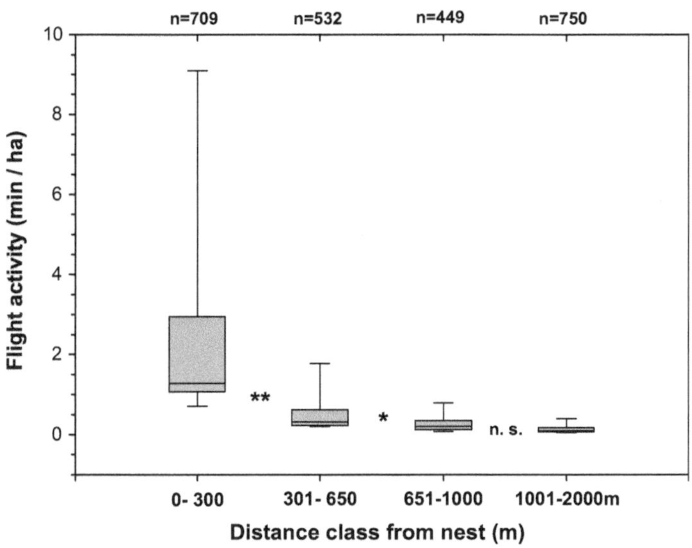

Fig. 4.19 Flight activity (recorded flight minutes/ha) depending on the distance from the nest site. Significance levels of the pair-wise comparison of flight intensities: ** = $p < 0.01$, * = $p < 0.05$, n. s. = not significant, Mann-Whitney-U tests

bird did not use the site of the wind farm itself, but the adjacent extensive sheep pasture at the dike. Several successful stoops for prey were observed in this area. Hunting flights always passed wind turbines close to the ground at heights of 2–5 m and at partly very small distances of 5–10 m.

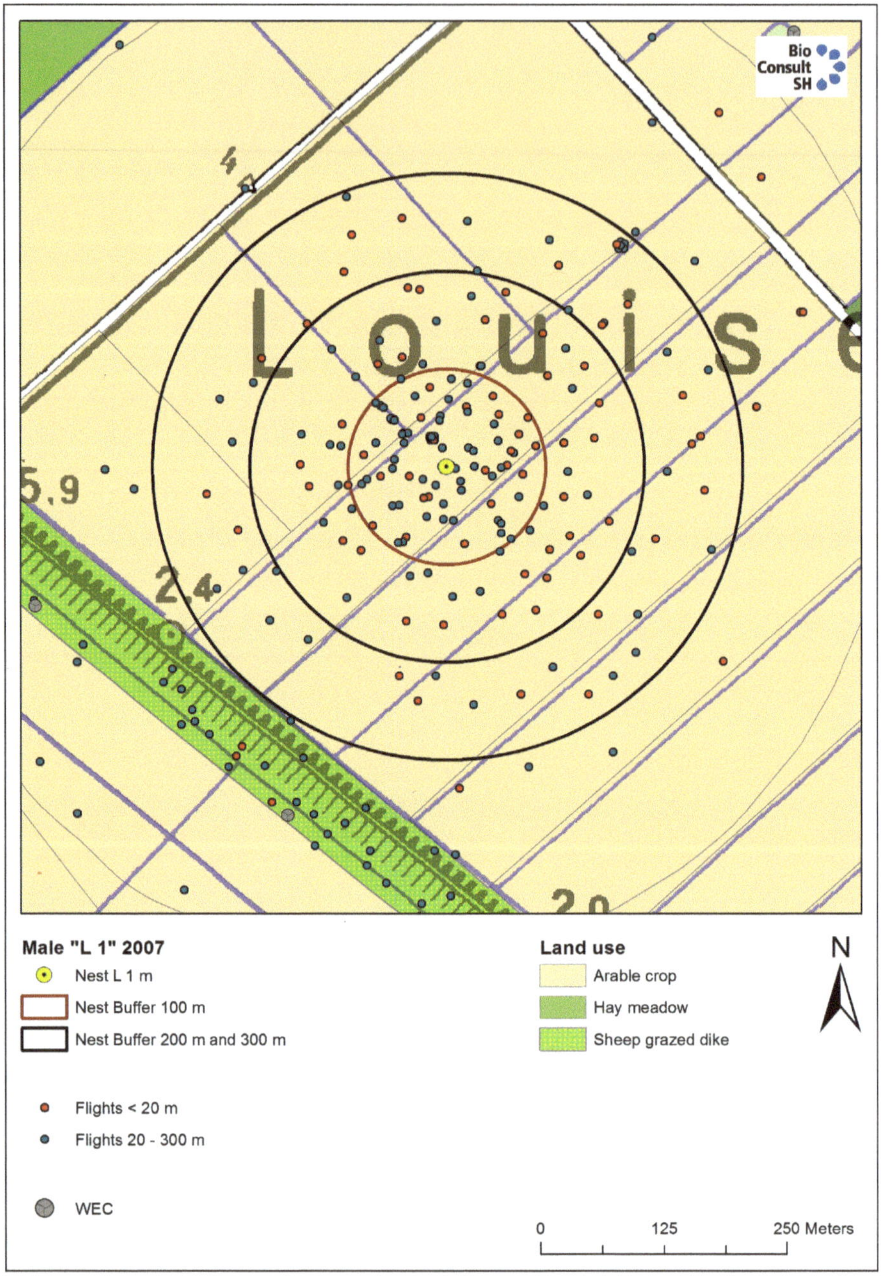

Fig. 4.20 Recorded locations and distribution of flight heights >20 m in the nesting area of two male birds in the southern study area. The 100, 200 and 300 m radii are given

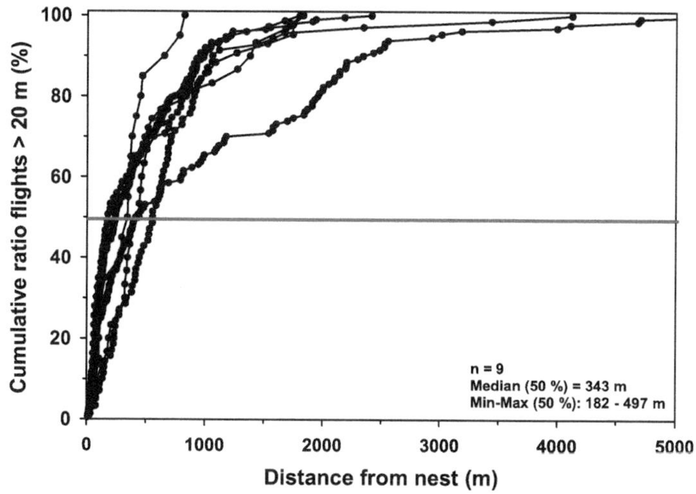

Fig. 4.21 Cumulative representation of the distances of the recorded locations at flight heights between 20 and 100 m from the nest sites. The distance radii with 50% of the total number of flight minutes are indicated for reference

Table 4.4 Proportion of locations recorded within a distance of 100 m around wind turbines and number of wind turbines within the home ranges

Bird/area	Gender	Locations (n)	Proportion within the 100 m around wind turbines (%)	Number of wind turbines within the 95% home range
L1m 2007	m	1271	15.3	11
FWL2w 2007	f	439	12.1	15
L3m 2008	m	1967	10.2	14
L3w 2007	f	537	10.2	9
L1w 2007	f	208	1.3	0
FWL2m 2007	m	455	1.3	5
L1m 2008	m	329	1.1	16
FWL1m 2008	m	329	0.8	2
FWL 2m 2008	m	213	0	0

Both partners were using sources of food temporarily available for a short time like freshly mowed meadows or harvested fields, which were used for up to 2 or 3 days. With 12% of the locations recorded within a radius of 100 m around wind turbines, the rate of contact with wind turbines of the female was relatively high (turbine row used as guidance). Due to the flight paths in mainly northern direction towards the wetland areas, the male did not get into conflict with the row of turbines in the western part of the home range.

Detailed analyses of habitat use and preferences used as a basis for habitat management measures for the Montagu's harrier are presented in Sect. 4.11.

4.6.3 Behaviour of Hunting Birds Towards Wind Turbines

A verification range of 300 m around wind turbines was observed to analyse the behaviour of foraging Montagu's harriers in the close-up area or impact zone of the turbines (see Sect. 4.3.5.3). Within this radius, spatial distribution and distances from wind turbines of single recorded locations were determined and compared to a random distribution (expected distribution) for the home ranges. The statistical comparison of the mean distances of both distributions is a measure of the degree of attraction or avoidance behaviour in hunting birds. Fife birds with more than 30 approximated independent locations while hunting were included in the analysis.

These five Montagu's harrier (three male, two female birds) approached the base of wind turbines up to a distance of 4 m while hunting for prey. Significantly lower mean distances than expected were determined for three of the five birds ($p < 0.01$, $df = 1$, chi^2 test), indicating an attracting effect of wind turbines (Table 4.5). Two of the birds maintained distances within the expected range, showing indifference towards wind turbines. Therefore, there is no indication of avoidance behaviour of hunting Montagu's harriers towards wind turbines.

Motives for approaching wind turbines may be found in the preference of Montagu's harriers for certain habitat structures within a wind farm site or the close vicinity thereof. The access routes lined with low grass and the often low growing grassland around the base of wind turbine towers are of particular importance. Extensively used grassland adjoining wind farms may lead to a concentration of hunting activity in the close-up area of wind turbines. In the northern study area, female FWL2 (2007) used an extensive sheep pasture directly adjoining the row of wind turbines intensely to hunt for prey (see Sect. 4.11, Fig. 4.22), resulting in the largest mean approach to wind turbines of the investigated birds (91.4 m, Table 4.5).

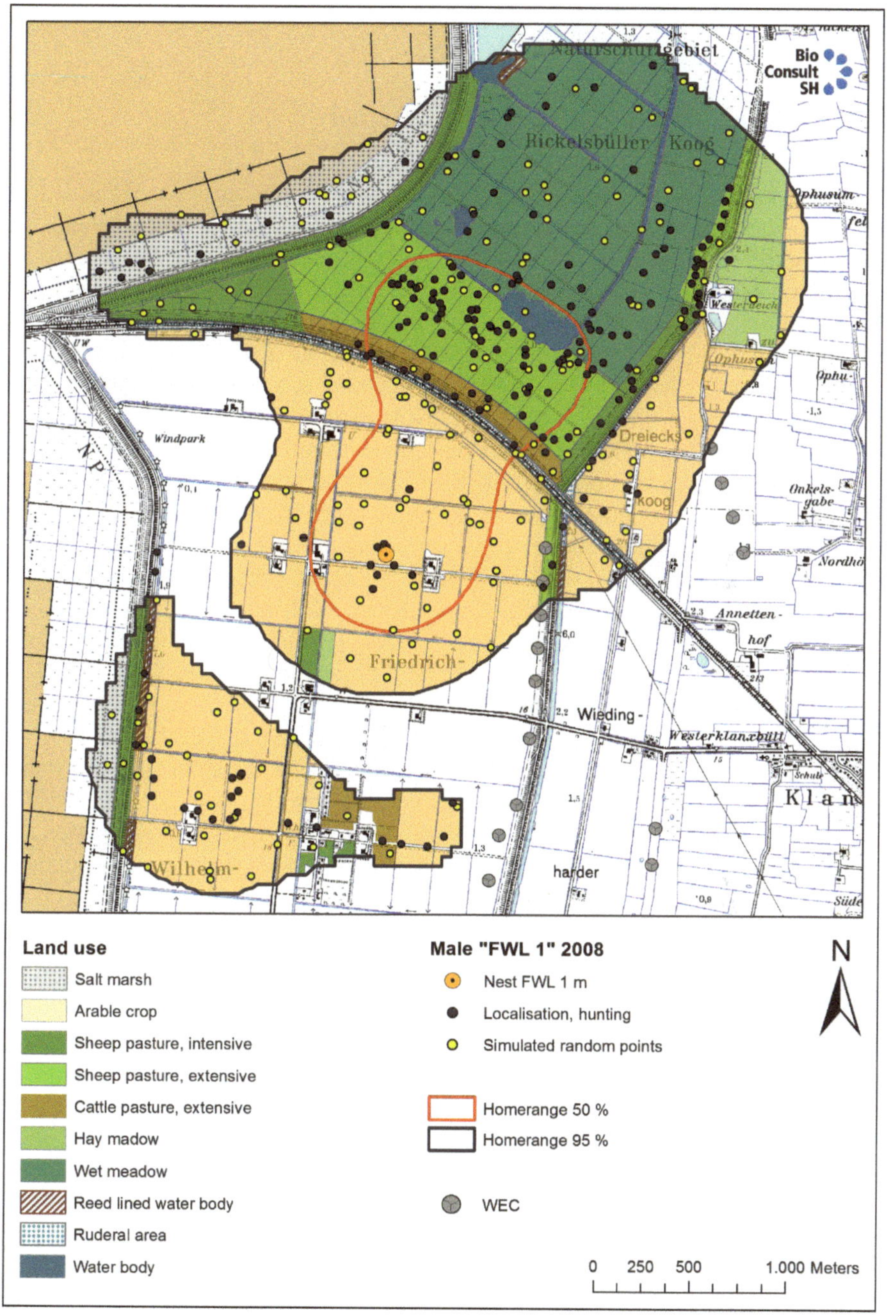

Fig. 4.22 Location of the nest, recorded locations while hunting, simulated random locations and habitat types within the home range of male FWL2 (2008, northern study area)

4.7 Home Range, Displacement and Barrier Effects—Discussion

4.7.1 Home Range and Centre of Activity

The large range and individual variability of home range sizes is confirmed by results for different European regions. Comparable data published so far indicate that home range sizes depend on habitat structure and quality. In the large scale agricultural landscape of the Netherlands provinces of Flevoland and Groningen, very large home ranges of on average 3490 ± 380 ha (90%) were determined for 10 male birds using the kernel method (Trierweiler et al. 2010). The results comply with earlier publications for the same regions. Visser et al. (2006) calculated home ranges for two male birds breeding in Flevoland of 5800 and 5200 ha using the MCP method. After excluding unused areas this was reduced to 2200 and 4000 ha. Near Groningen, two territories of 6900 and 3300 ha were determined (De Voogd 2004) including larger unused areas in the home range. Comparing the home range sizes determined by Trierweiler et al. (2010) using the same method as the present study, the home ranges of Dutch male birds are about three times as large as the ones of male birds on the west coast of Schleswig-Holstein.

In the small scale agricultural landscape of western France, flight distances of a maximum of 5.0 km from the nest and a mean home range size of 1423 ± 647 ha were determined for five tagged male birds (Salomolard 1997). Sizes varied between 854 and 2346 ha covering the size range calculated for Schleswig-Holstein.

Home ranges in the Extremadura (Spain) were between those in the Netherlands and France. Arroyo et al. (2009) calculated kernel home ranges of 2410 ± 1630 ha for 19 male birds in agricultural land with varying proportions of arable crops and grassland. Ranges of female birds were like in Schleswig-Holstein considerably smaller (670 ± 350 ha, 11 birds).

Own results and comparable data indicate that the habitats investigated in Schleswig-Holstein are relatively rich in structure, offering in comparison to

Table 4.5 Recorded locations of hunting Montagu's harriers within a radius of 300 m around wind turbines and comparison of mean distances with expected values using a random distribution (chi^2 tests)

	Locations (n)	Distance from wind turbine (m)	Chi^2	*p*	
		Observed	Expected		
L1m 07	91	136.2	187.8	3284.82	<0.01
L3w 07	84	146.5	143.3	903.54	n/a
FWL2w 07	135	91.4	125.9	3663.25	<0.01
L1m 08	36	150.1	151.6	793.94	n/a
L3m 08	96	133.5	160.4	1455.26	<0.05

average agricultural landscapes additional high-quality foraging grounds (salt marshes, extensive grassland and wetland). The proximity of this kind of habitat to breeding sites especially within the northern study area (flight distances of between 300 to a maximum of 1400 m) results in relatively short flight distances between nest site and foraging grounds and thus smaller home ranges.

4.7.2 *Displacement Effects Caused by Wind Turbines*

Within the framework of the telemetric studies, no displacement effects or avoidance behaviour of Montagu's harriers in the close-up area of wind turbines could be detected.

The monitored nest sites were located between 76 and 990 m from wind turbines. Six of the nests were closer than 300 m to wind turbines. The distribution of breeding sites within the study area indicated that the selection of habitats depends on the location and growth of arable crops resulting in an annually varying distribution of nests. Winter barley with a proportion of fewer than 5% within the study area is of particular importance and at least in some years strongly preferred by breeding birds. In 2007, occupation of single isolated winter barley crops resulted in a concentration of nest sites in the close-up area of wind turbines. Preference of Montagu's harriers for winter barley as a breeding location in agricultural landscapes is well documented for different regions in southern and central Europe (e.g. Arroyo et al. 2009; Liminana et al. 2006; Millon et al. 2002; Pürkhauer et al. 2009) and is confirmed for the present study areas by a recent study based on historic data on the distribution of breeding sites and land use (Gahrau 2011). Which small structures or habitat parameters within the crop are decisive for the selection of the nest site was not investigated so far. The structure of growth (height, density) depending on weather conditions at the time of occupying a territory will probably result in the annually varying use and selection of different crops (wheat, oilseed rape) in some years. Possible displacement effects of wind turbines on Montagu's harriers in the selection of the nest site may be superimposed by habitat preferences. The low availability of winter barley crops (<5% within the study area), typical for Schleswig-Holstein, may cause the birds to accept a closer proximity to wind turbines. In the breeding area of the Hellweg Börde (NRW) with a higher proportion of winter barley of 20–30% offering more suitable locations, nests were built further away from wind turbines than in the present study areas as derived from the GIS assisted analysis of the distribution of breeding sites (see Sect. 5.5.5, Joest et al. 2013). There, the minimum distance was 170 m and the mean distance 500 m. In a multi-factor analysis, influences of further habitat parameters like metres above sea level, slope gradient and the presence of extensive vertical structures on the breeding distribution could be proven (Joest et al. 2013). The distance of nests from electricity pylons and the total number of point vertical objects including wind turbines had a significant influence on the selection of the breeding site. Wind turbines alone had no verifiable influence on the distribution of breeding sites. These results show that wind turbines may well cause displacement effects, but measurable impacts only occur along with further vertical structures in open landscapes. For the two Schleswig-Holstein study areas, the observed smaller distances between nests and wind turbines may be due to lack of further influences (low height above sea level, no slope, far-reaching absence of additional vertical structures). It can therefore be assumed that wind turbines have no measurable impact on the selection of the nest site in the existing local constellation.

In general, observations and results show that impacts in form of displacement effects and avoidance behaviour towards wind turbines play a minor role in the selection of the breeding habitat for Montagu's harriers, but possible displacement effects caused by wind turbines may be intensified when overlapping with further factors (elevation of the area, slope gradient, additional vertical structures, for hen harriers see O'Donoghue et al. 2011). In the investigated landscapes in Northern Germany, which lack any of the further known disruptive factors, no measurable displacement effects caused by the current wind turbine set-up are to be expected. To a large extent, results and evaluation comply with the sensitivity classifications of the RSPB bird sensitivity map for hen harriers in Scotland (Bright et al. 2006) expecting possible breeding site displacement by wind turbines for the close-up area of up to 100 m only (Whitfield and Madders 2006b).

Results on habitat use and flight activity within the home range gave no indication of displacement effects or avoidance behaviour caused by existing wind turbines on Montagu's harriers hunting or carrying prey. Wind farms or wind turbines are part of the home range of Montagu's harriers as soon as they are located within the breeding area. They were traversed on the way to and from foraging and breeding grounds and were even used for foraging. For single birds, wind farms took up an area of between 1 and 12% of the home range. Intensity of use of wind farms and approach of wind turbines is determined by the distance from the breeding site. For the individual birds investigated in the present study, the centres of activity and were concentrated around the nesting location for both male and female birds. If wind turbines are located within areas of high flight activity, approaches are common. The influence of the location of the nest site on habitat use and approach of wind turbines is most apparent in the habitat use of the polygynous male. The surrounding area of the three nest sites of the female birds to be supported, were centres of activity within the home range. Because of the wind turbines within the centres of activity and in the flight paths between nest sites, about 25% of the locations for this bird were recorded in the close-up area of wind turbines (100 m verification range around wind turbines). Comparative data on the flight activity of Montagu's harriers in the vicinity of wind farms is not available so far.

The spatial distribution of recorded locations and the determination of flight paths of foraging Montagu's harriers underline that they fly close to wind turbines on a regular basis. Evaluation of frequently flying birds showed that no avoidance behaviour can be observed in the close-up area. Some birds showed even higher rates of approach than expected, implying an attracting effect of wind turbines due to accompanying structures. Important motives to use wind farms or to approach wind turbines are additionally available structures suitable for foraging found near wind farm sites (gravel roads, grassland) or adjacent high-quality foraging habitats (e.g. extensive grassland). Edge effects may be equally important due to an increased food supply (see Sect. 4.11.3). For birds investigated in the present study, wind farm sites were no important foraging habitat (<5% of the recorded time spent on searching for food) because high-quality foraging habitats (salt marshes, large-scale grassland) were available within the home range. In large-scale, featureless agricultural areas, wind farm sites will be of greater importance to Montagu's harriers due to the additional structures and edges, which may lead to a more intense use. In Lower Saxony, Baum and Baum (2011) found that Montagu's harrier use wind farms as resting or roosting grounds. The breeding birds visited the edges of access routes in the vicinity of the nest. The authors assume that the birds prefer open areas in the vicinity of breeding sites at night, which are not available in intensive cropland. Overall, use of wind farms of Montagu's harriers highly depends on the structure of agricultural landscape, allowing for regulation and direction by means of habitat management measures (see Sect. 4.13).

4.7.3 Summary of the Results

The home ranges determined for Montagu's harriers in Schleswig-Holstein vary between 340 ha and 1560 ha (kernel). Males have larger home ranges than female birds. Centres of flight activity are in the surrounding area of nests and comprise between 41 ha and 252 ha. The salt marshes of the intertidal zone and large extensive grassland rich in ecotones, are important foraging grounds. Wind farms within the home range are used without avoidance of the close-up area of wind turbines. The intensity of use depends on location and distance of wind turbines from the centre of activity (nest) and available foraging grounds. The results give no hint of displacement effects in the selection of the nest site or avoidance behaviour in the use of the habitat.

4.8 Collision Risk—Results

4.8.1 Flight Activity

The calculation of the activity budgets of the tagged birds is based on a total of 13,531 valid minute intervals. Records cover the entire period of diurnal activities. The main proportion of time records fell within the daily intervals between 9–13 and 13–17 (CEST, a total of 71%). For presentation purposes, the proportions of time used for certain activities were normalised to a mean day of activity of 16 h (daylight from sunrise to sunset during the study period).

Male and female birds differ considerably in proportions of time used for foraging flights and sitting on the ground. With on average 38% of their daily activity, the six male spent more time foraging (6.1 h out of 16) than females (27.5% equalling 4.4 h). Females on the other hand, spent most of the day sitting on the ground (9.57 h). Sitting included being present at the nest, guarding and brooding of the young, waiting for prey to be passed by the male as well as roosting and plumage care. With 5.5 h, the proportion of time spent sitting on the ground is relatively large for male birds as well. After passing prey they roosted in the surrounding area of the nest in the roost area on a regular basis. Longer hunting flights were interrupted for occasional short rests.

All further types of behaviour were observed in flight. The investigated male birds spent on average 6.4 h of the day in flight, just under two hours more than females. The higher flight activity of male birds may have an influence on the gender-specific collision risk (see Sect. 4.8.2).

4.8.2 Flight Altitudes

A height band between 20 and 100 m was used as a reference value, defining the rotor swept risk area, as the dimensions of most wind turbines within the study areas and as well Schleswig-Holstein were within this range at the time of the study.

In telemetric monitoring the visually observed flight heights of a total of 7800 min of flight were recorded for the Montagu's harriers. All recorded flight sequences showed that Montagu's harriers typically fly low. Almost 90% of the flight minutes occurred at a height of up to 20 m with the majority of time spent in the height band of up to 5 m (63%). On average, 7% of the flight time was recorded within the height band between 20 and 100 m (marked in red in Fig. 4.15). Flight heights above 100 m were rarely used with a total proportion of time of only 2.9%.

The small variability of individual birds is especially obvious for low flight heights (Fig. 4.16). Within the height band of 0–5 m, the proportions of time vary between 48 and 67%. Thus, individual birds showed broad agreement in flight heights.

Comparison between male and female birds shows hardly any differences in the use of flight heights (Fig. 4.16). The slightly higher proportion of time females spent within the band between 20 and 100 m can be attributed to aerial food passes usually performed in this height band in the vicinity of the nest. Flight heights for food passes are more heavily weighted for females, as these hunt much less in lower heights.

4.8.2.1 Comparison of the Years 2007 and 2008

For the five birds monitored in 2008, the mean flight heights were significantly higher than for the six birds in 2007 (mean: 7.6 $\pm$ 2.1 m as opposed to

6.4 ± 2.3 m, $p < 0.05$, Mann-Whitney-U test). The difference was almost entirely caused by a larger proportion of flight minutes in the height band of 6–10 m. Flights in these heights mainly were foraging flights. Behavioural observations in the field showed that flight heights differing between the years were related to differing food supply related hunting strategies. 2007 was a year with high abundance of small mammals. Foraging flights mainly occurred at heights below 5 m. In the following year, with less small mammals available, the birds hunted small birds (recognisable from plucking sites) at flight heights between 5 and 10 m, which were captured when flying up.

4.8.2.2 Comparison of the Study Areas

Comparing the flight heights of birds from the southern study area (Sönke-Nissen-Koog/Louisenkoog, n = 5362 flight minute intervals) and northern study area (Friedrich-Wilhelm-Lübke-Koog, n = 3438 flight minute intervals) no differences were found (df = 1, $p > 0.05$, Mann-Whitney-U test).

4.8.2.3 Influence of the Season

No seasonal effect was found in the distribution of flight heights. Flight heights did not significantly differ for incubation, nestling and fledged young stages (Mann-Whitney-U tests, p always > 0.05).

4.8.3 Flight Behaviour and Flight Heights

Different behaviours of Montagu's harriers had a large influence on the distribution of flight heights. The following important types of flight behaviour can be distinguished:

- Hunting flight
- Cross-country flights
- Flights with prey
- Interaction (intra- and interspecific)
- Food passing
- Courtship display and circling.

These types of flight can be summarised into three groups of behaviour by characteristic flight heights and thus potential risks (see Fig. 4.17).

4.8.3.1 Hunting and Cross-Country Flights

Hunting flights differed substantially from other behaviours, by low flight heights and only few variance (median: 3.0 m, mean 4.07 ± 2.65 m; n = 4175 min; df = 1 in each case, $p < 0.01$, Mann-Whitney-U tests). They were characterised by a low variance of flight heights. 93.8% of the recorded hunting minutes were spent at flight heights below 10 m (Fig. 4.17).

Due to the high proportion of time the birds spent hunting for prey during the day (male birds 6.1 h, female birds 4.3 h, see Fig. 4.14), the distribution of flight heights of Montagu's harriers is shifted towards a height band clearly below 10 m. While hunting, only few flights at a maximum height of 20 m were observed.

Cross-country flights served to cover short or longer distances with no obvious intention. With a median of 10 m and a proportion of 90% of flights below 20 m, these flights showed wider variance in flight height but still belong to the group of flight behaviours with the lowest potential for risks.

4.8.3.2 Flights with Prey and Interaction

This group covers flights to the nest with prey, subsequent food passes as well as interaction, aggressively pursuing individuals of the same or different species in the area around the nest. These behaviours mainly occur within a band of flight heights between 15 and 20 m with high variance and maximum heights of between 50 and 80 m. Flights above a height of 20 m occur on a regular basis (Fig. 4.17).

These flights bear a potential risk for collision, as they may reach heights within the rotor swept area of most currently operated wind turbines.

4.8.3.3 Courtship Displays and Circling

The highest flight heights and largest variance was determined for courtship displays and circling (median 29 m, maximum heights > 300 m). Both behaviours often were linked and sky dances of courtship displays often start with high circling. Flight heights of these behaviours were found to be highly dependent on temperature, as the birds use thermal streams to soar up high (Fig. 4.18). Flight heights of 50 m were observed in temperatures of at least 17 °C, heights of at least 100 m were reached from 19 °C onwards. Considering the distribution of flight heights, this flight behaviour bears the highest potential of risk, although flight heights well beyond the rotor swept area will be reached where thermals occur (Fig. 4.17).

4.8.4 Spatial Distribution of Flight Activity

In the evaluation of the collision risk, the spatial distribution of flight activity needs to be investigated in addition to flight heights. The spatial occurrence of potentially conflicting behaviour and the corresponding flight heights are of particular interest.

The spatial distribution of flight activity of the nine investigated individuals confirms the location of the centre of activity in the vicinity of the nest site during incubation and nestling stage determined using kernel analysis (Fig. 4.19). A strong concentration of flight activity occurs within a radius of 300 m around the nest mainly due to flights with prey, food passes, feeding and strong presence of the male in the nesting area. With increasing distance from the nest, the area-related flight activity was clearly and significantly decreasing. At a distance of between >300 and 650 m, only 16% of the flight minutes/ha occurring within a radius of 300 m around the nest were recorded.

Especially flight behaviour in critical heights of over 20 m is clearly nest-related. Figure 4.20 gives the example of the flight activity of two male birds and illustrates the concentration of flight heights above 20 m in the nesting area. The distinct decrease in flight intensity and distribution of recorded locations in areas outside the 300 m radius is obvious (see also Fig. 4.19).

Occurrence of critical flight heights depending on distance from the nest, becomes obvious in the cumulative representation of flight activity (Fig. 4.21). Distances with 50% of the total flight time at heights between 20 and 100 m of the individual Montagu's harriers are located within 182 and 497 m around the corresponding nest. The 50% median of eight of the birds was at 343 m (Table 4.6). Beyond these radii, the frequency of critical flight heights decreased considerably and was distributed throughout the area up to distances of several kilometres (Fig. 4.21).

Table 4.6 Distance radii with 50% of the total flight time at heights between 20 and 100 m around the nest sites of eight birds

Bird/area	Gender	Radius with 50% of the flight activity (m)
L1m 2007	m	311
L3m 2008	m	221
L1m 2008	m	497
FWL2m 2007	m	343
FWL1m 2008	m	182
L3w 2007	f	343
FWL2m 2008	m	404
FWL2w 2007	f	370
	Median:	343

4.9 Flight Behaviour and Collision Risk—Discussion

No published literature on flight behaviour and flight heights of Montagu's harriers is available so far. Descriptions of the behaviour in Clarke (1996) and species-related chapters in Cramp and Simmons (1980), Glutz von Blotzheim and Bauer (1971) as well as Mebs and Schmidt (2006) generally confirm the basic principles of flight activity determined in the present study. Hunting predominantly occurs at flight heights of <10 m. Greater heights are usually reached in courtship displays ("sky dancing"), food passes and by soaring up in circles. Standardised comparative data is available for the closely related hen harrier (*Circus cyaneus*) showing flight behaviour and hunting strategies similar to those of the Montagu's harrier (Mebs and Schmidt 2006, Whitfield and Madders 2006a, b). The flight height distribution of hen harriers was among others determined for eight different wind farm projects in the US (overview in Whitfield and Madders 2006a). Similar to the Montagu's harrier, the hen harrier showed a distinct preference for low heights. For all investigated project sites, 80% of the flight sequences were recorded within the height band of up to 25 m. Whitfield and Madders (2006a) investigated the flight heights of hunting hen harriers in Scotland. These hunted in 70–80% of the observations below 2 m. The flight heights determined for the Montagu's harrier widely correspond to these findings.

To assess the collision risk, the level of flight activity within the rotor swept area needs to be determined (Band et al. 2007, Whitfield and Madders 2006a). In the present study, 7% of the recorded flight minutes occurred within the risk area of wind turbines with a rotor swept area reaching from 20 to 100 m used as a reference value. 63% of the flight minutes were recorded in heights up to 5 m, and just below 3% at heights above 100 m. Accordingly, most of the flights of Montagu's harriers occur below the rotors of currently installed wind turbines, leading to the assumption that the risk of collision is generally low. Detailed investigation of the flight behaviour and distribution of flight heights in the area showed however, that the collision risk depends on the location of wind turbines in relation to the nest. Potentially risky flights at rotor blade height (courtship displays, food passes, circling, interaction) took place in the surrounding area of the nest. 50% of the recorded flights above 20 m recorded for the individual birds, were within an average radius of 343 m (median) around the nest. The flight activity per area is decreasing considerably with increasing distance from the nest and mainly consists of hunting flights at low heights. Flight activity in the vicinity of the nest causing a possible risk, account for a relative low proportion of time units, but may reach high frequencies over the whole breeding season (e.g. about 1000 food passes per breeding pair per season). Due to the concentration of flight activity on the area of the nest, the distance between wind turbine and nest site is an important factor in the determination of the collision risk. The minor avoidance behaviour of Montagu's harriers towards wind turbines therefore results in an increased collision risk in the nesting area. At distances above 300 m, flight activity and mean flight height are clearly decreasing, resulting in a low collision risk in the periphery of the home range or spatially separated foraging grounds. In the assessment of the collision risk of individual breeding pairs, the high variability of single birds as well as the relative position of the breeding site, wind turbines and preferred flight paths need to be taken into account.

A calculation of the collision risk of Montagu's harriers is not possible due to variations in behaviour and the locally differing habitat parameter. The two collision fatalities registered in the central German database for collision victims by now (one of them being a male bird ringed and tagged for the present study in the northern study area in 2008, Dürr 2013), do not allow to draw conclusions on the actual collision rate, as this long-distance migrant is only present in the breeding territories in the growing season (May to August) and collision victims can hardly be found in high growing cereal crops. The number of unreported casualties may therefore be significant. No monitoring of mortality due to collisions accounting for the abundance and the period of presence of Montagu's harriers is carried out so far. The two birds bound dead and the three birds listed in the central German register as probable collision victims (Dürr 2013) were found randomly outside ongoing monitoring programmes.

With the detailed results of telemetric monitoring for this study on flight behaviour, reliable data is available for realisation of collision models (Band model, Band et al. 2007). Only relative comparison of risks for different site conditions are possible though, as the main factor to calculate victim rates, avoidance behaviour in the rotor swept area (avoidance rate, Chamberlain et al. 2006), cannot be determined with absolute certainty (see Rasran et al. 2013). This would only be possible on the basis of a standardised search for collision victims, which for the Montagu's harrier is limited by the conditions in the growing season (see above) and will require very long investigation periods to yield reliable data for this rare species.

The circumstances under which the so far reported collision victims and the probable collision victims were found (Dürr 2013) underline the assessment of the collision risk based on telemetric monitoring. According to the present results, especially male birds are at risk due to the considerably higher flight activity in comparison to female birds. Locations and time of birds found dead confirm that collisions will mainly occur during the breeding season, if nests are located in the vicinity of wind turbines. One of the as yet documented collision victims is a male bird in the northern study area (Friedrich-Wilhelm-Lübke-Koog), VHF-tagged in 2008 for this project. In 2010, the year the bird was found dead, the nest was located 250 m from the next wind turbine. Similar spatial relations were true for the male bird found dead near Petjenburg in Lower Saxony in the same year. According to the autopsy report of the IZW Berlin, the bird died of barotrauma induced by moving turbine blades (Dürr 2013).

On the basis of the autopsy report by the IZW Berlin, the tagged polygynous male from the southern study area of the present study, which we found dead on 22.06.2008, is listed by Dürr (2013) as probable collision victim. With a distance of 300 m from the next wind turbine, the nest of this bird again was located in relative proximity to a wind turbine. Possible contacts with the nearest wind turbine were documented by means of telemetric locations recorded around the time of the probable collision.

4.9.1 Influence of Wind Turbine Dimensions on the Collision Risk

Apart from the location of the wind turbine, its dimensions have an influence on the collision risk for Montagu's harriers. As the distribution of flight heights is clearly concentrated on height ranges below 50 m for this species, the lower area of the rotor and especially the distance between lower rotor blade tip and ground are decisive for the degree of collision risk.

By choosing higher towers for the same rotor diameter, the clearance below the blade tip may be increased thus reducing the potential for conflicts. According to the distribution of flight heights of the Montagu's harriers investigated for the present study, an increase of clearance from 20 to 30 m would already lead to a reduction of flight activity (length of stay) in the rotor swept area by about 50%. An increase of the overall height of a wind turbine (usually combined with an enlargement of the rotor swept area) does not result in an increased collision risk, as heights above 100 m are not used by Montagu's harriers for a significant proportion of time in relation to the total number of flight movements in the breeding season even under favourable conditions (e.g. in case of strong thermals). In repowering projects where small turbines are replaced by turbines with a greater clearance below the blade tip, a lower collision risk can be expected despite an increase of the rotor swept area. Whether such differences will result in a measurable change in collision risk is doubtful. Collisions of Montagu's harriers are still relatively rare and quite often random events, which may occur at any wind turbine. In East Frisia for example, a male bird collided with a wind turbine directly after small wind turbines were replaced by larger turbines with more favourable dimensions for Montagu's harriers in a repowering project (Baum and Baum 2011).

In many areas of the central and southern German inland, wind turbines of different dimensions than those in Schleswig-Holstein are installed. Here wind turbines with heights of 60–80 m below the rotor swept area are typically installed. Under these conditions, only a very small proportion of the flight movements of Montagu's harriers in Schleswig-Holstein would occur within the rotor swept area (mean values for courtship displays, food passes, interaction with other birds of prey and prey transfers at heights of around 20 m). Only in courtship displays and circling under favourable thermal conditions, Montagu's harriers used higher flight heights on a regular basis, which made up only a small part of the total flight activity. Single observations suggest that Montagu's harriers fly higher in inland areas (e.g. Hellweg Börde, Mainfranken) than in coastal areas (except for hunting flights) so that the collision risk with turbines of harrier-friendly dimensions still needs to be verified by reliable data on flight behaviour in these areas.

4.10 Flight Behaviour and Collision Risk—Overall Conclusion

60–70% of the flight heights of Montagu's harriers are below 10 m rarely reaching the risk area of currently installed wind turbines (rotor swept area 20–100 m or 50–150 m). Flight activities in critical heights are concentrated on the nest site with 50% of the recorded locations within a radius of 182–497 m (median 343 m) around the nest. Due to the high level of flight activity and the minimal avoidance behaviour towards wind turbines, an increased collision risk occurs in the vicinity of the nest site. With increasing distance from the nest, the flight activity per area is decreasing considerably and is distributed over distances of several kilometres throughout the home range. Due to hunting flights mainly below 10 m, low collision risk occurs in the periphery of the home range. The distance between nest and wind turbine is a decisive factor for the collision risk. In the assessment of the possible collision risk, flight paths between breeding and foraging grounds need to be taken into consideration. Here, a higher number of flights at conflicting heights may be concentrated in areas further away from the nest site (e.g. prey transfers).

4.11 Habitat Use and Selection

The boundaries of the 95% kernel home ranges were used as a reference for habitats available within reach.

To analyse the habitat use, only hunting-related behaviours were included, as only these depend on habitat type. Other behaviours have a spatial relation to the nest site (e.g. resting, courtship displays, circling) but are independent of the type of habitat or land use. Eight birds with more than 50 valid location records were included in the analysis.

4.11.1 Areal Habitats

Arable crops made up the largest proportion of all eight investigated home ranges (between 43 and 91%). Important foraging areas within the home ranges (95% kernel) were salt marshes and/or grassland of different use or intensity of use. In the analysis of the habitats, the following habitats or types of habitat use were established on the basis of the respective proportion of land use and total number of recorded locations:

- Cropland
- Intensive pasture
- Extensive pasture

- Hay meadow
- Wet meadow (extensive)
- Salt marsh.

Figure 4.22 gives an example of the home range of a male bird with recorded locations and random locations used for comparison for different types of habitats.

The analysis of the Euclidian distance ratio (following Conner and Plowman 2001, Table 4.7) shows that the distances of the recorded locations differ from the expected distribution of distances to the habitats (mean of all habitats with a ratio $\neq$ 1.0). Values <1.0 indicate a lower distance ratio and association of the animal with the habitat, values >1.0 indicate a lack of habitat association or avoidance due to a higher distance ratio. The investigated habitats were used disproportionately relative to availability (Table 4.7).

On average, all Montagu's harriers used extensive pastures ($\rho_{extensive\ pasture} = 0.766 \pm 0.23$, $t_8 = -6.18$, $p < 0.01$) and hay meadows ($\rho_{hay\ meadow} = 0.904 \pm 0.54$, $t_8 = -5.48$, $p < 0.01$) significantly more often.

Salt marshes returned on average higher distances to locations than expected ($\rho_{salt\ marsh} = 1.55 \pm 0.671$, $t_8 = 1.63$, $p = 0.128$) as they were not or to a small proportion within the home range of five of the eight birds. The three remaining birds showed a very strong association with this type of habitat (L1m, L3m, FWL2m with values of <0.5 each, Table 4.7). Representative for all birds with a preference for salt marshes, home range and spatial distributions of the recorded locations while hunting of male FWL2 are presented in Fig. 4.22. Typically, salt marshes make up a smaller proportion of the area within the home range than agricultural land, but return a much larger amount of locations than to be expected in a random distribution (simulated points). In case of a male bird it was obvious that the wind turbines east of the breeding site were not approached due to a strong focus on the salt marshes in the west, which were therefore not within the centre of activity.

Wet meadows only occurred in two of the eight investigated home ranges located in the northern study area within the nature reserve Rickelsbüller Koog. Again, the low distance ratio resulted in a strong association of the birds with this type of habitat.

A considerably higher distance ratio than expected was determined for crops and intensive pastures. Only the recorded locations of female L1w (2007) showed lower distances than expected (see Table 4.7), using only cropland in the vicinity of the nest for foraging. The example of habitat use in Fig. 4.22 confirms that arable crops and intensive pastures were used disproportionately despite the high proportion of these habitats within the home ranges (between 44 and 69%).

4.11.2 Temporary Available Habitats

The habitat of type "hay meadow" was summarized in Table 4.7 without consideration of the state. For seven of the nine investigated birds, strong association with this habitat was temporary and only occurred directly after mowing. Selecting only data for freshly mown meadows from all data available for this type of habitat, the distances determined for four of the birds with sufficient data available (>30 locations) resulted in even lower distances and thus stronger association with this type of habitat directly after mowing (Table 4.8). For three of the four birds the increase in preference as well as the lowered ratio (stronger association) was significant.

Mowing of hay meadows prompts Montagu's harriers to specifically visit and intensely hunt these areas. In times of mowing, hunting flights of several kilometres towards areas sometimes far outside the kernel home ranges occurred. This intensity of use only lasts for one to three days. These observations correspond to results found for Montagu's harriers in the Netherlands (Trierweiler et al. 2010). Red kite show the same behaviour (see Sect. 3.3.3; Mammen et al. 2013).

A similar reaction was caused by harvesting of cereal crops starting with winter barley during raising the nestlings. Here too flights were aimed towards areas partly several kilometres away after threshing. The data available for this temporary preference is too small to carry out statistical analyses, but the reaction of the birds can be seen from a noticeable concentration of recorded locations in such areas, which are partly located outside hunting grounds or paths.

Table 4.7 Preferences (distance ratio between used locations and random points) for selected habitat types

Bird	Arable crop	Extensive pasture	Intensive pasture	Hay meadow	Wet meadow	Salt marsh
L1m 07	1.554	0.913	1.111	0.855	–	0.339
L1w 07	0.892	–	–	0.639	–	–
L3w 07	2.544	0.533	1.037	0.681	–	1.088
L1m 08	1.878	0.736	0.977	0.939	–	–
L3m 08	1.653	1.208	1.430	1.087	–	0.395
FWL2m 07	1.620	0.876	1.872	0.865	0.643	–
FWL2w 07	3.346	0.351	2.891	1.233	–	3.130
FWL1m 08	3.754	0.468	1.222	0.968	0.894	1.108
FWL2m 08	1.265	1.188	2.106	0.872	–	0.411
Medium	2.056	0.766	1.648	0.904	–	1.550

Bold values: distances lower than expected, <1.0 preference. – habitat not available in the home range

4.11.3 Linear Habitats or Structures

In habitat mapping, all visible boundary and fringe structures within the investigated home ranges were included as far as possible. Observations of the radio-tagged Montagu's harriers showed that they follow boundary structures especially while hunting. To verify, if birds will show a preference for boundary structures within their home range, the distance ratio between recorded locations and random points was investigated for all available linear habitats or structures using the methods described in Sect. 4.11. These may be boundary lines and fringe structures between habitats or partial areas of the same type or structure, reed lined ditches or banks as well as ruderal edges.

On average, locations of hunting Montagu's harriers (n = 1785) within the home range were considerably closer to edge structures than expected ($\rho_{\text{boundaries}} = 0.686 \pm 0.211$, $t_8 = -5.36$, $p < 0.01$). The low ρ-ratio indicates a strong preference for boundary lines or fringe structures.

Selecting only arable crops of all types of habitat available within the home ranges, this effect is intensified by a further decrease of distances as compared to expected values ($\rho_{\text{cropland boundaries}} = 0.504 \pm 0.262$, $t_8 = -8.08$, $p < 0.01$). Therefore, birds hunting near arable crops mainly depend on the use of boundaries. The cropland itself is largely unsuitable for hunting due to the growth of the crop with the exception of crops with lower coverage (beet, potatoes, etc.).

For the habitat type "extensive pasture", preferred by six of the birds, an edge effect could be proven, which was less pronounced than for cropland ($\rho_{\text{extensive parsture boundaries}} = 0.815 \pm 0.322$, $t_5 = -4.31$, $p < 0.01$). An especially strong association with boundaries was found for female FWL2 (2007) in the northern study area, intensely hunting an extensive sheep pasture directly adjoining a row of wind turbines, resulting in frequent approaches of the wind turbines (Fig. 4.23). The extensive pasture made up 1.5% of the home range area of this bird, but 39.5% (n = 332) of all locations were recorded in this area.

For the four birds using the salt marsh hunting habitat, no association with boundaries could be found ($\rho_{\text{salt marsh boundaries}} = 1.304 \pm 0.262$, $t_3 = -0.09$, $p <$ 0.19). This is probably due to the areal distribution of the nests or young birds of breeding open-area birds, demanding extensive search while hunting.

Table 4.8 Preferences (distance ratio between used locations and random points) for hay meadows before and after mowing of four birds

Bird	Before mowing	After mowing	Difference (*p*, t-test)
L1m 07	1.082	0.433	0.023
L3w 07	0.811	0.334	0.015
L3m 08	1.087	0.803	0.042
FWL1m 08	1.086	0.964	n/a

Bold values: distances lower than expected, <1.0 preference

4.12 Habitat Use and Selection—Discussion

Analysis of habitat use within the Montagu's harriers home ranges resulted in a strong preference for extensively grazed grassland. In both study areas, the extensive sheep pastures were most important, as extensively grazed cattle pastures were not or at larger distances from the breeding site available within the home ranges. Depending on availability, single birds showed a preference for further types of extensive use e.g. hay meadows, fallow grassland and wetland, which only occurred in the northern study area (nature reserve Rickelsbüller Koog). One male bird flew several times into a military training area over 10 km away, to hunt on semi-natural and dry grasslands.

Salt marsh habitats are exceptional due to spatial limitation on coastal areas and the resulting limited availability. This habitat differs from other preferred types of habitat substantially in structure. The harriers prefer coastal and not or only partially grazed areas of the intertidal zone, which are part of the Wadden Sea National Park. Here vegetation is clearly higher, denser and more heterogeneous than in intensively short grazed pastures. The areas are traversed by an extensive system of tideways and ditches adding to the structural diversity. This type of vegetation, which in terms of structure can be compared to arable land left fallow for several years, usually is unsuitable for birds hunting visually for small mammals and are otherwise avoided by Montagu's harriers. On the other hand, small mammals do not occur in zones with a high salinity level and Montagu's harrier instead focus on hunting young birds of grassland breeders (common redshank, oystercatcher, lapwing, skylark, meadow pipit), which are extremely abundant in these areas. The habitat offers a rich supply of prey over the whole breeding season and was even used after fledging the young in the beginning of August on a regular basis. Availability of these areas depends on distance from the nest site, competition with other Montagu's harriers and individual or gender-specific preferences for certain hunting habitats. None of the female birds showed a preference for salt marshes and only few locations of females were recorded in these areas. Hints on intra-specific feeding competition and niche differentiation of Montagu's harriers can be found in studies carried out by Kitowski (2003) on sub-adult and adult male birds. Possibly different hunting techniques are used by male and female birds (e.g. due to the difference in size, see Garcia and Arroyo 2005) leading to partitioning of habitats. Location and distance from suitable hunting grounds probably play an equally important role in differences of use. The minimum distance between nest site (centre of activity) and salt marshes was between 436 and 2060 m for the eight birds. For the three territorial birds showing a preference for salt marshes, these areas were located, with the exception of one bird, closer to the nest (436, 1205 and 1469 m) than for birds preferring different types of habitat. As female birds show stronger attachment to the nest over the whole breeding season than male birds and usually stay closer to the nest, they mainly use habitats in the vicinity of the nest for foraging. In the late nestling stages distances over 2 km will only occur, if favourable food sources are temporarily available (mown meadows, harvested cereal crops).

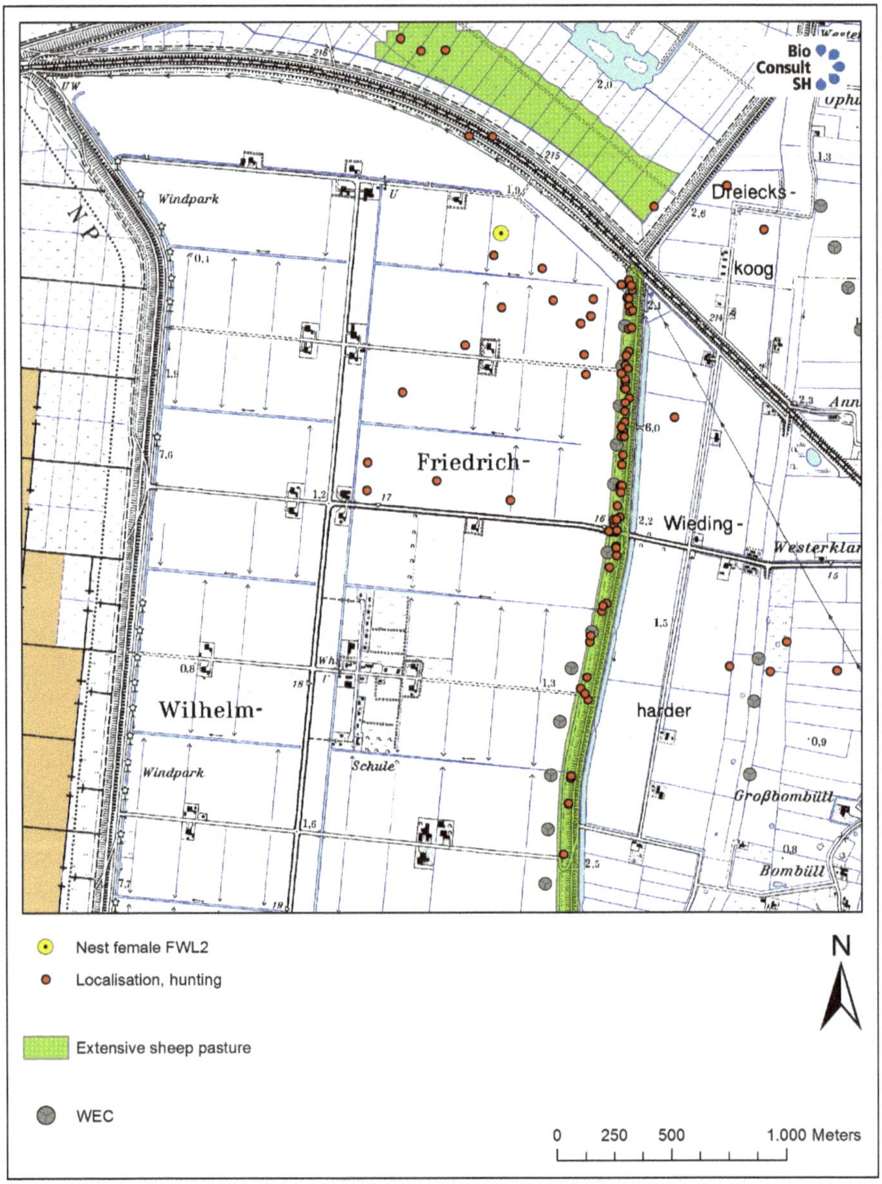

Fig. 4.23 Edge effect of an extensive sheep pasture adjoined by a line of wind turbines (northern study area): Locations of female FWL2 (2007) while hunting for prey

Apart from salt marshes, Montagu's harriers showed a clear preference for short pastures, meadows and fallow grassland. Preference of low vegetation was especially obvious, when freshly mown meadows were used intensely and hunted for a short period of one to three days, approached over partly large distances. The intensively used cattle and sheep pastures present in all territories, appear to structurally suit the hunting strategy of Montagu's harriers due to the likewise low vegetation, but were used by none of the birds disproportionally. These areas probably do not offer a sufficient amount of small mammals or breeding bird species. The habitat preferences of Montagu's harriers in agricultural landscapes (apart from salt marshes) determined in this study are mostly identical with the results of telemetric monitoring in the Netherlands, where the abundance of small mammals was investigated for different types of habitat by means of trap catches (Trierweiler et al. 2010). This proves that the present study may well differ from many other breeding ranges in Germany and the Netherlands due to the presence of salt marshes, but results on habitat selection and the derived measures for habitat protection in an agricultural landscape still are transferable to other breeding ranges.

4.13 Habitat Management and Preventive Measures

On the basis of the results of the telemetry study, BioConsult SH has developed a model concept for habitat management of Montagu's harriers in the northern part of the southern study area (Sönke-Nissen-Koog, Louisenkoog) to be applied in the planning of future wind farm projects in areas of conflict. Aim of the concept is to relocate breeding sites to locations outside the range of impact of wind turbines (avoidance of collision risk) and to channel flight activity along suitable created foraging grounds or edge strips towards the important foraging grounds of the salt marshes and the extensive grassland. The required pooled space is made available within a spatially defined area (study area) for preventive measures (cereal crops used as breeding habitat) and compensation measures (grassland compensation areas, foraging habitats). Over a period of 20 years, management measures adapted to the variable breeding behaviour of Montagu's harriers shall be carried out for varying parts of the project area. Tests of the suitability of different alternative cereals and types of vegetation as a breeding habitat are planned for. The management concept comprises the following key components based on the breeding and hunting behaviour of Montagu's harriers:

- No cultivation of crops suitable for breeding (mainly winter barley) in the vicinity of wind turbines (within a radius of 300 m due to the determined concentration of flight activity in this area around nests).
- Targeted cultivation of barley and other structurally suitable cereals serving as breeding habitat on varying agricultural land areas away from wind farm sites (>300 m from wind turbines) and in the vicinity of existing or created suitable foraging grounds.
- Creation of suitable small-scale grassland areas serving as foraging habitats including extensification and maintenance measures in the surrounding area of nest sites (maximum distance: 2 km).

- Creation and expansion of buffer strips or linear water systems with edge strips (ditches, sluices with banks) between breeding and main foraging habitats.

Due to the annually distinct preference of winter barley as a nest site as well as of linear structures as foraging habitat or guidance in foraging flights, breeding sites and foraging grounds will effectively be moved out of the impact area of wind turbines by applying this management concept. An improvement of the quality of Montagu's harriers habitats can be expected as a result of spatial connection of compensation measures within the breeding area (foraging habitats).

References

Arroyo B, Mougeot F, Pinilla A, Crystal F, Guerrero A (2009) Home Range Sizes and Foraging Habitat Selection in the Montagu's Harrier (*Circus pygargus*): Implications for Conservation Management. Proceedings of Raptor Research Foundation Annual Conference, Pitlockry, 29 Sep–4 Oct 2009.

Band W, Madders M, Whitfield DP (2007) Developing field and analytical methods to assess avian collision risk at wind farms. In: de Lucas M, Janss G, Ferrer M (eds) Birds and Wind Power. Quercus, Madrid, p 259–275.

Baum R, Baum S (2011) Wiesenweihen in der Falle. Beobachtungen in einem ostfriesischen Windpark. Falke 58:230–233.

Belting C, Krüger RM (2002) Populationsentwicklung und Schutzstrategien für die Wiesenweihe *Circus pygargus* in Bayern. Ornithologischer Anzeiger 41:87–92.

BirdLife International (ed) (2004) Birds in Europe: Population estimates, trends and conservation status. BirdLife Conservation Series No. 12, Cambridge.

Bright JA, Langston RHW, Bullman R, Evans RJ, Gardner S, Pearce-Higgins J, Wilson E (2006) Bird Sensitivity Map to provide locational guidance for onshore wind farms in Scotland. RSPB Research Report No 20, Sandy, pp 136.

Chamberlain DE, Rehfisch M, Fox A, Desholm M, Anthony SJ (2006) The effect of avoidance rates on bird mortality predictions made by wind turbine collision risk models. Ibis 148:198–202.

Clarke R (1996) Montagu's Harrier. Arlequin Press, Chelmsford.

Conner LM, Plowman BW (2001) Using Euclidean distances to assess nonrandom habitat use. In: Millspaugh JJ, Marzluff JM (eds) Radio Tracking and Animal Populations. Academic Press, New York, pp 474.

Conner LM, Smith MD, Burger LW (2003) A comparison of distance based and classification-based analysis of habitat use. Ecology 84:526–531.

Cramp S, Simmons KEL (1980) Handbook of the birds of Europe, the Middle East and North Africa. Vol II, Oxford University Press, Oxford, pp 695.

De Voogd M (2004) Hunting Yield and Habitat Use in the Montagu's Harrier. Master thesis, University of Groníngen.

Dürr T (2007) Vogelverluste an WKA in Deutschland und Fledermausverluste weltweit. Summe der Funde seit 1989. Stand April 2007. Daten aus dem Archiv der Staatlichen Vogelschutzwarte im Landesamt für Umwelt, Gesundheit und Verbraucherschutz Brandenburg. Accessed 26 April 2007.

Dürr T (2013) Vogelverluste an WKA in Deutschland und Fledermausverluste weltweit. Summe der Funde seit 1989. Stand April 2013. Daten aus dem Archiv der Staatlichen Vogelschutzwarte im Landesamt für Umwelt, Gesundheit und Verbraucherschutz Brandenburg. http://www.lugv.brandenburg.de/cms/detail.php/bb1.c.321381.de. Accessed 30 April 2013.

Gahrau C (2011) Untersuchungen zur Brutbiologie der Wiesenweihe (*Circus pygargus*) in Schleswig-Holstein. Bachelor thesis, Christian-Albrechts-University of Kiel.

Gahrau C., Schmüser H (2012): Artenschutzprojekt Wiesenweihe (Circus pygargus) des Landes Schleswig-Holstein. Abschlussbericht zur Brutperiode 2011. Wildtierkataster Schleswig-Holstein.

Garcia JT, Arroyo BE (2005) Food-niche differentiation in sympatric Hen *Circus cyaneus* and Montagu's Harriers *Circus pygargus*. Ibis 147:144–154.

Glutz von Blotzheim UN, Bauer KM (1971) Handbuch der Vögel Mitteleuropas. Vol 4 Falconiformes. Aula Verlag, Wiebelsheim.

Hertz-Kleptow C, Schmüser H (2013) Wiesenweihe. In: Ministerium für Energiewende, Landwirtschaft und ländliche Räume Schleswig-Holstein (ed) Jagd- und Artenschutz Jahresbericht 2013, p 103–105.

Höntsch K (2004) Der Kleinspecht. Autökologie einer bestandsbedrohten Vogelart im hessischen Vordertaunus. Dissertation, Johann Wolfgang Goethe-Universität of Frankfurt/ M. M.-G.-Schmitz-Verlag, Kelkheim.

Joest R, Griesenbrock B, Illner H (2013) Auswirkungen von Windenergieanlagen auf den Bestand und die Nistplatzwahl der Wiesenweihe *Circus pygargus* in der Hellwegbörde, Nordrhein-Westfalen. In: Hötker H, Krone O, Nehls G (eds) Greifvögel und Windkraftanlagen: Problemanalyse und Lösungsvorschläge. Schlussbericht für das Bundesministerium für Umwelt, Naturschutz und Reaktorsicherheit. Michael-Otto-Institut im NABU, Leibniz-Institut für Zoo- und Wildtierforschung, BioConsult SH, Bergenhusen, Berlin, Husum.

Kenward RE (2001) A Manual for Radio Tagging. Academic Press, New York, pp 311.

Kernohan BJ, Millspaugh JJ, Jenks JA, Naugle DE (1998): Use of home range estimators in a GIS environmentto identify habitat use patterns. J. Environmental Management 53: 83–89.

Liminana R, Soutullo A, Urios V, Surroca M (2006) Vegetation height selection in Montagu's Harriers *Circus pygargus* breeding in a natural habitat. Ardea 94:280–284.

Looft V, Busche G (1980) Vogelwelt Schleswig-Holsteins 2 - Greifvögel. Karl Wachholtz Neumünster, pp.83–90.

Mammen K, Mammen U, Resetariz A (2013) Rotmilan. In: Hötker H, Krone O, Nehls G (eds) Greifvögel und Windkraftanlagen: Problemanalyse und Lösungsvorschläge. Schlussbericht für das Bundesministerium für Umwelt, Naturschutz und Reaktorsicherheit. Michael-Otto-Institut im NABU, Leibniz-Institut für Zoo- und Wildtierforschung, BioConsult SH, Bergenhusen, Berlin, Husum, http://www.nabu.de/themen/energie/erneuerbareenergien/windkraft/17417.html, Accessed 30 June 2014.

McNay RS, Morgan JA, Bunnell EL (1994) Characterizing independence of observations in movements of Columbian black-tailed deer. Journal of Wildlife Management 58: 422–429.

Mebs T, Schmidt D (2006) Die Greifvögel Europas, Nordafrikas und Vorderasiens. Kosmos, Stuttgart, pp 495.

Millon A, Bourrioux JL, Riols C, Bretagnolle V (2002) Comparative breeding biology of Hen Harrier and Montagu's Harrier: an 8-year study in north-eastern France. Ibis 144:94–105.

Millspaugh JJ, Marzluff JM (eds) (2001) Radio Tracking and Animal Populations. Academic Press, New York, pp 474.

Mohr CO, Stumpf WA (1966): Comparison of methods for calculating areas of animal activity. Journal of Wildlife Management30: 293–304.

Neu CW, Byers CR, Peek JM (1974) A technique for analysis of utilization availability data. J Wildlife Manage 38:541–545.

O'Donoghue B, O'Donoghue TA, King F (2011) The hen harrier in Ireland: Conservation issues for the 21st century. Biol Environ 111B: 83–93.

Pürkhauer C, Hoh E, Klein H, Krüger R, Leuchs O, Pauldrach G (2009) Einsatz für die Wiesenweihe. Falke 56:419–425.

Rasran L, Grajetzky B, Mammen U (2013) Berechnungen zur Kollisionswahrscheinlichkeit von territorialen Greifvögeln mit Windkraftanlagen. In: Hötker H, Krone O, Nehls G (eds) Greifvögel und Windkraftanlagen: Problemanalyse und Lösungsvorschläge.

Schlussbericht für das Bundesministerium für Umwelt, Naturschutz und Reaktorsicherheit. Michael-Otto-Institut im NABU, Leibniz-Institut für Zoo- und Wildtierforschung, BioConsult SH, Bergenhusen, Berlin, Husum, http://www.nabu.de/themen/energie/erneuerbareenergien/windkraft/17417.html, Accessed 30 June 2014.

Salomolard M (1997) Utilisation de l'espacepur le Busard cendre *Circus pygargus*: Superficie et distribution des zones de chasse. Alauda 65:307–320.

Schmüser H, Gahrau C (2011) Artenschutzprojekt Wiesenweihe (*Circus pygargus*) des Landes Schleswig-Holstein. Abschlussbericht zur Brutperiode 2010 im Auftrag Ministerium für Landwirtschaft, Umwelt und ländliche Räume des Landes Schleswig-Holstein, Kiel.

Schoener TW (1981) An empirically based estimate of home range. Theoretical Population Biology 20: 281–325.

Swihart RK, Slade, NA (1985) Influence of sampling intervalon estimatesof home range size. Journal of Wildlife Management 49:1019–1025.

Swihart RK, Slade NA (1997) On testing for independence of animal movements. J Agr Biol Envir St 2:48–63.

Südbeck P, Bauer HG, Boschert M, Boye P, Knief W (2009) Rote Liste und Gesamtartenliste der Brutvögel (Aves) Deutschlands. Naturschutz und Biologische Vielfalt 70(1): 159–227.

Trierweiler C, Drent RH, Komdeur J, Koks BJ (2010) Home range size and habitat selection of the endangered Montagu's harrier *Circus pygargus* in NW-Europe: implications for conservation. In: Trierweiler C (ed) Travels to feed and food to breed: The annual cycle of a migratory raptor, Montagu's harrier, in a modern world. Dissertation, University of Groningen.

Visser E, Koks B, Trierweiler C, Arisz J, van der Leij RJ (2006) Grauwe Kiekendieven *Circus pygargus* in Nederland in 2006. De Takkeling 15: 116–132.

White GC, Garrott RA (1990) Analysis of Wildlife Radio-Tracking Data. Academic Press, Inc., San Diego, New York, Boston, London, Sydney, Tokyo, Toronto.

Whitfield DP, Madders M (2006a) Flight height in the Hen Harrier *Circus cyaneus* and its incorporation in wind turbine collision risk modelling. Natural Research Information Note 2, Natural Research, Banchory.

Whitfield DP, Madders M (2006b) A review of the impacts of wind farms on the Hen Harrier *Circus cyaneus* and an estimation of collision avoidance rates. Natural Research Information Note 1. Natural Research, Banchory.

Worton BJ (1987) A rewiew of models of home range for animal movement. Ecological modelling 38:277–298.

Worton BJ (1989) Kernel methods for estimating the utilization distribution in home range studies. Ecology 70:164–168.

Chapter 5
Impacts of Wind Turbines on the Population and Nest Site Selection of the Montagu's Harrier in the Hellweg Börde, North Rhine-Westphalia

Ralf Joest, Bastian Griesenbrock and Hubertus Illner

5.1 Introduction

Apart from the Mainfränkische Platten and the East and North Frisian Marshes, the Hellweg Börde is one of the most important traditional breeding areas of Montagu's harriers in Germany. Since the beginning of the 1990s, between 17 and 44 pairs breed in this area per year. Due to the particular importance of the Hellweg Börde for the breeding population of the Montagu's harrier and other birds breeding in open agricultural landscapes, such as the corncrake, as well as for migratory birds as red kite, lapwing or golden plover, the area was designated as Special Protection Area (SPA) by the North Rhine-Westphalia federal state in 2004. At the same time, the Hellweg Börde is one of the most suitable regions for wind energy use due to the open landscape and the favourable location on the northern edge of the Central German Uplands. Since the beginning of the 1990s, about 270 wind turbines were erected in the Hellweg Börde, including single wind farms with more than 20 turbines, partly located in the vicinity of core areas of the breeding distribution of Montagu's harriers. As obvious from several planning approval procedures for wind turbines in recent years, there is still a potential for conflicts between wind energy use and the protection of Montagu's harriers in the Hellweg Börde.

Aim of this study was to investigate potential impacts of the construction of wind turbines on the population and the nest site selection of Montagu's harriers in the Hellweg Börde. Data collected between 1993 and 2007 was analysed. Due to

R. Joest (✉) · B. Griesenbrock · H. Illner
Arbeitsgemeinschaft Biologischer Umweltschutz im Kreis Soest e.v.
(Soest District Working Group for Nature Conservation) ABU,
Biological Station Soest, Teichstraße 19,
59505 Bad Sassendorf Lohne, Germany
e-mail: r.joest@abu-naturschutz.de

H. Hötker et al. (eds.), *Birds of Prey and Wind Farms*,
DOI 10.1007/978-3-319-53402-2_5

comprehensive annual mapping, number and location of Montagu's harriers nests are known for this period. The underlying question was, whether the construction of wind turbines caused a loss of suitable breeding areas for the Montagu's harrier during this period. Further possible impacts of wind turbines on the Montagu's harrier, i.e. on the choice of hunting grounds, on the hunting behaviour and on the collision risk, could not be investigated within the framework of the present study. The influence of wind turbines on the spatial distribution and the nest site selection of Montagu's harriers were investigated by overlaying nest sites with locations of wind turbines using a geographic information system. Different complementing methods were used.

In a first step, the development of wind energy use and the Montagu's harrier population within the Hellweg Börde will be described. Subsequently, a possible relation between the number of wind turbines and the number of Montagu's harrier nests within the study area will be investigated. In selected areas with wind farms constructed in the vicinity of previous nesting areas of Montagu's harriers, the distribution of nest before and after construction of the turbines will be compared to determine to what extent the construction had an influence on the establishment of territories. A further methodological approach serves to investigate, whether Montagu's harriers will avoid the vicinity of wind turbines while selecting a nest site and which additional landscape parameters will have an influence on the selection.

5.2 Protection and Conservation Status of the Montagu's Harrier in the Hellweg Börde

The Montagu's harrier is a species protected under European law (Annex I, EU Birds Directive, 79/409/EWG). In the current German Red List, the species is listed as "critically endangered" (Südbeck et al. 2007) and in the current Red List of North Rhine-Westphalia as "threatened with extinction" (Sudmann et al. 2009). The nationwide population of about 450 breeding pairs (as from 2005: 410–470, Südbeck et al. 2007) is concentrated on the Mainfränkische Platten in Bavaria, the coastal lowlands of Lower Saxony and Schleswig-Holstein, the Hellweg Börde in North Rhine-Westphalia as well as the federal states of Brandenburg, Mecklenburg Western-Pomerania and Saxony-Anhalt (Boschert 2005; Mebs and Schmidt 2006).

In North Rhine-Westphalia, the breeding population is concentrated in the Hellweg Börde, an intensively farmed Börde landscape on the southern edge of the Westphalian Lowland (NWO 2002). This population in the Hellweg Börde is well documented since the 1920s (Peitzmeier 1969). From only few pairs in the 1960s, the population was constantly growing since the 1970s due to extensive protection measures for birds nesting in cereal crops (avoidance of breeding failure due to harvesting activities, see below). Towards the end of the 1980s, about 20 breeding pairs could be counted in the Hellweg Börde. In the 1990s, the population was still increasing to an average of about 35 breeding pairs due to the continuing conservation efforts and a larger availability of prey due to the creation of numerous fallows in the course of the set-aside scheme of the EU. Since the beginning of the millennium, a declining trend of the breeding population of the Montagu's harrier can be observed in the Hellweg Börde coinciding with a decline in set-aside farmland.

Natural breeding sites of the ground-nesting Montagu's harrier are the moorland areas and reed beds of the lowlands (Bauer et al. 2005; Mebs and Schmidt 2006). After loss of larger wetland areas in Central Europe, the Montagu's harrier is more and more nesting in agricultural farmland throughout Europe (Bauer et al. 2005; Mebs and Schmidt 2006). Nests are built in winter cereal crops, mainly winter barley. Many broods are endangered by harvesting machinery, as young birds are often not fully fledged at the time of harvest. This circumstance requires the implementation of special protection measures, which have a long tradition in the Hellweg Börde. Protection of the broods by voluntary ornithologists started in the 1960s. Since 1993, protection is implemented by the biological station of the Arbeitsgemeinschaft Biologischer Umweltschutz im Kreis Soest e.V. (ABU) within the framework of the Montagu's harriers protection programme initiated by the North Rhine-Westphalia state government (Glimm et al. 2001).

5.3 Study Area

5.3.1 Location

The study area Hellweg Börde covers an area of 1013 km^2 and is situated between the edge of the North German Plain and the northern edge of the Central German Uplands in North Rhine-Westphalia (coordinate of the north-eastern edge: 51°46N, 8°46E). With a width of 5–30 km, it extends over a length of 77 km from Unna in the west up to Paderborn in the east (Figs. 5.1 and 5.2). Landscape in the study area is characterized by intensive farming along today's federal highway B1 (Fig. 5.2). The elevation above sea level is evenly rising from 70 m in the north to a maximum of 410 m in the south-east. The Münsterland and the Lippetal lowlands mark the northern edge. Towards the west, the study area is delimited by the Ruhr conurbation. Natural boundaries of the study are the valleys of the rivers Ruhr and Möhne in the south and in the east the valley of the river Alme. Politically, the major part of the study area is located within the municipality of Arnsberg. With 75%, the Soest district makes up the largest part followed by the Unna district in the west with 17% and the Paderborn district in the east with 7%. The proportion of the Hamm municipal area in the north-west accounts for less than 1%.

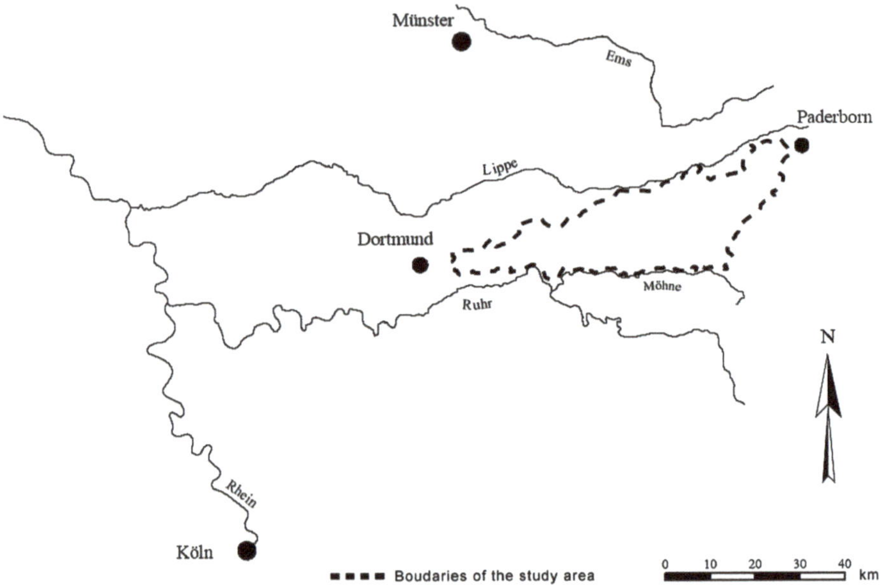

Fig. 5.1 Location of the Hellweg Börde study area

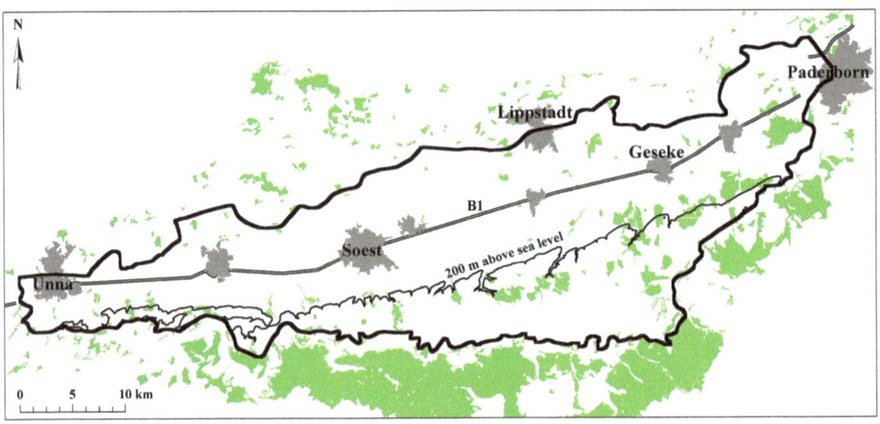

Fig. 5.2 Boundaries of the Hellweg Börde study area

5.3.2 Classification of Natural Landscape Regions

The study area follows the natural boundaries of the German natural regions (Bürgener 1963a, b; Meisel 1959, 1960) and covers the main area of distribution of Montagu's harriers in North Rhine-Westphalia (Glimm et al. 2001; NWO 2002).

The study area is located in the natural area of the Westphalian Lowland (54) and is largely consistent with the major unit Hellweg Börde (542) (Fig. 5.2).

The Hellweg Börde stretches between the Kernmünsterland (541) and the East Münsterland (540) in the north and the Nord Sauerland Oberland (334) und Lower Sauerland (337). The Paderborn Plateau (362) is the neighbouring major unit to the east, and Emscherland (543) and West Hellweg (545) follow in the west.

Within the Hellweg Börde major unit (54), the study area comprises the Soest Lower Börde (542.12), the Geseke Lower Börde (542.13), the Werl-Unna Börde (542.21), the Soest Upper Börde (542.22) and the Geseke Upper Börde (542.23) followed by the Haar ridges in the south (542.30). The Kamener Hügelland (542.0), the Dortmunder Hellwegtal (542.10), the Dortmund Börde (542.20) and the Süderhaar (543.31) were excluded, as they are not used by Montagu's harriers on a regular basis.

The Lower Hellweg (542.1) is a slightly undulating to even and 1–7 km wide lowland running in west to east direction with an elevation of 80–100 m above sea level. The lowest parts can be found in the Lippe valley, which forms the northern boundary. The relief of the at most slightly undulating Lower Hellweg (542.1) is formed by streams running northwards. These have their source along the southern boundary, which follows the Hellweg, the trading route of the Middle Ages, today's B1. Bordering in the south are the boundaries of the Upper Hellweg (542.2) gently sloping up to the Haarstrang. The natural structures of the Upper Hellweg (542.2) with an elevation of 100–320 m above sea level mainly differ in their north to south expansion. The Werl-Unna Börde (542.21) is between 3 and 4 km wide in the west and rises steeply towards the Haarstrang. The Soest Upper Börde (542.22) follows further to the east. The north-facing edge is getting wider and reaches its maximum width of up to 12 km in the area of the Geseke Upper Börde (542.23) with the lowest slope gradient of the Upper Hellweg, but relatively large elevations of up to 320 m above sea level. The transition between Upper Hellweg (542.2) and Haarstrang (542.3) is fluent. According to height, exposition and soil surface, the Haar (542.3) is divided into the Haarhöhe (542.30) and the Süderhaar (542.31), which is descending into the Mittelruhrsenke. The Haarhöhe (542.30) forms the even hill ridge. The at most gently inclined crest stretches from Dortmund-Aplerbeck eastwards to Büren. Its elevation varies between 200 and 300 m in the western part of the study area and is ascending to a maximum of 400 m above sea level towards the east. The steeper southern slope, descending into the valleys of Ruhr and Möhne, forms the sharp, geological southern boundary of the Haarhöhe (542.30), the Westphalian Lowland (54) and at the same time of the entire study area.

5.3.3 Present Use

In the Hellweg Börde, the historical open agricultural landscape formed by natural conditions is still intact today (LWL 2007). Open agricultural landscape accounts for the major part of the study area (Table 5.1).

Table 5.1 Landscape structure in the Hellweg Börde (spatial database LVERMA NRW 1999, 2000; LDS 2003)

Land use	Area (km^2)	Proportion (%)
Open farmland	765	76
Settlement	136	13
Forest and woodland	75	7
Infrastructure	36	4
Total	1013	100

Table 5.2 Proportion of crops cultivated on arable land (77,523 ha) in the Soest district in 2003 (Chamber of Agriculture (LWK) 2004)

Crop	Proportion [%] of farmland within the Soest district
Winter wheat	32
Winter barley	21
Winter triticale	3
Winter rye	2
Summer cereals	7
Maize	7
Oilseed rape	12
Sugar beet	4
Potatoes	2
Other	4
Fallow	6
Total arable crops	100

Crop farming prevails in the open landscape. Cultivation of winter cereals, especially wheat, barley, oilseed rape and different root crops is typical for the region. As an example, Table 5.2 shows the current crops in the Soest district, which takes up the major part of the study area. Pastures and meadows make up a proportion of only 9%. They are restricted to village outskirts, the moist lowlands of the Lower Hellweg and the dry valleys ("Schledden") of the Upper Hellweg.

Next to mainly agricultural use with a small proportion of woods, the open character of the Hellweg Börde landscape can be attributed to the small degree of urban sprawl in comparison to the Münsterland in the north and the Ruhr area in the west. Villages and towns scattered over the Hellweg Börde cover a total of only 13%. Built-up areas are mainly restricted to settlements. Only single human structures can be found in the open landscape of the study area. Over the last decades though, the use of open spaces for industrial estates, roads, mining (cement industry), dumps and wind turbines is increasing.

With 7%, forests and woods take up only a small proportion of the total area of the study area. They are restricted to parts of the Upper Hellweg, where they are concentrated along the dry valleys, which cannot be farmed, as well as on low-quality soils in the Geseke Upper Börde. The crest of the Haarstrang is virtually treeless. Smaller patches of forest or woodland are evenly distributed over the entire study area. An almost treeless strip can only be found in the transition zone between Lower and Upper Hellweg.

5.3.4 Conservation Status

About half of the study area became the Natura 2000 site "Hellwegbörde" (Code: DE-4415-401) in 2004. In the designation of the area, existing concentration areas for wind turbines were excluded. Protective purpose of this area is to protect the breeding populations of Montagu's harriers, marsh harriers, corncrakes and other open landscape species listed in Annex I of the EU Birds Directive, occurring in numbers of international importance in the area as well as of internationally important numbers of birds resting on migration as red kite, lapwing and golden plover (www.naturschutz-fachinformationssysteme-nrw.de/natura2000/meldedok/, 12.10.2008). The major part of the protected area is situated in the Soest district. Here, the conservation obligations are mainly carried out within the framework of an agreement on the protection of the Montagu's harrier and other species of the open landscape in the Hellweg Börde, which aims to account for the interests of politics, trade, industry, agriculture and nature conservation.

5.4 Materials and Methods

5.4.1 Underlying Data

5.4.1.1 Nest Sites of Montagu's Harriers

Between 1993 and 2007 including, nest sites of the Montagu's harrier within the Hellweg Börde were comprehensively mapped on an annual basis by the biological station of the Arbeitsgemeinschaft Biologischer Umweltschutz (ABU) im Kreis Soest on behalf of the state North Rhine-Westphalia. An almost complete annual coverage of all breeding sites can therefore be assumed (Glimm et al. 2001; Hölker 1994–2006; Illner 2007, 2008). Data on 595 Montagu's nests with evidence of nesting were included in the present study. As the number of replacement clutches only made up a small proportion of the total broods and does not show a distinct temporal trend (Illner 2007), the number of nests built per year can be used as a measure for the breeding population and its long-term development.

The maximum spatial error in the localisation of nest sites in the field and their transfer into topographical maps is 50–100 m for the years 1993–2005 and 10 m for the years 2006–2007 (verification by means of GPS). The digitised nest sites are available at a scale of 1:25,000 with an accuracy of up to 25 m (LVERMA NRW 2008). This results in a maximum total error of 75–125 m for the location data recorded between 1993 and 2005 and of approximately 35 m for the years 2006 and 2007.

5.4.1.2 Wind Turbines

Locations of wind turbines within the study area as well as turbine specifications and time of commissioning were researched and included in the present evaluation. These data were made available by the district councils of Arnsberg, Unna, Soest and Paderborn, the municipalities Rüthen, Anröchte, Möhnesee, Ense and Fröndenberg as well as the German Wind Energy Association (BWE, regional association of Southern Westphalia) and Enercon GmbH. Site, date (or at least month) of commissioning, nominal capacity, rotor diameter and hub height are known for all wind turbines. It is assumed that the date of commissioning does not differ significantly from the date of construction. As no dates on the start of the construction were available, the impacts of the construction phase could not be determined. Turbine- and operation-related impacts could not be evaluated separately, as no further information was available on the duration of operation. It is assumed that most of the turbines were in operation. The resulting uncertainty may have an influence on an additional visual impact caused by the moving rotor but not on the impact of the turbine as a vertical structure within the landscape.

All turbines put into operation until May of the breeding season were assumed to be relevant for the nest site selection of Montagu's harriers from that year onwards. Wind turbines put into operation at a later date were considered to be relevant from the following year. The choice of this date is due to the breeding phenology of the Montagu's harrier. As the selection of the nest site ends with egg-laying, wind turbines put into operation at a later date will not have an influence on the selection for the current breeding season. Between 1993 and 2005, egg-laying usually started between 16 and 27 May (Hölker 1994–2006).

In the definition of single wind turbines as opposed to wind farms, the locations given by the operator as well as the spatial relation between wind turbines and their distance from each other were used. Groups of at least three wind farms were defined as wind farm, as this is the minimum number to define an area around these turbines. The area of wind farms was defined by a minimum polygon connecting the outermost wind turbines (minimum convex polygon).

Due to possible impacts on the study area, all wind turbines outside the study area within a radius of 2 km were included in the evaluation. Wind turbines with a capacity of below 100 kW were not included, as these single do-it-yourself constructions are relative small in comparison to state-of-the-art turbines. As this kind of turbine can usually be found in the vicinity of residential areas, their impact cannot be clearly separated from impacts caused by the residential area.

5.4.1.3 ATKIS

Data on land use and landscape structure were taken from the German federal topographic information system (ATKIS-OK, as from 1999). In a nationwide standardized digital data pool, this information system makes geometrical and thematic information of topographical maps available (Bartelme 2005). Digitisation is based on the German basic map and the aerial map on a scale of 1:5000 allowing for accuracy to 3 m (Dickmann and Zehner 1999). Within the ATKIS feature catalogue (ATKIS-OK) 170 features can be found as point, line and area objects (Dickmann and Zehner 1999).

5.4.1.4 Digital Terrain Model

The ATKIS digital terrain model (DGM5) describes the natural terrain of the earth's surface using grid points geo-referenced by location and altitude with a regular spacing of 5 m. These models achieve an accuracy of ±0.5 m (LVERMA NRW 2008).

5.4.2 Method

5.4.2.1 Preliminary Remark

The present study is an evaluation of existing data, searching for possible correlations of the construction of wind turbines with the number and distribution of Montagu's harrier's nests in the Hellweg Börde. This is no experimental setup, investigating the impacts of wind turbines on the breeding population in a before and after comparison accounting for reference areas under controlled conditions.

It should be noted that already in the planning of new sites for wind turbines attempts were made to avoid conflicts with the spatial distribution of the Montagu's harrier and other open landscape breeders. In 1994, a map of areas suitable for wind energy use from a nature and species conservation point of view was compiled by the advisory board of the landscape authority of the Soest district, including the main distribution area of the Montagu's harrier in the Hellweg Börde. Especially the protection of breeding areas of the Montagu's harrier against possible impacts by wind turbines was taken into consideration (Bunzel-Drüke and Schulze-Schwefe 1994; Kampik 1995), ensuring a far-reaching spatial separation of breeding areas and wind turbine locations. In the interpretation of the results, these limitations to the existing data and the descriptive or correlative evaluation methods need to be taken into consideration, as only impacts of wind turbines with far-reaching effect distances can be determined on the distribution of Montagu's harrier's nests. The distance within which an effect on the selection of the nesting or resting site can be assumed, has been determined to fall within a range of below 1 km in studies on other onshore bird species (e.g. Leddy et al. 1999; Larsen and Madsen 2000; Bergen 2001; Kruckenberg and Borbach-Jaene 2001; Hötker 2006). In the present evaluation, the sample size within this distance range was rather small.

5.4.2.2 Number of Montagu's Harriers Breeding in the Hellweg Börde in Relation to the Number of Wind Turbines

This approach aimed to determine a possible correlation between the long-term development of the Montagu's harrier population and the number of wind turbines. The study area was overlaid with a grid of 2 km side length. This grid size was chosen to achieve a sufficiently large number of gird cells containing breeding Montagu's harriers as well as wind turbines. This relatively coarse spatial resolution needs to be taken into account in the interpretation of the results. This approach is only to a limited extent suitable to prove avoidance behaviour of Montagu's harriers towards wind turbines, as merely the number of nests and wind turbines within the grid cells covering 4 km^2 each is included in the evaluation but not their spatial arrangement. The aim is to determine whether the number of Montagu's harrier's nests was influenced by the number of wind turbines during the investigation period.

For each of the grid cells, the number of nests (between 1993 and 2007) and the number of wind turbines as well as the total capacity of the turbines were noted. For grid cells with at least one brood or at least one wind turbine put into operation, the relation between the number of Montagu's harrier nests and the number of wind turbines was determined by means of an analysis of variance (ANOVA) with repeated measures (software R). This method serves to determine whether the number of nests within a grid cell is influenced by the number of wind turbines within the same grid cell. Repeated measures take account of the fact that a time series within a grid cell is no independent sample. At the same time, the relation between the number of Montagu's harrier's nests and the total capacity of wind turbines within the same grid cell were investigated per cell. As the capacity of wind turbines is closely related to rotor diameter and hub height (Spearman rank correlation: capacity–hub height r: 0.655 $p < 0.001$; capacity–rotor radius r: 0.969 $p < 0.001$), the total capacity of the wind turbines within the grid cell was used as an index of the spatial effect of the turbines, comprising the factors number, height and rotor diameter.

5.4.2.3 Establishment of Territories of Montagu's Harriers in the Hellweg Börde Before and After Construction of Wind Farms

In the individual review of case studies, farmland areas with Montagu's harriers breeding within a radius of 1 km around the maximum area of a later constructed wind farm, i.e. within the polygon around the outermost turbines of the farm, were included in the evaluation. Comparison of the number of Montagu's harrier broods before and after the construction of a wind farm aimed to determine the influence of the construction of the wind farm on the use of the area as breeding ground. As only single wind turbines were constructed in some cases, the date of construction was defined as the year with at least three wind turbines in operation for the first time, as this is the number of turbines required to define an area for the wind farm. Wind farms with construction of wind turbines around the middle of the investigation period were chosen for comparison, to ensure that a similar amount of years was available for a before and after comparison.

The number of nests before and after construction of the wind turbines was compared to the number of expected nests based on annual average occupation of the area. The average probability of a brood in an area in a given year can be derived from the total number of nests in this area divided by the number of investigation years (15 years). Expected values for the average occupation of an area in the years prior to or after the construction and the wind farm can then be calculated as the product of the average number of nests over the entire period and the number of nests in the years prior to or after construction (total number of nests/total number of years * number of years before or after). As further factors influencing the occupation of single farmland areas could not be looked at individually and only a small number of valid cases was available, statistical significance testing of the before and after comparison would not have been useful.

5.4.2.4 Nest Site Selection of Montagu's Harriers in the Hellweg Börde in Relation to Wind Turbines and Other Habitat Parameters

In the analysis of the impacts of wind turbines on the nest site selection of Montagu's harriers in the Hellweg Börde, the minimum distances between nest sites and the next wind turbine were compared to randomly selected reference points and the next wind turbine. Apart from wind turbines, the distribution of nest sites may be influenced by additional vertical landscape structures. Results of a diploma thesis for the Institute for Landscape Ecology of the University of Münster (Griesenbrock 2006) showed that Montagu's harriers avoided the proximity of other vertical structures within the landscape as settlements, individual buildings, forests and woodland in the selection of a nest site in the Hellweg Börde. This already resulted in a considerable reduction of suitable breeding areas for Montagu's harriers. The combined effect of wind turbines and other landscape structures on the distribution of nest sites of Montagu's harriers in the Hellweg Börde was therefore investigated in a multi-factor analysis. This analysis was limited to the years 2003–2007, as only few additional wind turbines were erected during that period and only a single wind farm was developed around an already existing wind turbine. According to present knowledge, all existing wind turbines with only a few exceptions (technical failure of single turbines for a few months, one turbine near a village was dismantled, no repowering) were constantly in operation over the investigation period. A nearly stable as-is situation (status quo) can therefore be assumed for the period of investigation between 2003 and 2007. While determining the distances, only those wind turbines present in the respective breeding season were taken into consideration.

Using a geographic information system (ArcView3.2 software) further habitat parameters with possible influence on the nest site selection of Montagu's harriers

Table 5.3 Overview of habitat parameters with possible impact on the nest site selection of Montagu's harriers in the Hellweg Börde included in the present study

Habitat parameter	Abbreviation
Terrain height (metres above NN)	Height NN (m)
Slope gradient (°)	Slope (°)
Distance from next areal vertical object (m)	Dist. areal vert. obj. (m)
Distance from next point vertical object (m)	Dist. point vert. obj. (m)
Distance from next power pylon (m)	Dist. power pylon (m)
Distance from next wind turbine (m)	Dist. WEC (m)
Distance from next wind turbine within a wind farm (m)	Dist. wind farm (m)
Distance from the next neighbour (m)	Dist. neighbour (m)

(Table 5.3) were determined for nesting locations in the years 2003–2007 (n = 162). These were chosen based on preliminary studies within the framework of the diploma thesis by Griesenbrock (2006) on the nest site selection of Montagu's harriers in the Hellweg Börde. As within this evaluation the main focus was on the impacts of vertical structures, roads were not included as a further habitat parameter, especially since the sample size was too small to include a larger number of independent variables in multi-factor analysis.

The influence of land use and especially the location of winter barley as a preferred crop for breeding were assumed to be minor, as winter barley and other winter cereals suitable for breeding made up a rather large proportion of the total area. At least one winter barley crop was available in almost all farmland areas or grid cells of 4 km^2 per year.

Here, vertical areal objects are settlements and forests causing an occlusion effect within the landscape, thus reducing the openness of the landscape. These areal vertical objects are avoided by Montagu's harriers in the selection of a nest site. Features including areas with buildings, forests and woodland were chosen from the ATKIS feature catalogue and summarised for use in the analyses (see Table 5.4). Impacts of single features as forests or settlements were not assessed separately, because they will not differ significantly in the spatial effect relevant for the nest site selection of Montagu's harriers (Griesenbrock 2006).

Point vertical objects are upright, higher structures influencing the openness of the landscape and thereby possibly the nest site selection of Montagu's harriers. These consist of the ATKIS feature mast (Mast, No. 3541), power pylons only in the study area, as well as wind turbines present in the respective breeding season. The few radio masts present in the open landscape (about 20 in 2005) were disregarded, as they have a much smaller spatial effect than the considerably larger power pylons and wind turbines.

In statistical analysis, habitat parameters were determined for a number of randomly selected reference locations (random points) corresponding to the number of nests within areas suitable for nesting. For each year of the study, reference points corresponding to the total number of nests were randomly distributed within the areas suitable for Montagu's harriers using the ArcView Random Point Generator

Table 5.4 Habitat types defined in the ATKIS landscape model—feature types

Open landscape, suitable for breeding (agricultural land i. a.) Area: 80,000 ha
4101 Ackerland (arable crop); 4102 Grünland (grassland); 4104 Heide (heath); 4105 Moor, Moos (moorland, mosses); 4106 Sumpf, Ried (swamp, reed bed)
Open landscape, unsuitable for breeding (sealed and uses unsuitable for occupation) Area: 3000 ha
2121 Bergbaubetrieb (mining operation); 2122 Deponien, in Steinbrüchen (landfills in quarries); 2222 Sportplatz (sports ground); 2230 Golfplatz (golf course); 2301 Tagebau (open pit mine); 2314 Absetzbecken (settling basin); 2345 Schwimmbecken (swimming pool); 3103 Platz (place); 3301 Flughafen, ohne Gebäude (airport, excluding buildings); 3302 Flugplatz, ohne Gebäude (airfield, excluding buildings); 3402 Hafenbecken (harbour basin); 4101 Ackerland (innerorts) (arable farmland, build-up areas; 4102 Grünland (innerorts) (grassland, build-up areas); 4103 Gartenland (gardening land); 4109 Sonderkultur (special crop); 4110 Brachland (fallow); 4120 vegetationslose Fläche (area without vegetation); 4199 Fläche, z. Z. unbestimmbar (currently indefinable); 5100 Wasserflächen (water bodies)
Areal vertical features (areas including buildings, forests and woodland) Area: 20,000 ha; buildings and facilities: 10,000 ha; forest and woodland: 9000 ha
2111 Wohnbaufläche (residential building area); 2112 Industrie- und Gewerbefläche (industrial and commercial areas); 2113 Fläche gemischter Nutzung (mixed use); 2114 Fläche besonderer funktionaler Prägung (areas of special functional character); 2201 Sportanlage (sports facility); 2202 Freizeitanlage (recreational facility); 2213 Friedhof (cemetery); 2227 Grünanlage (park); 2228 Campingplatz (campsite); 3301 Flughafen (airport); 3302 Flugplatz (airfield); 3501 Bahnhofsanlage (railway station); 3502 Rastanlage (service station); 3514 Brücke und Überführung (bridge, footbridge); 4107 Wald, Forst (forest); 4108 Gehölz (woodland)

1.3 tool (Jeness 2005). A total of 162 random points were generated and used in all comparisons with nest sites.

ATKIS-OK was used in the selection of suitable breeding areas for Montagu's harriers (Table 5.4). Features included were arable crops (Ackerland—4101), grassland (Grünland—4102), heathland (Heide—4104), moorland, mosses (Moor, Moos—4106). With about 90%, arable crops made up the major proportion of these types of feature. These forms of land use were assumed to be suitable for nesting Montagu's harriers. Smaller areas, totally enclosed by buildings, were assumed to be not suitable. The same is true for features of unsuitable open landscapes including all sealed surfaces as well as other land uses unsuitable for breeding (Table 5.4).

The fact that Montagu's harriers breed in loose colonies (Arroyo 1995; Clarke 1996) needed to be taken into consideration in the evaluation. Nesting groups mainly occur under favourable conditions, especially under good feeding conditions (Newton 1979; Krogulec 1993). In traditionally occupied farmland areas, part of the Hellweg Börde Montagu's harrier population is annually nesting semi-colonially (referred to as colonies in the following). Single nests outside these colonies are equally common (Glimm et al. 2001). The formation of colonies is presumably based on social attraction by other birds of the same species already occupying the area, thereby resulting in a possible overlay or weakening of effects of the landscape structure on the nest site selection of later arriving breeding pairs. In a strict statistical sense, pairs breeding in colonies will therefore not be independent of each other. Analyses were therefore carried out for the total number of nests (n = 162) as well as separately for the total number of single nests and single colonies (n = 81) to obtain samples largely independent from effects of social attraction. In this reduced sample, habitat parameter values determined for nests belonging to colonies were averaged so that a colony of several broods is a data point in the calculation (Table 5.5). Following Arroyo (1995) and Griesenbrock (2006), nests at a distance of below 600 m from the next nest are considered to be part of a colony. At larger distances from the next neighbour, the nest is assumed to be a single nest. Analogous to the total number of nests, a corresponding number of a total of 81 random points was generated for the reduced but largely independent sample as a reference for the nest sites.

5.4.3 Statistical Evaluation

In the analysis of the influences of the investigated habitat parameters on the nest site selection of Montagu's harriers it is assumed that a significant difference in a habitat parameter between nest sites as compared to randomly selected reference locations (random points), indicates preference for or avoidance of the respective parameter.

Table 5.5 Distribution of Montagu's harrier nests in the Hellweg Börde on single nests and colonies for the years 2003–2007

	2003	2004	2005	2006	2007	2003–2007
Total number of nests	40	46	26	17	33	162
Number of single nests	8	8	11	8	14	49
Number of nests in colonies	32	38	15	9	19	113
Number of colonies	8	11	4	4	5	32
Nests per colony (averaged)	4.0	3.5	3.8	2.3	3.8	3.5
Total single nests and colonies	16	19	15	12	19	81

Differences in habitat parameters between nest sites and random points were described using medians and quartiles, which are robust towards deviations from the normal distribution, and tested for significance using the non-parametric Wilcoxon rank test (Software R).

As comparisons of habitat parameters for nest sites and random points are based on the assumption that single parameters are independent of each other, disregarding possible combined effects, binary logistic regressions of the dependent variables "nest site" (nest = 1 and no nest (random point) = 0) and the investigated habitat parameters as independent variables were tested for significance (Backhaus et al. 2006, SPSS software). These provide information on the influence of the selected parameters on the probability of breeding in a given location. As the parameters "terrain height" and "slope gradient", "distance from next power pylon" and "distance from next vertical point object", "distance from next wind turbine" and "distance from next wind farm" were significantly and strongly related, respectively (Spearman rank correlation, $r > 0.5$), they were used alternatively in the models.

All analyses were carried out for the overall sample, including the total number of nest sites, as well as for the reduced sample taking account of breeding colonies by averaging the number of nests per colony. A two-way approach was used in all statistical analyses. P values smaller than or equal to 0.05 were assumed to be significant and p values >0.05 and up to 0.10 as potentially significant.

5.5 Results

5.5.1 Development and Spatial Distribution of Wind Energy Use and the Breeding Population of Montagu's Harriers in the Hellweg Börde

5.5.1.1 Wind Energy Use

The number of wind turbines and wind farms within the study area has continuously increased over the period of investigation. Different phases can be identified in the development of wind energy use (Fig. 5.3).

The first wind turbine within the study area was constructed in 1989. Whereas the number of turbines increased slowly until 1993, the construction of wind turbines was clearly promoted from 1994 and the first wind farms developed. Up until 1997, 100 wind turbines were erected in the study area. 56 wind turbines (56%) were distributed on eight wind farms. Between 1998 and 2000, the further development of wind energy use within the study area continued at a slower pace. In 2000, 140 wind turbines were located within the study area with 98 turbines (70%)

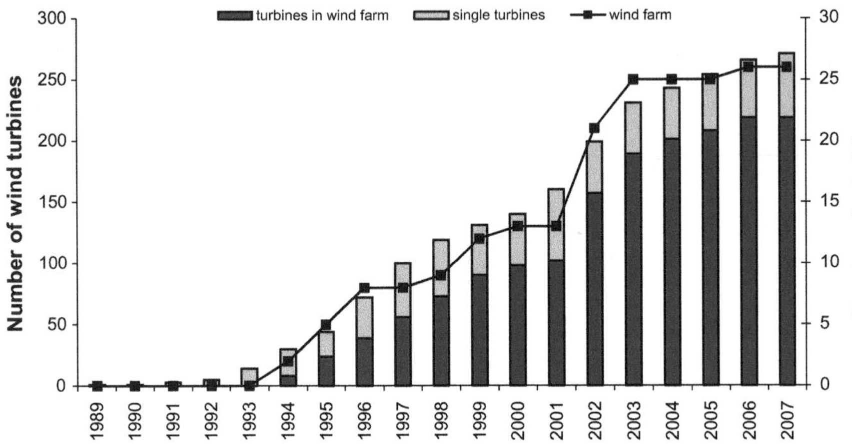

Fig. 5.3 Number of wind turbines within and outside wind farms and total number of wind farms in the Hellweg Börde between 1989 and 2007

in 13 wind farms. From 2001, less favourable sites could be used due to the development of larger and more efficient wind turbines. This resulted in a further rapid increase of the number of wind turbines in the Hellweg Börde. Up until 2003, a further 71 wind turbines were constructed and 12 new wind farms went into operation. A total of 189 (82%) of the then 231 wind turbines were distributed on 25 wind farms. In the following, the further development of wind energy use slowed down again. Up until 2007, a total of 271 wind turbines were constructed within the study area. 219 (81%) of these turbines were located in 26 wind farms. Thus, an average of 0.27 wind turbines per square kilometre was in operation in the study area in 2007.

Corresponding to technical development and the increasing number of wind turbines, the total installed nominal capacity of wind turbines within the study area has increased considerably over the period of investigation (Fig. 5.4). Significant increases in nominal capacity can be found for the periods 1993–1997 and 2001–2003. The increasing nominal capacity was accompanied by a considerable increase in turbine heights and rotor diameter (Fig. 5.5).

5.5.1.2 Development of the Breeding Population of the Montagu's Harrier

Between 1993 and 2007, the number of nests including replacement broods of the Montagu's harrier varied around a median of 41 (a minimum of 17 in 2006 and a maximum of 56 in 1994) within the study area. In the mean, a slight decline in the

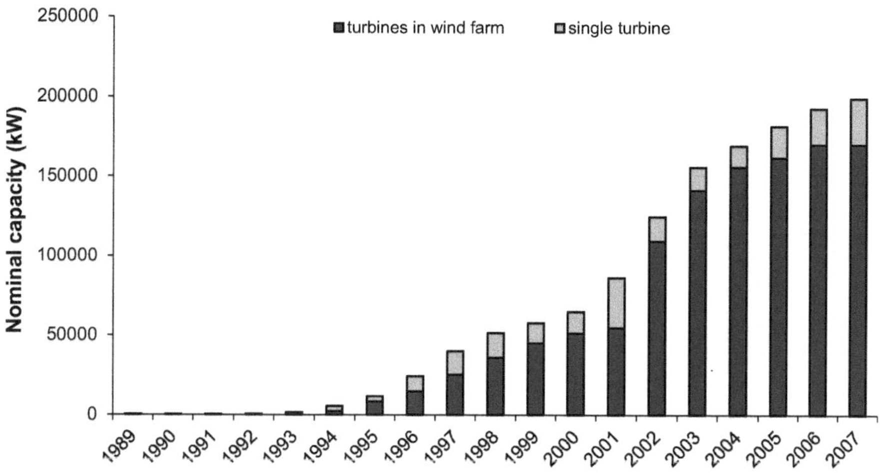

Fig. 5.4 Nominal capacity of single wind turbines and wind turbines within wind farms in the Hellweg Börde between 1989 and 2007

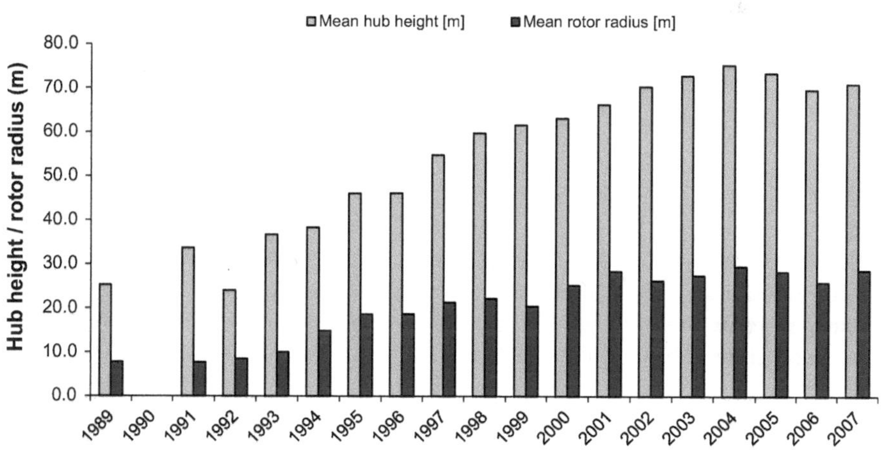

Fig. 5.5 Average hub height and rotor radius of the annually newly erected wind turbines in the Hellweg Börde between 1989 and 2007

number of Montagu's harrier's nests can be identified over the investigation period (1.7 nests less/year, linear regression, $R^2 = 0.58$) with partly strong annual fluctuations (Fig. 5.6) without a necessary causal correlation with the number of wind turbines. The negative trend became obvious after a severe decrease in population (from 46 down to 25 nests) in 2005.

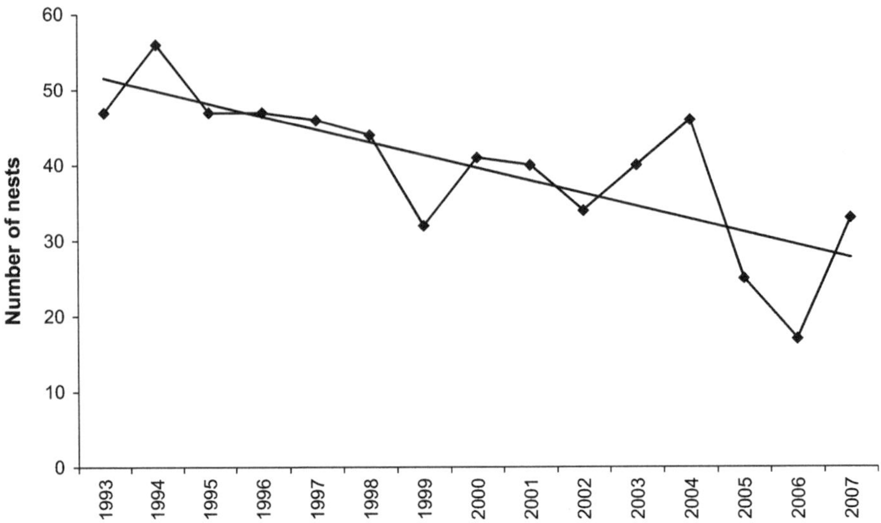

Fig. 5.6 Number of Montagu's harrier nests (including replacement clutches) in the Hellweg Börde between 1993 and 2007

5.5.2 Spatial Distribution of Wind Energy Use and Montagu's Harrier Nests

Figure 5.7 summarises the distribution of wind turbines and Montagu's harrier nests within the study area for the years 1993–2000 and 2001–2007.

In the spatial distribution of wind turbines within the study area, two phases of the construction of wind turbines can be identified. Between 1993 and 2000, wind turbines were, with only a few exceptions, erected in the higher elevations of the Haarstrang above 200 m. On the one hand, this was due to better wind conditions in the higher ranges of the hill ridges but on the other hand a result of spatial planning for the purpose of nature protection (map for wind energy uses of the advisory board of the landscape authority). This resulted in a far-reaching spatial separation of wind turbines and regular breeding areas of Montagu's harriers within the study area, which are concentrated in the area of the Lower Hellweg at heights of below 200 m. An exception was the Ruhne area, an regularly used breeding ground for Montagu's harriers in the higher ranges of the Haarstrang, where the first wind turbines were already constructed in 1994 in the proximity of a Montagu's harriers breeding area. From 2001, an increasing number of new wind turbines was constructed in the lower parts of the Lower Hellweg Börde at heights of below 200 m and therefore closer to breeding areas of the Montagu's harrier. Up until 2007, more and more new wind turbines and wind farms developed, some of these near Montagu's harrier's breeding areas.

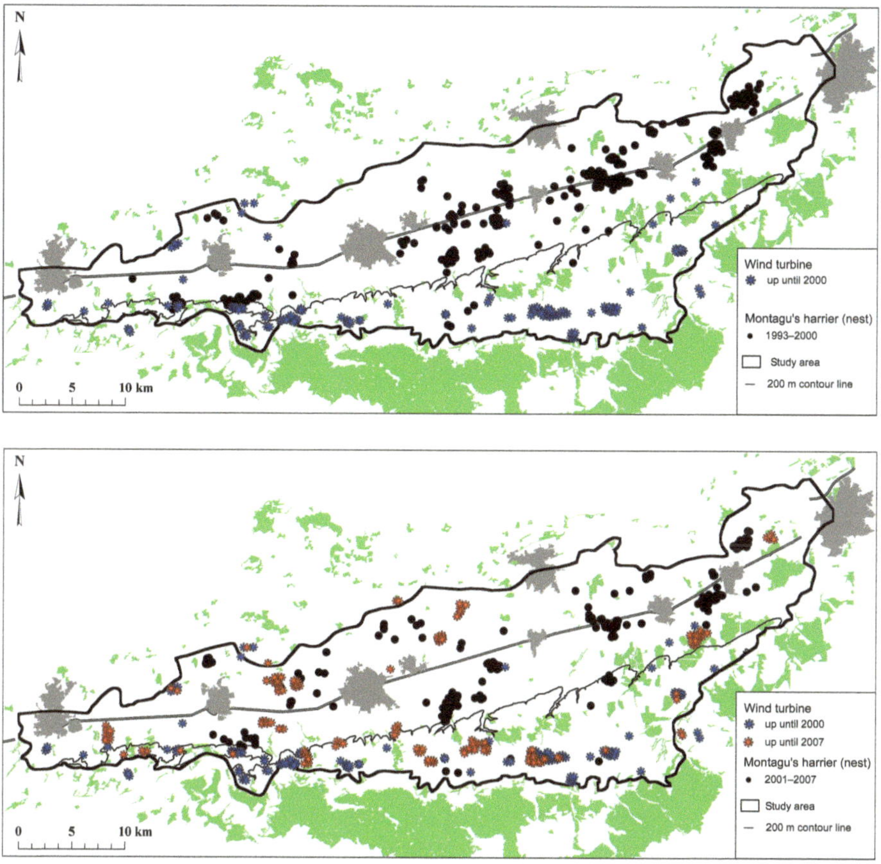

Fig. 5.7 Spatial distribution of wind turbines and Montagu's harrier nests in the Hellweg Börde between 1993 and 2000 (*above*) and 2001 and 2007 (*below*)

5.5.3 Number of Montagu's Harriers Breeding in the Hellweg Börde in Relation to the Number of Wind Turbines

Figure 5.8 shows the distribution of Montagu's harrier nests in the Hellweg Börde between 1993 and 2007 and the wind turbines put into operation until 2007 in grid cells of 4 km^2.

In 127 of the total of 316 grid cells, at least one wind turbine or at least on Montagu's harrier brood could be found between 1993 and 2007. Montagu's harriers were breeding in 63 of the grid cells, without a wind turbine being constructed during the investigation period. In 48 of these grid cells, wind turbines

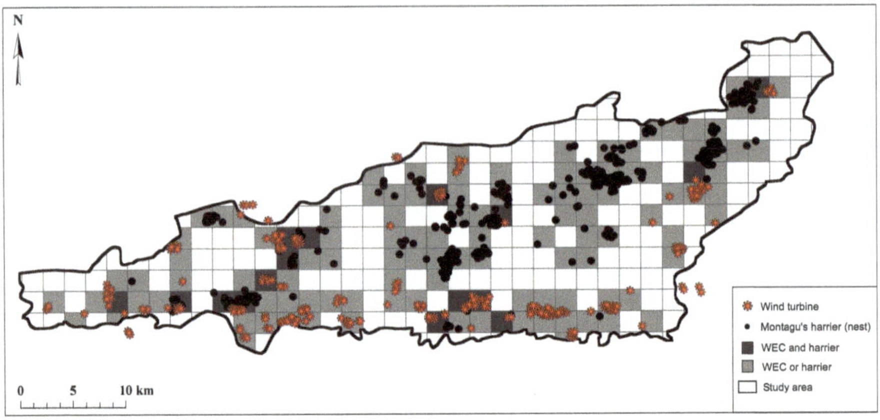

Fig. 5.8 Distribution of Montagu's harrier nests (total 1993–2007) and of wind turbines put into operation until spring 2007 per grid cell (side length: 2 km) in the Hellweg Börde

were constructed but no Montagu's harrier brood was found. With 16 grid cells, the number of cells with breeding Montagu's harriers and construction of a wind turbine is relatively low. This small number reflects the large degree of separation between breeding areas of the Montagu's harrier and areas used for wind turbines within the study area. This is mainly due to a concentration of wind turbines on higher areas of the southern parts and avoidance of traditional breeding areas of the Montagu's harrier in lowland areas as a result of spatial management in the construction of wind turbines.

To investigate a possible influence of the construction of wind turbines on the number of Montagu's harrier nests, the relation between the annual number of wind turbines and the number of nests was determined in an analysis of variance with repeated measures for the 127 grid cells which were either occupied by Montagu's harriers or had wind turbines.

The analysis showed that the year but not the number of wind turbines had an influence on the number of Montagu's harrier nests per grid cell (Table 5.6). Likewise, no significant influence of the total capacity of wind turbines per grid cell (index of the combined effect of number, height and rotor diameter) on the number of nests per grid cell could be found. Accordingly, no influence of the construction of wind turbines on the frequency of use of Montagu's harriers of the 4 km^2 grid cells could be determined for the total Hellweg Börde area during the investigation period between 1993 and 2007 (Table 5.6).

Table 5.6 Results of the analysis of variance with repeated measures on the influence of number and nominal capacity (Number_WEC, KW_WEC) of installed wind turbines on the number of Montagu's harrier nests (Number_nests) in 4 km^2 grid cells in the Hellweg Börde between 1993 and 2007

	Model	df	AIC	BIC	logLik	L.Ratio	*p*
H0: Number_nests	1	4	4,111,209	4,133,417	−2,051,604		
H1: Number_nests~year	**2**	**5**	**4,108,646**	**4,136,407**	**−2,049,323**	**4,562,969**	**0.0327**
H2: Number_nests ~ year + Number_WEC	3	6	4,110,461	4,143,774	−2,049,230	0.184811	0.6673
H3: Number_nests ~ year + KW_WEC	4	6	4,110,561	4,143,874	−2,049,280	0.085028	0.7706

Parameters with significant influence are highlighted in bold font (Number of observations: 1905, Number of cells: 127)

5.5.4 *Establishment of Territories of Montagu's Harriers in the Hellweg Börde Before and After Construction of Wind Farms*

Between 1993 and 2007, Montagu's harriers bred at distances of below 1 km around (later) wind farms in a total of nine farmland areas (Table 5.7). Seven of these wind farms were constructed around the middle of the investigation period, thus enabling a comparison of the number of Montagu's harrier broods before and after construction of the wind farm. The Ruhne and Rüthen wind farms were already erected in the beginning of the investigation period, making a comparison with the situation prior to the construction of the wind farm difficult.

Table 5.7 Number of Montagu's harrier nests (MH) and wind turbines (WEC) in selected farmland areas with construction of a wind farm in the Hellweg Börde between 1993 and 2007; nests at a distance of <1 km around (later) wind farms only

		1993	1994	1995	1996	1997	1998	1999	2000	2001	2002	2003	2004	2005	2006	2007	Years before	Years after	Nests before	Nests after	Nests expected before[a]	Nests expected after[a]
Ostbüren	WEC			1	1	1	1	1	1	1	1	7	7	7	7	7	10	5				
	MH										1								1	0	0.7	0.3
Schlückingen	WEC				1	1	1	1	4	4	4	4	5	5	5	5	7	8				
	MH	3	3	1	3	1													11	0	5.1	5.9
Mawicke	WEC										5	5	6	6	6	6	9	6				
	MH						1												1	0	0.6	0.4
Merklingsen[b]	WEC		1	1	1	1	1	1	1	2	3	16	22	23	24	24	9	6				
	MH						1				1	1				1			1	3	2.4	1.6
Bettinghausen	WEC										8	8	8	8	8	8	9	6				
	MH					1													1	0	0.6	0.4
Haarhöfe[c]	WEC		1	2	2	2	2	2	2	4	14	16	16	17	17	17	9	6				
	MH						2												2	0	1.2	0.8
Geseke	WEC					1	1	1	1	1	12	15	15	16	16	16	9	6				
	MH														1				0	1	0.6	0.4
Total																			17	4	11.2	9.8
Ruhne	WEC		4	9	9	9	9	9	9	10	10	10	10	10	10	10	1	14				
	MH	6	3	4	4		1	3	1	1		1		2	2							
Rüthen	WEC		4	7	15	15	15	15	15	15	16	16	16	16	16	16	1	14				
	MH										1											

Years after the construction of the wind farm are shaded grey

[a]The number of expected nests is derived from an assumed average distribution of nests over the period prior to and after the construction of the wind farm: total number of nests/total number of years * number of years prior to or after

[b]The Merklingsen wind farm consists of two spatially separate partial areas

[c]The wind farm was in operation since 2002

Within a radius of 1 km around these seven wind farms, a total of 17 broods were counted prior to and four after the construction. This ratio deviates from an assumed average distribution for the years before and after of 11.2–9.8 nests. This effect is mainly causes by the Schlückingen wind farm, where 11 broods (between one and three per year) were found before the construction but none afterwards. The abandonment of this farmland area as a breeding site started in 1998 already, well before the significant decline in Montagu's harrier population in the Hellweg Börde from 2005. For five of the seven wind farms no broods were counted within a radius of 1 km after construction. Only in the surrounding areas of the Geseke and Merklingsen wind farms Montagu's harriers still established breeding territories after the construction. Within a radius of 1 km around the Rüthen wind farm only one Montagu's harrier nest was found at a distance of 920 m from the next wind turbine after the construction of the farm in 14 years.

The wind farm near Ruhne is an exception, as it was already constructed in the beginning of the investigation period (1994), making a comparison with the situation prior to the construction nearly impossible. Within a radius of 1 km around this wind farm, 22 broods were counted in 14 years (one to four broods per year at a declining rate). The spatial constellation of wind turbines and Montagu's harrier nests is quite special for this wind farm. Whereas the wind turbines are linearly aligned along the crest of the Haarstrang, Montagu's harrier nests were almost exclusively located on a slope of the Haarstrang following towards the north. Therefore, the wind farm is situated on the southern edge of the Montagu's harrier area oriented towards the closed forest areas of the Sauerland, whereas the nests are located in north-western to north-eastern direction within the open landscape of the Hellweg Börde and the Westphalian Lowland. Overlap of the row of wind turbines with the close-up area of the nest sites of the Montagu's harrier is therefore only minor. None of the six breeding pairs of the year 1993, the year prior to the construction of the first wind turbine, was breeding closer than 220 m to later erected wind turbines.

No brood occurred within any of the nine investigated wind farms. The minimum distances of breeding Montagu's harriers from the next wind turbine within one of the above wind farms was between 170 and 9710 m. Broods in the close-up area occurred at a mean distance of 500 m (Table 5.8).

5.5.5 Nest Site Selection of Montagu's Harriers in the Hellweg Börde in Relation to Wind Turbines and Other Habitat Parameters

5.5.5.1 Habitat Parameters

Table 5.9 shows the habitat parameter investigated in this study for all nest sites of the Montagu's harrier between 2003 and 2007 and the randomly selected reference points. Significant differences occurred in the distance from the next nest, the slope

Table 5.8 Minimum distances between Montagu's harrier nests and the nearest wind turbine after construction of a wind farm (more than three wind turbines) in the Hellweg Börde between 1993 and 2007 (values rounded to 10 m)

	Year of the construction of the wind farm	Year of the brood	Minimum distance in the close-up area of below 1 km (m)	Minimum distance for areas >1 km (m)
Ostbüren	2003	2007		9710
Schlückingen	2000	2007		3410
Mawicke	2002	2002		1520
Merklingsen	2002	2007	170	
Bettinghausen	2002	2004		1460
Haarhöfe[a]	2002	2005		1950
Geseke	2002	2006	710	
Ruhne	1994	2000	290	
Rüthen	1994	2002	920	
Median			500	1950
Mean			523	3610

[a]The wind farm was in operation since 2002

Table 5.9 Comparison of habitat parameters for nest sites of the Montagu's harrier (N) and random points (R)

		n	Min	First quartile (25%)	Median	Third quartile (75%)	Max	W	*p*
Height NN (m)	N	162	77	105	118	160	274	12,064.0	0.210
	R	162	74	95	127	202	365		
Slope (°)	**N**	**162**	**0.1**	**0.5**	**0.8**	**1.3**	**3.2**	**8649.5**	**<0.001**
	R	**162**	**0.0**	**0.6**	**1.4**	**2.6**	**7.3**		
Dist. areal vert. obj. (m)	**N**	**162**	**125**	**281**	**402**	**532**	**752**	**22,325.0**	**<0.001**
	R	**162**	**0**	**75**	**168**	**268**	**903**		
Dist. point vert. obj. (m)	N	162	100	527	848	2190	4537	14,144.5	0.225
	R	162	35	429	917	1549	4283		
Dist. power pylon (m)	N	162	100	543	952	2679	8684	12,539.5	0.490
	R	162	35	588	1372	2585	8086		
Dist. WEC (m)	**N**	**162**	**170**	**2267**	**3527**	**5007**	**8263**	**17,599.0**	**<0.001**
	R	**162**	**88**	**1100**	**2125**	**3557**	**9856**		
Dist. wind farm (m)	**N**	**162**	**170**	**2912**	**3840**	**6913**	**9925**	**17,269.5**	**<0.001**
	R	**162**	**147**	**1586**	**2697**	**4702**	**10,519**		
Dist. neighbour (m)	**N**	**162**	**49**	**188**	**344**	**962**	**9386**	**3706.0**	**<0.001**
	R	**162**	**174**	**1400**	**2773**	**3976**	**10277**		

Included are nests occupied between 2003 and 2007 as well as nests in colonies. Wilcoxon rank sum test (W). Parameters with significant influence are highlighted in bold font. A description of the habitat parameters can be found in Table 5.3

gradient, the distance from areal vertical objects as forests, settlements and buildings, the distance from the next wind turbine and the distance from the next wind farm. On average, nest sites of Montagu's harriers were located in less sloping areas at larger distances from areal vertical objects, wind turbines and wind farms than to be expected in a random distribution. Additionally, they preferred smaller distances from the next neighbour than to be expected in a random distribution.

Nests of Montagu's harriers were located at a median distance of 3527 m from the next wind turbine or 3840 m from the next wind farm. The minimum distance of a Montagu's harrier nest from a wind turbine or a wind farm was found to be 170 m. More than 75% of all Montagu's harriers bred at a distance of at least 2267 m or 2912 m from wind turbines or wind farms, respectively. During the five years, included in the evaluation, the mean minimum distance of Montagu's harrier nests from wind turbines or wind farms was found to be 500 m (Table 5.10). The mean minimum distance from areal vertical objects as forests or settlements equalled 350 m.

The pattern of a smaller number of nests within the immediate surrounding area of wind turbines is confirmed by the evaluation of the frequency of Montagu's harriers nests and random points for different distance classes from the next wind turbine or wind farm (Fig. 5.9). A smaller number of nests was located within the distance classes 2000 m or up to 3000 m from the next wind turbine or wind farm than to be expected in a random distribution based on area availability.

For the reduced sample, taking account of the effect of breeding in colonies, significant differences in habitat parameters were found between the selected nest sites and random points (Table 5.11). Nests or colonies of Montagu's harriers were located in areas lower above sea level and in less sloping areas at larger distances form areal vertical landscape structures and wind turbines than to be expected in a random distribution.

For this sample, the median distances from the next wind turbine or wind farm equalled 3054 m or 3471 m, respectively. Possibly due to the small sample size, the difference determined for wind farms was potentially significant with $p = 0.053$. Again for this sample, a smaller number of nests was located within the distance classes up to 2000 m or up to 3000 m from the next wind turbine or wind farm than to be expected in a random distribution (Fig. 5.10).

The determination of distances of Montagu's harrier nests from wind turbines and wind farms was once again carried out for nests and random points located at distances of below 1 km from the next wind turbine or wind farm only, to verify an

Table 5.10 Minimum distances of Montagu's harrier nests from areal vertical objects, power pylons, wind turbines and wind farms between 2003 and 2007 in the Hellweg Börde (values rounded to 10 m)

	2003	2004	2005	2006	2007	Median	Mean
Dist. areal vert. obj. (m)	450	480	300	350	150	350	346
Dist. power pylon (m)	150	150	1940	1800	630	630	934
Dist. WEC (m)	530	590	320	500	170	500	422
Dist. wind farm (m)	5900	5840	320	500	170	500	2546

A description of the habitat parameters can be found in Table 5.3

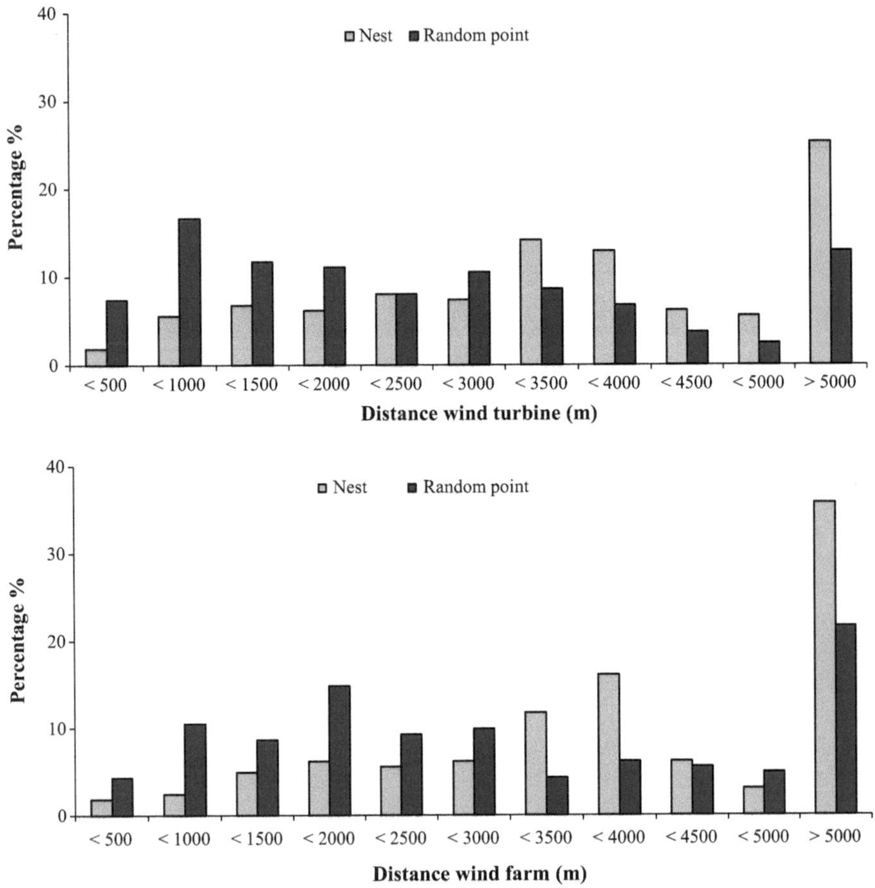

Fig. 5.9 Percentage distribution of Montagu's harrier nests and random points in distance classes from the next wind turbine (*above*) or from the next wind farm (*below*). Included are all nests occupied between 2003 and 2007 (n = 162)

overlap of large-scale effects of the investigated habitat parameters, especially the elevation, and the large degree of separation of wind turbines and breeding areas with the effects of wind turbines in the close-up area. For this sample, including Montagu's harriers breeding in the close-up area of wind turbines, no significant differences were found between nest sites and random points. This result cannot be assumed to be reliable due to the small sample size (Table 5.12).

It can be assumed that colonies of Montagu's harriers are mainly occupying especially suitable locations, whereas solitary breeding pairs will also choose less suitable locations. For this reason, the selection of a nest site was evaluated separately for solitary breeding Montagu's harriers and Montagu's harriers breeding in colonies. Solitary and colonially breeding Montagu's harriers showed significantly

Table 5.11 Comparison of habitat parameters for nest sites of the Montagu's harrier (N) and random points (R)

		n	Min	First quartile (25%)	Median	Third quartile (75%)	Max	W	*p*
Height NN (m)	**N**	**81**	**81**	**103**	**114**	**161**	**274**	**2507.0**	**0.010**
	R	**81**	**76**	**99**	**154**	**231**	**369**		
Slope (°)	**N**	**81**	**0.2**	**0.5**	**0.8**	**1.4**	**3.2**	**2324.5**	**0.001**
	R	**81**	**0.0**	**0.8**	**1.6**	**2.2**	**8.0**		
Dist. areal vert. obj. (m)	**N**	**81**	**125**	**271**	**384**	**503**	**671**	**5226.0**	**0.000**
	R	**81**	**25**	**106**	**190**	**301**	**637**		
Dist. point vert. obj. (m)	N	81	100	552	971	2024	4446	3378.0	0.745
	R	81	79	576	1055	1651	3936		
Dist. power pylon (m)	N	81	100	620	1110	2579	8684	2915.5	0.222
	R	81	79	800	1616	3143	8436		
Dist. WEC (m)	**N**	**81**	**170**	**1821**	**3054**	**4102**	**8263**	**4198.0**	**0.002**
	R	**81**	**106**	**1166**	**1742**	**3289**	**8031**		
Dist. wind farm (m)	N	81	170	2035	3471	4545	9925	3859.0	0.053
	R	81	106	1258	2671	4473	10853		

Included are nests occupied between 2003 and 2007; the average number of nests was used per colony. Wilcoxon rank sum test (W). Parameters with significant influence are highlighted in bold font. A description of the habitat parameters can be found in Table 5.3

different distances from areal vertical objects, wind turbines and wind farms (Table 5.13). Nests of Montagu's harriers breeding in colonies were located significantly further away from these structures than those of solitary breeding Montagu's harriers.

5.5.5.2 Landscape Parameters of Montagu's Harrier Nest Sites and Wind Turbines

Comparison of the habitat parameters of Montagu's harrier nest sites in the Hellweg Börde with parameters of the wind turbine sites again confirmed the previous, far-reaching spatial separation of nest sites and wind turbine sites. On average, Montagu's harrier nests were located at lower heights above sea level and in areas with less slope and at larger distances from areal vertical objects than wind turbines (Table 5.14). It can be assumed that wind turbines are constructed at larger distances from settlements than from forests due to planning guidelines resulting in comparably large distances between settlements and wind turbines, which can also be observed in Montagu's harrier nests.

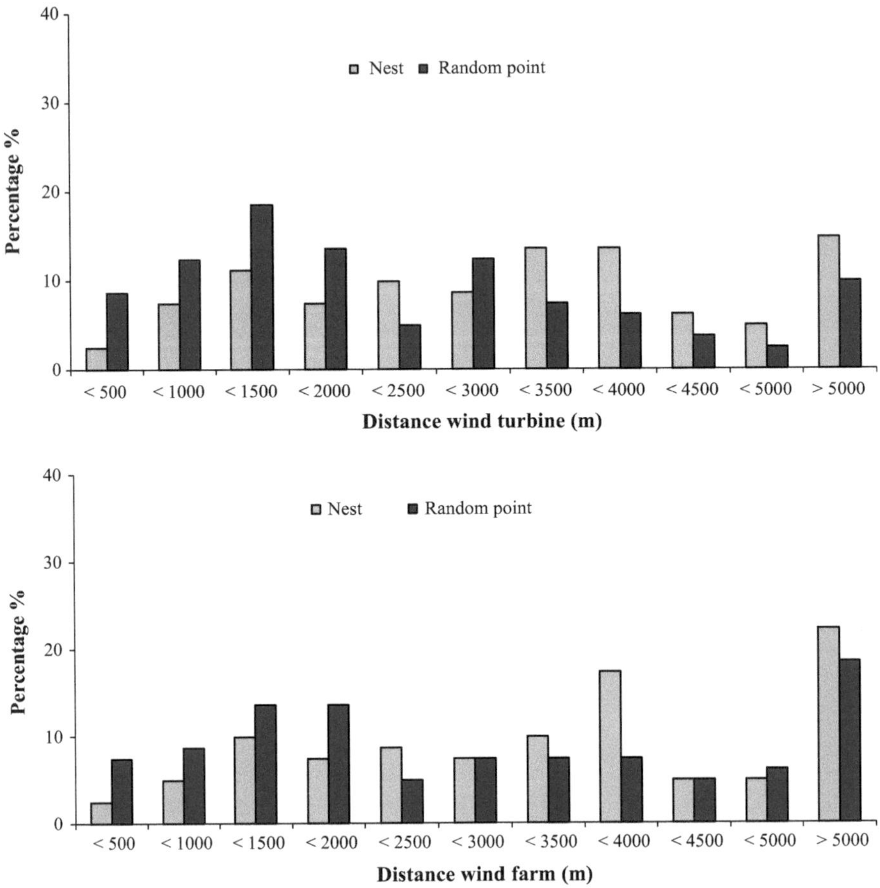

Fig. 5.10 Percentage distribution of Montagu's harrier nests and random points in distance classes from the next wind turbine (*above*) or from the next wind farm (*below*). Included are nests occupied between 2003 and 2007; the average number of nests per colony was used (n = 81)

5.5.5.3 Multi-factor Evaluation

The above evaluation showed that the nest site selection of Montagu's harriers in the Hellweg Börde is, apart from the tendency to breed socially, influenced by habitat parameters as height above sea level, slope gradient and distance from the next areal vertical object and from the next wind turbine or wind farm. It cannot be excluded that parameters have a combined effect or that correlation between habitat parameters causes results not found in an independent evaluation of single parameters. For this reason, logistic regressions of the combined effect of the investigated habitat parameters as independent variable on the dependent binary variable "nest site" (1: nest, 0: random point) were carried out. Strongly intercorrelated variables were used alternatively in the models.

Table 5.12 Comparison of the distances of Montagu's harrier nests (N) and random points (R) from the next wind turbine or wind farm

		N	Min	First quartile (25%)	Median	Third quartile (75%)	Max	W	*p*
Total nests									
Dist. WEC (m)	N	12	170	515	615	720	833	213.5	0.657
	R	39	88	388	596	824	986		
Dist. wind farm (m)	N	7	170	407	520	678	801	54.0	0.163
	R	24	147	457	703	854	988		
Single nests and colonies									
Dist. WEC (m)	N	8	170	461	678	798	839	69.0	0.977
	R	17	106	286	586	853	925		
Dist. wind farm (m)	N	6	170	364	577	695	801	32.0	0.579
	R	13	106	266	723	853	925		

Included are nests occupied in area of below 1 km from the next wind turbine or wind farm between 2003 and 2007. The sample size varies depending on the number of nests within this distance class. Wilcoxon rank sum test (W). A description of the habitat parameters can be found in Table 5.3

Table 5.13 Comparison of habitat parameters for nest sites of Montagu's harriers breeding solitary (S) and in colonies (C) in the Hellweg Börde

		n	Min	First quartile (25%)	Median	Third quartile (75%)	Max	W	*p*
Height NN (m)	C	49	77	111	118	152	205	2961.5	0.483
	S	113	81	97	113	165	274		
Slope (°)	C	49	0.1	0.5	0.8	1.1	2.7	2444.0	0.236
	S	113	0.2	0.5	0.9	1.5	3.2		
Dist. areal vert. obj. (m)	**C**	**49**	**127**	**304**	**437**	**551**	**752**	**3603.0**	**0.002**
	S	**113**	**125**	**250**	**313**	**461**	**671**		
Dist. point vert. obj. (m)	C	49	125	527	813	2158	4537	2650.0	0.667
	S	113	100	552	1107	2200	4317		
Dist. power pylon (m)	C	49	125	527	836	2158	5930	2431.0	0.219
	S	113	100	627	1179	2712	8684		
Dist. WEC (m)	**C**	**49**	**499**	**2720**	**3755**	**5193**	**7761**	**3671.0**	**0.001**
	S	**113**	**170**	**1460**	**2979**	**3834**	**8263**		
Dist. wind farm (m)	**C**	**49**	**499**	**3383**	**4296**	**7081**	**9710**	**4028.0**	**0.000**
	S	**113**	**170**	**1545**	**2979**	**4102**	**9925**		

Included are nests occupied between 2003 and 2007. Wilcoxon rank sum test (W). Parameters with significant influence are highlighted in bold font. A description of the habitat parameters can be found in Table 5.3

Table 5.14 Comparison of habitat parameters for Montagu's harrier nest sites (N, nests between 2003 and 2007) and wind turbine sites (as from 2007) in the Hellweg Börde

		N	Min	First quartile (25%)	Median	Third quartile (75%)	Max	W	*p*
Height NN (m)	**Wind turbine**	**258**	**78**	**152**	**229**	**306**	**390**	**33,262.5**	**0.000**
	N	**162**	**77**	**105**	**118**	**160**	**274**		
Slope (°)	**Wind turbine**	**258**	**0.0**	**1.1**	**2.0**	**2.8**	**9.3**	**32,342.0**	**0.000**
	N	**162**	**0.1**	**0.5**	**0.8**	**1.3**	**3.2**		
Dist. areal vert. obj. (m)	**Wind turbine**	**258**	**35**	**181**	**259**	**378**	**626**	**11,367.0**	**0.000**
	N	**162**	**125**	**281**	**402**	**532**	**752**		

Wilcoxonrank sum test (W). A description of the habitat parameters can be found in Table 5.3

In the evaluation of all single nests, the parameters “terrain height” and “slope gradient”, alternatively used in modelling, as well as the distance from the next areal vertical object and from the next neighbour were found to have a significant influence on the nest site selection of Montagu’s harriers. Increasing terrain height, slope and distance from the next neighbour reduced the probability of a Montagu’s harrier brood, whereas an increasing distance from the next areal vertical object increased the probability of a Montagu’s harrier brood (Table 5.15). Taking into consideration the effect of these parameters, the distance from the next power pylon, which was insignificant in the single-factor evaluation, also had an influence on the probability of a Montagu’s harrier brood. The alternatively used distances from the next wind turbine and from the next wind farm had no significant influence in the logistic regression models. Combining the effect of the point vertical objects “wind turbine” and “power pylon” resulted in a significant influence with increasing probability of a Montagu’s harrier brood with increasing distance from the point vertical object.

The effect of breeding colonies (social attraction) was included in the models using the parameter “distance from next neighbour”. Still, corresponding to the above evaluation, logistic regressions were calculated for the reduced sample with averaged habitat parameters for Montagu’s harriers breeding in colonies. For the reduced sample, the alternatively used parameters “terrain height” and “slope gradient” had a negative influence on the probability of a Montagu’s harrier brood, whereas the distance from the next areal vertical object had a positive influence (Table 5.16). The alternatively used distances from the next power pylon, wind turbine, wind farm or point vertical object had no influence on the probability of a Montagu’s harrier brood.

In the multi-factor analysis of the combined effects of single habitat parameters, all models revealed a significant influence of the distance from the next neighbour,

Table 5.15 Summary of the binary logistic regressions of the effects of habitat parameters on the nest site selection of Montagu's harriers in the Hellweg Börde between 2003 and 2007. All nests (n=162)

Factor	Factor	Dist. neighbour (m)	Dist. neighbour (m)	Height NN (m)	Height NN (m)	Dist. areal vert. obj. (m)	Dist. areal vert. obj. (m)	Dist. point vert. obj. (m)	Dist. point vert. obj. (m)	Dist. power pylon (m)	Dist. power pylon (m	Dist. WEC (m)	Dist. WEC (m)	Dist. wind farm (m)	Dist. wind farm (m)	2 Log likelihood	Nagelkerke's R^2	Omnibus test
B	Sig.	B	Sig.	B	Sig.	B	Sig.	B	Sig.	B	Sig.	B	Sig.	B	Sig.			Sig.
−0.334	0.505	**−0.001**	**<0.001**	**−0.008**	**0.003**	**0.009**	**< 0.001**									263.522	0.582	<0.001
−0.867	0.168	**−0.001**	**<0.001**	**−0.006**	**0.017**	**0.008**	**<0.001**					0	0.154			261.494	0.586	<0.001
−0.455	0.461	**−0.001**	**<0.001**	**−0.007**	**0.006**	**0.009**	**<0.001**							0	0.736	263.409	0.582	<0.001
−0.64	0.224	**−0.001**	**<0.001**	**−0.009**	**0.001**	**0.008**	**<0.001**	**0.0004**	**0.009**							256.179	0.598	<0.001
−0.279	0.599	**−0.001**	**<0.001**	**−0.013**	**0.001**	**0.009**	**<0.001**			**0.0003**	**0.002**					252.393	0.607	<0.001
−0.729	0.266	**−0.001**	**<0.001**	**−0.012**	**0.001**	**0.009**	**<0.001**			**0.0003**	**0.002**	0.0001	0.242			251.026	0.61	<0.001
−0.48	0.457	**−0.001**	**<0.001**	**−0.013**	**0.001**	**0.009**	**<0.001**			**0.0003**	**0.001**			0.00004	0.584	252.093	0.608	<0.001
Factor	Factor	Dist. neighbour (m)	Dist. neighbour (m)	Slope (°)	Slope (°)	Dist. areal vert. obj. (m)	Dist. areal vert. obj. (m)	Dist. point vert. obj. (m)	Dist. point vert. obj. (m)	Dist. power pylon (m)	Dist. power pylon (m)	Dist. WEC (m)	Dist. WEC (m)	Dist. wind farm (m)	Dist. wind farm (m)	2 Log likelihood	Nagelkerke's R^2	Omnibus test
B	Sig.	B	Sig.	B	Sig.	B	Sig.	B	Sig.	B	Sig.	B	Sig.	B	Sig.			Sig.
−0.367	0.398	**−0.001**	**<0.001**	**−0.753**	**<0.001**	**0.008**	**<0.001**									251.304	0.609	<0.001
−0.564	0.318	**−0.001**	**<0.001**	**−0.714**	**<0.001**	**0.008**	**<0.001**					0	0.584			251.004	0.61	<0.001
−0.18	0.747	**−0.001**	**<0.001**	**−0.791**	**<0.001**	**0.008**	**<0.001**							0	0.588	251.009	0.61	<0.001
−0.722	0.122	**−0.001**	**<0.001**	**−0.777**	**<0.001**	**0.008**	**<0.001**	**0.0003**	**0.03**							246.367	0.62	<0.001
−0.682	0.133	**−0.001**	**<0.001**	**−0.907**	**<0.001**	**0.008**	**<0.001**			**0.0003**	**0.006**					242.976	0.628	<0.001
−0.842	0.144	**−0.001**	**<0.001**	**−0.872**	**<0.001**	**0.008**	**<0.001**			**0.0003**	**0.006**	0.00004	0.652			242.774	0.628	<0.001
−0.576	0.318	**−0.001**	**< 0.001**	**−0.928**	**< 0.001**	**0.008**	**< 0.001**			**0.0003**	**0.006**			−0.00002	0.766	242.887	0.628	<0.001

Parameters with significant influence are highlighted in bold font. A description of the habitat parameters can be found in Table 5.3

Table 5.16 Summary of the binary logistic regressions of the effects of habitat parameters on the nest site selection of Montagu's harriers in the Hellweg Börde between 2003 and 2007. Single nests and colonies (n=81)

Factor	Factor	Height NN (m)	Height NN (m)	Dist. areal vert. obj. (m)	Dist. areal vert. obj. (m)	Dist. point vert. obj. (m)	Dist. point vert. obj. (m)	Dist. power pylon (m)	Dist. power pylon (m)	Dist. WEC (m)	Dist. WEC (m)	Dist. wind farm (m)	Dist. wind farm (m)	2 Log likelihood	Nagelkerke's R^2	Omnibus Test
B	Sig.	B	Sig.	B	Sig.	B	Sig.	B	Sig.	B	Sig.	B	Sig.			Sig.
−0.712	0.229	**−0.01**	0.002	0.007	<0.001									169.836	0.382	<0.001
−0.913	0.198	**−0.009**	0.005	0.007	<0.001					0	0.604			169.567	0.384	<0.001
−0.455	0.52	**−0.01**	0.002	0.007	<0.001							0	0.507	167.394	0.385	<0.001
−0.903	0.142	**−0.01**	0.001	0.007	<0.001	0	0.168							167.888	0.394	<0.001
−0.525	0.395	**−0.014**	0.002	0.007	<0.001			0	0.128					167.403	0.397	<0.001
−0.682	0.357	**−0.013**	0.004	0.007	<0.001			0	0.138	0	0.703			167.257	0.397	<0.001
−2.64	0.721	**−0.014**	0.002	0.007	<0.001			0	0.129			0	0.518	166.982	0.399	<0.001
Factor	Factor	Slope (°)	Slope (°)	Dist. areal vert. obj. (m)	Dist. areal vert. obj. (m)	Dist. point vert. obj. (m)	Dist. point vert. obj. (m)	Dist. power pylon (m)	Dist. power pylon (m)	Dist. WEC (m)	Dist. WEC (m)	Dist. wind farm (m)	Dist. wind farm (m)	2 Log likelihood	Nagelkerke's R^2	Omnibus Test
B	Sig.	B	Sig.	B	Sig.	B	Sig.	B	Sig.	B	Sig.	B	Sig.			Sig.
−1.287	**0.009**	**−0.575**	**0.008**	**0.007**	**<0.001**									171.611	0.372	<0.001
−1.482	**0.015**	**−0.537**	**0.016**	**0.007**	**<0.001**					0	0.584			171.31	0.374	<0.001
−1.071	0.073	**−0.622**	**0.007**	**0.007**	**<0.001**							0	0.527	171.208	0.374	<0.001
−1.454	**0.007**	**−0.574**	**0.008**	**0.007**	**<0.001**	0	0.425							170.967	0.376	<0.001
−1.329	**0.008**	**−0.611**	**0.008**	**0.007**	**<0.001**			0	0.637					171.386	0.373	<0.001
−1.518	**0.013**	**−0.572**	**0.017**	**0.007**	**<0.001**			0	0.648	0	0.594			171.1	0.375	<0.001
−1.119	0.066	**−0.654**	**0.008**	**0.007**	**<0.001**			0	0.669			0	0.548	171.023	0.375	<0.001

Parameters with significant influence are highlighted in bold font. A description of the habitat parameters can be found in Table 5.3

the height above sea level, the slope gradient as well as of the distance from the next areal vertical object on the nest site selection of Montagu's harriers. In the evaluation of the complete data, both models indicated a significant influence of the distance from the next power pylon or the total number of point vertical objects, including wind turbines, on the nest site selection.

5.6 Discussion

5.6.1 Method

In the interpretation of the results, it should be taken into consideration that no experimental approach was used in the investigation of impacts of wind turbines on the distribution of Montagu's harrier nests in a before and after comparison with reference areas. In this study, a breeding area of the Montagu's harrier, where topography (wind conditions), political guidelines (German Renewable Energies Act (EEG), Wind Energy Ordinance of the North Rhine-Westphalia federal state) as well as nature conservation issues have determined the distribution of wind turbines. Already in the planning phase for wind turbine sites, attempts were made to avoid potential conflicts with the spatial distribution of Montagu's harriers and other open landscape breeders. In 1994, a map of areas suitable for wind energy use from a nature and species conservation point of view was compiled by the advisory board of the landscape authority of the Soest district, including the main distribution area of the Montagu's harrier in the Hellweg Börde. Especially the protection of traditional breeding areas of the Montagu's harrier in lowland areas against possible impacts of wind turbines was taken into consideration (Bunzel-Drüke and Schulze-Schwefe 1994; Kampik 1995). This precautionary approach was based on possible impacts of wind turbines found for some bird species studied so far. Following the results of these studies, Montagu's harriers may be exposed to a risk of displacement or collision caused by wind turbines as well. For this reason, the construction of wind turbines was to a large extent already avoided in the traditional breeding areas of Montagu's harriers in the Hellweg Börde even before the area became a Natura 2000 area. After the area was designated as a Special Protection Area under the Bird's Directive in 2004, administrative barriers to erect wind turbines in the Hellweg Börde were even higher, as the Wind Energy Ordinance of the North Rhine-Westphalia federal government does not permit the construction of wind turbines in EU Special Protection Areas.

The geographical characteristics and especially the more favourable wind conditions in higher ranges led to spatial separation of wind turbines concentrated on the higher elevated southern part of the area from the main breeding areas of the Montagu's harrier in the lower central and northern part (Fig. 5.7). The inhomogeneity of both distributions was intensified by a concentration of most wind turbines in wind farms and a relatively large proportion of Montagu's harriers nesting in colonies. In the analyses based on a comparison with random points, this inhomogeneity may have caused that possible relations between wind turbines and the location of Montagu's harrier nests were found to be statistically insignificant. Within an area of a maximum of 1 km around the next wind turbine, a range within which an effect on the selection of the breeding or resting site can be assumed as determined in studies on other onshore bird species (e.g. Leddy et al. 1999; Larsen and Madsen 2000; Bergen 2001; Kruckenberg and Borbach-Jaene 2001; Hötker 2006), a small sample size of only 12 or eight Montagu's harrier nests, respectively, was available for statistical analysis.

Furthermore, the positional inaccuracy of the nesting locations of a maximum of 100 m for the years 1993–2005, and 35 m for the years 2006 and 2007 needs to be taken into account. These errors are especially relevant in the evaluation of the determined minimum distances between Montagu's harrier nests and wind turbines. The maximum error of the determined values falls within the range of 1% (evaluation of all nesting locations) and a maximum of 10% (evaluation of nest sites within an area of 1 km around the next wind turbine) depending on the investigated parameter. Especially for small sample sizes, these inaccuracies may blur possible differences between two distributions, which can therefore remain undiscovered.

The present study is bases on ATKIS data as from 1999. Changes occurring in the following years and especially the period between 2003 and 2007 could therefore not be included. This may result in errors in the determination of the distances between nests and areal vertical objects. According to present knowledge, only few buildings (mainly single agricultural outdoor facilities) were added after 1999. Smaller barns, livestock shelters and the like not included in the ATKIS data could not be taken into account. The resulting rather small error may have had an influence on the mean distances of nests from these structures but not on possible differences in the distances from random points, which should have been effected in the same way. No such errors occur in the determined distances from wind turbines, as these were included exact to the month of commissioning for each of the years. To present knowledge no changes in number and location of power pylons occurred between 2003 and 2007 as compared to 1999.

Influences of agricultural land use and the location of hunting grounds as well as differences in food availability (Montagu's harriers prefer areas with a higher density of voles for breeding, Arroyo et al. 2002) on the selection of a nest site could not be included in the evaluation. Even in a sedentary pair, the preference for winter barley as a nest site in the Hellweg Börde (Glimm et al. 2001; Illner 2007, 2008) would result in small-scale relocation of the nest site between years, as the same preferred crop will not be cultivated in the same place for two years in a row in the Hellweg Börde due to crop rotation (Illner and Steinwarz 2003). Due to the comparably large proportion of winter barley of about 20%, it can be assumed that in larger farmland areas within the Hellweg Börde more than one winter barley crop was available per year during the investigation period. In cannot be excluded though that with only one sufficiently large winter barley crop available within an farmland area this was then chosen as a site for nesting, even though other habitat characteristics were not ideal. Overlap of the results for the parameters investigated in the present study with the distribution of crops are assessed to be negligible for large-scale evaluation. In small-scale evaluations, the effect is assumed to be relatively small, as due to the three-field crop rotation prevailing in the investigation area (Illner and Steinwarz 2003), winter barley was cultivated in each of the fields twice or more times during the investigation period of five (2003–2007) or 15 years (1993–2007).

Broods in the data pool for the 15 years or 5 years of investigation, respectively, cannot be assumed to be statistically independent due to the relative longevity of the Montagu's harrier and the strong site fidelity of especially male birds (Arroyo et al. 2004), which would have been a prerequisite for the applied statistical tests. The same is true for the few replacement clutches and for nests of Montagu's harriers breeding in colonies, where the social attraction of already breeding birds may overlap effects of landscape structures. This was compensated for by parallel evaluation of the complete date and a reduced data set by averaging the values of nests in colonies. It can still be assumed that every nest is the result of a free decision of a breeding pair for a particular nest site, influenced by a multitude of factors. Individual factors (site fidelity, experience, individual behaviours), social factors (colonies) and habitat characteristics must be weighed up in the selection. The above general methodological restrictions play a differently important role in the different approaches of evaluation. These and further specific restrictions need to be taken into consideration in the following discussion of the results.

5.6.2 Development and Spatial Distribution of Wind Energy Use and the Breeding Population of Montagu's Harriers in the Hellweg Börde

In the Hellweg Börde wind energy started to be used in 1989 and has continuously increased between 1993 and 2007. This resulted in an increasing number of wind turbines, wind farms and total installed nominal capacity. Wind turbines were increasingly concentrated in larger wind farms. Average hub height, rotor diameter and nominal capacity of the single wind turbines rose substantially. Two phases of the construction of wind turbines can be identified in terms of the spatial distribution wind turbines. In the 1990s, a major part of the turbines was erected in the higher elevations of the Haarstrang outside the main area of distribution of the Montagu's harrier in the Hellweg Börde lowlands. On the one hand, this spatial separation results from the better suitability of higher elevations for wind energy use and on the other hand from the spatial management of wind energy development as a precautionary bird protection measure. From 2001, an increasing number of new wind turbines was constructed in the lower parts of the Lower Hellweg Börde and therefore closer to traditional breeding areas of the Montagu's harrier.

During this period, with its increasing number of wind turbines, the average number of Montagu's harrier nests showed a negative trend despite the annual fluctuations, which became very clear from 2005. The decline occurred, although measures to protect all birds nesting in cereal crops from damage by farming activities continued until 2007 (Glimm et al. 2001; Illner 2008). These measures significantly improved the breeding success, and were according to Glimm et al. (2001) a major reason for the earlier increase in population in the Hellweg Börde since the 1960s.

It was analysed, whether such a negative relation could be confirmed for 4 km^2 partial areas. On this small-scale level, no impact of wind turbines on the temporal development and the number of Montagu's harrier nests could be proven. This result can probably be attributed to the far-reaching spatial separation of wind turbine sites and breeding areas of Montagu's harriers. It is therefore unlikely that the decline of the Montagu's harrier's population in the Hellweg Börde between 1993 and 2007 is primarily due to large-scale displacement effects of wind turbines. Still, based on the coarse grid and the above described far-reaching separation of wind turbines and Montagu's harriers nests, it cannot be excluded that a displacement effect occurs in the close-up area of about 1 km around a wind turbine.

The long-term population decline is likely to have mainly ecological reasons. In Central Europe, breeding Montagu's harriers highly depend on availability and accessibility of voles and small birds (Arroyo et al. 2004; Koks et al. 2007). Fallows are a preferred hunting habitat of Montagu's harriers during the breeding season (Glimm et al. 2001), especially in years with lower vole abundances (Illner 2007), as they host considerably higher densities of farmland birds and voles than crops in the Hellweg Börde (Sayer et al. 2003; Braband et al. 2006). The number of fallows has strongly decreased in the Hellweg Börde over the last years (Illner 2007), a fact presumably contributing to a large extent to the decline in population by causing a shortage of prey (Illner 2007) together with at the same time intensified cereal crop cultivation (Arroyo et al. 2002). This overall pattern is confirmed by an especially strong decline of the Montagu's harrier population in years with low vole densities over the last years (Illner 2007, 2008).

5.6.3 *Establishment of Territories of Montagu's Harriers in the Hellweg Börde Before and After Construction of Wind Farms*

Looking at single farmland areas used by Montagu's harriers as a breeding ground with wind farms constructed during the investigation period, may give hints on the impacts of wind farms on the nest site selection of Montagu's harriers. In five of seven cases, where a comparison was possible, the respective farmland area was not used by Montagu's harriers for breeding after the construction of a wind farm. The overall results show that more Montagu's harriers nested in the respective farmland area before construction of the wind farm than afterwards. In the peripheral area of only one wind farm, several nests were found over several subsequent years (minimum distance from the next wind turbine: 290 m), possibly due to the special spatial constellation of the wind turbines and the specific geographical conditions. With the relatively small sample size and a lack of information on further parameters influencing the occupation of wind farm areas prior to and after construction of the wind farm, the determined differences were not tested for significance. The cautious conclusion that farmland areas within which wind farms are constructed will lose their value as a breeding ground for Montagu's harriers and may be completely avoided in nesting is supported by the fact that Montagu's harriers did not nest within wind farms in the Hellweg Börde so far. Apparently, most Montagu's harriers will observe a mean minimum distance of about 500 m from the next wind turbine, if they breed in the vicinity of a wind farm.

5.6.4 *Nest Site Selection of Montagu's Harriers in the Hellweg Börde in Relation to Wind Turbines and Other Habitat Parameters*

Looking at the complete data without taking into consideration the social attraction of colonies, the nest site selection of Montagu's harriers was found to be significantly influenced by landscape parameters such as slope gradient, distance from the next vertical landscape feature, distance from the next wind turbine or wind farms as well as from the distance from the next neighbouring nest. On average, nests of Montagu's harriers were located in less sloping areas at larger distances from areal vertical objects as forests or settlements as well as from wind turbines and wind farms than to be expected in a random distribution. Furthermore, they preferred smaller distances from the next neighbour as compared to a random distribution, reflected by the frequent colonial nesting pattern.

Similar results were found for a reduced sample using a mean value for nests in colonies. Differing from the complete sample, a significant difference was found for the height above sea level and nests were now located significantly lower than random points. Not too much importance should be attached to this, as the median values differed only slightly. The less frequent selection of steeply sloping areas as a nest site than to be expected in a random distribution can on the one hand be explained by the fact that the slope gradient is correlated to the height above sea level within the study area, which in turn influences the selection of a nest site (see below). On the other hand, the slope gradient, reducing the clear visibility of the terrain as much as vertical landscape structures, may impact the suitability of the area as a nest site due to an increased risk of predation (see below).

The comparison of solitary broods to broods in colonies showed that breeding colonies are on average located further away from areal vertical structures, wind turbines or wind farms than solitary broods. Differences in the distances from wind turbines and wind farms may have partly methodological reasons, as colonies of Montagu's harriers are located in lowland areas, whereas wind turbines are concentrated on the higher ranges. Based on the available data it cannot be excluded that a larger distance from wind turbines will lead to a higher quality of broods in colonies (measure: number of nests and breeding success) or that this quality criterion is still true for the traditional locations of breeding colonies in the Hellweg Börde. Concentration of Montagu's harrier colonies as well as of the total number of nests on lowland areas was already observed prior to the construction of wind turbines in the Hellweg Börde (Glimm et al. 2001). Among other things this could be due to faster growing winter cereals in lowland areas for climatic reasons, reaching a suitable height for breeding around one to two weeks earlier in these areas than in the higher elevated southern part of the study area (Glimm et al. 2001; Illner, unpublished). Furthermore, higher ranges always have more steep areas, which are less suitable for occupation by Montagu's harriers. The particular significance of breeding of colonies can be derived from the higher breeding success in colonies as compared to solitary broods (Arroyo et al. 2001) a fact which was confirmed for the Hellweg Börde (Illner, unpublished).

The minimum distances from vertical structures as forests and settlements as well as wind turbines and wind farms observed by most Montagu's harriers in the Hellweg Börde, fell within the range between 300 and 500 m. Only twice a nest was built at a distance of below 300 m from the next wind turbine at a distance of at least 170 m. No Montagu's harriers were observed breeding within a wind farm up to the year 2007. As wind turbines in wind farms have a mean spacing of 220 m (±75 m) in the Hellweg Börde, occupation of a territory between two wind turbines would lead to an average distance of 110 m from the next wind turbine. The overall results give an indication of a far-reaching avoidance of the close-up area of a median of 500 m around wind turbines by Montagu's harriers nesting in the Hellweg Börde. However, due to the small number of investigated years and the influence of other parameters, this value is not suitable as a guideline for planning practice.

Some of the investigated landscape parameters are correlated. Slope gradient and proximity to the next wind turbine are increasing with increasing height above sea level, as most wind turbines were constructed in higher elevations of the Hellweg Börde. These and further correlations complicate the interpretation of the results. For this reason, multi-factor analyses based on logistic regression models were carried out. In these models, the combined effects of several habitat parameters on the nest site selection of Montagu's harriers can be considered statistically independent of each other. In these models, the combined distance from the next neighbour and height above sea level or the slope gradient and the distance from the next areal vertical structure had a significant influence on the nest site selection of Montagu's harriers. Avoidance of settlements as well as colonial breeding attributed to social attraction was also found for a French study area (De Cornulier and Bretagnolle 2006). This trend towards the formation of colonies is particularly pronounced in years with good feeding conditions (Arroyo et al. 2004), profiting from the reduced risk of predation in colonial breeding (Arroyo et al. 2001) without strong competition for food.

Additionally, a significant influence of the distance from the next power pylon and the distance from the next point vertical structure (power pylons together with wind turbines) could be observed on the nest site selection. Looking only at wind turbines, no such significant effect occurs. As power pylons usually have dimensions similar to the less compact design of wind turbines within the study area and are mostly located further away from wind turbines, it can be assumed that the significant effect of the combined category "point vertical object" was caused by both of the structures. The lack of a significant influence of wind turbines alone on the nest site selection found in the regression models is not necessarily a contradiction, as power pylons are more frequently located within the traditional breeding areas of the Montagu's harrier in the lowland areas of the Hellweg Börde, whereas wind turbines are concentrated in the higher elevations with only few breeding Montagu's harriers. Therefore, the structure of the data may be a possible reason, why no effect of wind turbines alone could be determined in the regression models. Similar nesting distances from wind turbines of 200–300 m were found for the closely related hen harrier (Whitfield and Madders 2006) with similar habitat needs in terms of openness of the landscape. A more recent study of Pearce-Higgins et al. (2009) in Scotland proves significantly lower densities of hen harriers within a radius of 500 m around wind farms during the breeding season than in unaffected reference areas.

This leads to the overall conclusion that high vertical structures are influencing Montagu's harriers in the Hellweg Börde, as they breed at a certain distance from these structures. Openness in the field of vision seems to be an important habitat feature limiting the selection of possible breeding sites (see also De Cornulier and Bretagnolle 2006). The dimensions seem to play an important role, as the effect of areal vertical structures as forests or settlements proved to be more pronounced than that of point vertical structures including wind turbines. It can therefore be assumed that (larger) wind farms will have a stronger spatial effect on Montagu's harriers than single wind turbines and thus a stronger influence on the selection of a nest site. This question could not be answered conclusively in this study due to the large-scale separation of main breeding areas of the Montagu's harrier and areas with wind turbines and wind farms.

The differing results found for areal and point vertical structures may be partly due to the fact that peripheral zones around areal vertical structures are considerably more frequent and more evenly distributed within the study area than wind turbines and power pylons, resulting in a larger amount of data for nests and random points for these zones.

The determined larger distances of Montagu's harrier nests from vertical structures can be interpreted as avoidance behaviour with plausible biological reasons. An important factor may be that areal vertical structures are no suitable hunting ground for Montagu's harriers (Arroyo et al. 2004). Larger forests or settlement areas in the vicinity of a nest site will reduce the availability of potential feeding areas within reach without the need to cover larger distances involving a higher energy input. This will be of particular importance in years with a low abundance of prey and the resulting lower hunting success rate (Trierweiler et al. 2008).

Vertical structures may be used as a perching point by larger birds of prey or they may limit or prevent the detection of approaching avian predators causing an increased risk of predation. Montagu's harriers as a small ground-nesting bird of prey species are subject to a high risk of predation. In Central Europe predators mainly are carnivorous mammals, northern goshawk, marsh harrier and common buzzard. Predation of females on the nest by foxes, goshawks and marsh harriers was documented (Koks and Visser 2000; Kitowski 2003; K.D. Gierach, C. Pürkhauer and R. Krüger, personal communication). The high risk of predation shows indirectly in the pronounced alarm calls and attacks on carnivorous mammals or birds of prey approaching the breeding site (Arroyo et al. 2001, 2004; Kitowski 2003). Partly intense attacks of Montagu's harriers against red kite, marsh harriers, peregrine falcons and common buzzards were observed at breeding sites in the Hellweg Börde (Illner, unpublished). These species, especially common buzzards, using vertical structures (trees, hedges, masts, buildings) as a perching point while hunting, will be attacked and driven away by Montagu's harriers (repeated observations in the Hellweg Börde, Illner, unpublished). Rotating wind turbines with closed towers are only sporadically used as a perching point by birds of prey, but Montagu's harriers will probably not be able to anticipate this and may perceive wind turbines as a potential starting point for attacks of predators.

Finally it should be pointed out that the results of multi-factor analyses were determined using correlation analyses, which cannot always explain causal relations. This could only be done in an experimental approach, which was not intended in this study. To determine the effects of vertical structures on the nest site selection of Montagu's harriers, additional behavioural studies would be necessary. However, it can be assumed that some effects can be explained or made plausible based on the relatively well-known breeding biology of the Montagu's harrier (Arroyo et al. 2001).

5.6.5 Conclusions for the SPA Hellwegbörde

In this study, no clear, statistically significant influences of wind turbines on the development of the Montagu's harrier population in the Hellweg Börde or large-scale statistically significant avoidance of the surrounding area of wind turbines by nesting Montagu's harriers could be determined. However, due to the above described methodological limitations, the present analysis is only to a limited extent suitable to prove impacts of wind turbines on the distribution of Montagu's harrier nests in the close-up area. Form a scientific point of view it is therefore not possible to make reliable predictions for a scenario with additional wind turbines or repowering of small wind turbines in the vicinity of Montagu's harrier's breeding sites in the Hellweg Börde based on the results for the investigation period between 1993 and 2007. The same will be true for a prediction of impacts of wind turbines in other breeding areas of the Montagu's harrier, which would be statistically speaking an inadmissible extrapolation going beyond the investigated data. The results can therefore not be used to clearly or even to a sufficient degree of probability exclude long-term impacts of a further development of wind farms in the Hellweg Börde on the Montagu's harrier population.

Despite the far-reaching spatial separation of wind turbines and traditional breeding areas, the results indicate possible impacts on the habitat of the Montagu's harrier in the Hellweg Börde already for the current scenario. Montagu's harriers observed a minimum distance of on average (median) of 500 m from wind turbines. In single farmland areas occupied by Montagu's harriers, where wind farms were installed, only single broods occurred after the construction of wind turbines at a mean (median) distance of 500 m from the next wind turbine. No broods were found within wind farms. Avoidance of vertical landscape structures and the avoidance of power pylons and point vertical objects obvious from some of the logistic regressions provides evidence that wind turbines as a form of high vertical object may impair the habitat of Montagu's harriers. This is especially true in the cumulative evaluation of the combined effects of further parameters.

Against this background, the previously followed strategy to spatially separate areas for wind energy use from core breeding areas of Montagu's harriers proves to be justified. The results of the present evaluations show that possible conflicts can be largely avoided by this approach. This precautionary principle should still be followed and the construction of further wind turbines within the SPA is not acceptable. This is all the more true, as the long-term decline of the Montagu's harrier population in the Hellweg Börde obliges to take the utmost care to not further endanger the already affected conservation objectives of the European SPA. It needs to be taken into consideration that other relevant bird species of this European SPA are already impaired by the present development stage of wind energy use. In the Hellweg Börde, avoidance distances of about 300 m around wind turbines were determined for the corncrake and the common quail (Müller and Illner 2001). This also applies to resting waders, which were likewise found to show avoidance behaviour towards wind turbines (Hötker 2006).

Repowering will lead to an enhanced balance between wind energy use and the protection of the Montagu's harrier and other open landscape species, if the number of wind turbines is actually reduced and wind farms in problematic locations are decommissioned in the replacement of older wind turbines with larger and higher performance wind turbines. As far as Montagu's harriers are concerned, this mainly applies to wind farms in lowland areas. Other wind farms may be of concern for other open landscape species. A binding and subordinated planning accounting for all aspects of bird protection would be a prerequisite but involves considerable difficulties in the implementation. No statements on the impacts of larger wind turbines installed in repowering on the Montagu's harrier are possible in the present context.

5.6.5.1 Protection of the Open Landscape

It cannot be excluded that suitable breeding areas for Montagu's harriers in the Hellweg Börde will be reduced to a size falling below the required minimum size of available open landscape especially for colonial breeding in a further development of wind energy. Minimum distances of Montagu's harrier nests from wind turbines and vertical landscape structures as well as the avoidance of power pylons, often located in the vicinity of breeding areas, i.e. of structures similar to wind turbines, determined in some of the models point in this direction. This effect is especially to be feared in the flat open landscape of the Hellweg Börde lowland areas, which are particularly suitable for breeding. Especially in these areas, more and more wind farms were installed in recent times starting in 2000, e.g. near Merklingsen and Bettinghausen. The cumulative effect of wind energy use and other land uses as quarries, road construction, the expansion of residential and industrial areas as well as farm buildings in the open landscape needs to be taken into account.

It is already evident that the sufficiently open and flat lowland areas of the Hellweg Börde as a suitable breeding ground for the Montagu's harriers are considerably smaller than the total Hellweg Börde area (Griesenbrock 2006). A major proportion of Montagu's harrier nests is already now concentrated on the few remaining core breeding areas. In the reduction of the remaining open landscape areas suitable for breeding Montagu's harriers located at a sufficiently large distance from settlements due to other land use (mining, expansion of settlements, industry, agricultural projects), a concentration of wind turbines in these areas due to planning guidelines may further reduce these areas. Especially the construction of further large wind farms will lead to direct area competition with Montagu's harrier breeding grounds.

5.6.5.2 Collisions

Only indications of an avoidance of the close-up area of wind farms within a median radius of 500 m around wind turbines by nesting Montagu's harriers could be determined in the present study. If nesting Montagu's harriers do not actually avoid wind turbines, the risk that Montagu's harriers collide with wind turbines will be potentially higher, because they will use the air space in the surrounding area of the nest more frequently due to site fidelity lasting for months.

During the breeding season, many flight behaviours of Montagu's harriers like courtship displays, attacks on avian predators, food passes and practicing flights of young birds are performed within the height range of the rotor swept area of wind turbines in the surrounding area of the breeding site (Arroyo 1995; Clarke 1996; Arroyo et al. 2004). According to monitoring of 11 satellite-tagged Montagu's harriers in North German wind farms, about 5% of all Montagu's harrier flight movements occurred in heights between 30 and 100 m, i.e. within the height band of the rotor swept area, where several dangerous situations were observed (Grajetzky and Nehls 2013). The risk of collision does not only depend on the number of flights but on the duration of stay within the risk zone as well. The following descriptions are based on behavioural observations of Montagu's harriers made during breeding bird surveys in the Hellweg Börde between 2006 and 2010 (Illner, personal observations). During mating, courtship displays of male and female birds involving fast alternating vertical and horizontal flight movements (sky dancing) can be observed for several weeks mostly in heights above 30 m. 26.5% of these flights were performed near the prospective breeding site within a height range of between 30 and 120 m, the rotor swept area of today's wind turbines. A total of 944 flight minutes of 17 territorial pairs was evaluated. In one case, a Montagu's harrier was observed getting dangerously close to the rotor of a wind turbine in courtship display (rough estimate: 50 m) without showing a distinct startle response (breeding season 2007, Merklingsen wind farm, see Table 5.7). Birds of prey, crows and other large birds, especially red kites, marsh harriers, common buzzards, other Montagu's harriers and carrion crows, flying near the nest are often strongly and persistently mobbed, attacked and driven away in flight (height range: 5–100 m). Humans approaching the nest are mobbed mainly by female birds hovering persistently in one place above the intruder at heights of between 30 and 100 m. Especially in fine weather, female birds will take young birds older than one to two weeks for flights in heights above 30 m (circling considerably higher where thermals occur), moving up to 300 m or even more away from the nest. Practicing flights of young birds are often performed in the surrounding area of the nest in partly higher altitudes. Mainly male bird use rising thermal currents for energy-saving flight movements, soaring up in circles to heights above 1000 m before passing prey or returning to hunting grounds. Within the study area, Montagu's harriers spend a considerable proportion of time in higher altitudes being at risk of collision in flight movements requiring higher attention, possibly preventing proper estimation of the risks imposed by moving rotors of wind turbines.

Even if nesting Montagu's harriers actually avoid the close-up area of wind turbines, a collision risk can therefore not be excluded for the Hellweg Börde. Hunting flights are usually performed at heights of a few meters above the ground, but in years with a low abundance of voles, pursuits of passerines may occur, where Montagu's harriers rise to heights of about 40 m above the prey to strike from above (Illner, unpublished). Under favourable weather conditions, male birds catching prey further away from the breeding site will use thermal soaring to carry the prey, whereby they may get into the rotor swept area as proven for other bird of prey species (Barrios and Rodrìguez 2004).

The first two collisions of a Montagu's harrier with a wind turbine were documented in Spain, the Mediterranean country with the largest number of wind turbines. In 2006, a male bird was found dead near Bureta in the Saragossa province (A.C. Cardenal and M.F. Canevet, personal communication) and in 2009 a male bird in Navarra (J.M. Lekuona, personal communication). A further three suspected collisions of Montagu's harriers with wind turbines are documented in Germany. One male breeding bird was found dead at a distance of 300 m from the next wind turbine in a wind farm in Schleswig-Holstein within the framework of the study carried out by Grajetzy et al. (in: Hötker 2009). The injuries as well as the circumstances of the find suggested a collision with a wind turbine. In case of the two birds found dead in the Hellweg Börde, the external and internal injuries could not be attributed to a collision with a nearby wind turbine with absolute certainty, as both injured Montagu's harriers could initially still move.

As to present knowledge (as from 2010) breeding areas were not systematically searched for dead Montagu's harriers at wind turbines and taking into consideration all possible sources of error, the low number of documented collision victims cannot not be interpreted as an evidence that this rare species is not at risk. Montagu's harriers as a migratory species are only present in Central Europe between end of April and beginning of September, where breeding areas are mostly covered with high growing vegetation (mainly cereals and grassland), complicating the search for dead birds, which may have collided with wind turbines in these areas and thus the verification of collisions with wind turbines.

Up until 11 May 2010, a total of five collisions of marsh harriers, a larger and more abundant species, found dead at German wind turbines were reported (T. Dürr, central German database of collision victims of the Staatliche Vogelschutzwarte (State Bird Observatory and Field Centre) Brandenburg). Due to comprehensive investigations, several fatal collisions of hen harriers, a species, which is closely related and in many respects similar to the Montagu's harrier, with wind turbines could be verified for North America and Spain (Smallwood and Thelander 2008; Leukuona and Ursùla 2007). It can therefore be assumed that Montagu's harriers are also likely to collide with wind turbines.

5.7 Conclusions

Aim of the present study was to analyse possible impacts of the construction of wind turbines on the population and nest site selection of Montagu's harriers in the Hellweg Börde by overlaying nest sites with locations of wind turbines using a geographic information system. By evaluating existing data of the years 1993–2007, possible relations of the construction of wind turbines to number and distribution of Montagu's harrier nests in the Hellweg Börde were investigated. It should be noted that already in the planning phase for new wind turbine sites attempts were made to avoid conflicts with the spatial distribution of the Montagu's harrier and other open landscape breeders. In 1994, a map of areas suitable for wind energy use from a nature and species conservation point of view was compiled by the advisory board of the landscape authority of the Soest district, including the main distribution area of the Montagu's harrier in the Hellweg Börde. Especially the protection of breeding areas of the Montagu's harrier in lowland areas against possible impacts caused by wind turbines was taken into consideration. These measures were able to ensure a far-reaching spatial separation of wind turbine sites and breeding areas of the Montagu's harrier.

In this study, no clear, statistically significant influences of wind turbines on the development of the Montagu's harrier population in the Hellweg Börde or large-scale statistically significant avoidance of the surrounding area of wind turbines by nesting Montagu's harriers could be determined. However, due to the above described methodological limitations, the present analysis is only suitable to prove large-scale and distinct impacts of wind turbines on the distribution of Montagu's harrier nests. It is therefore not possible to make reliable predictions for a scenario with additional wind turbines or repowering of small wind turbines in the vicinity of Montagu's harrier's breeding sites in the Hellweg Börde or the impacts of wind turbines in other breeding areas of Montagu's harriers based on the results for the investigation period between 1993 and 2007. This would be an inadmissible extrapolation going beyond the so far investigated data. Despite the far-reaching spatial separation of wind turbines and traditional breeding areas of Montagu's harriers, the results indicate possible impacts on the habitat of the Montagu's harrier in the Hellweg Börde even for the current scenario. Montagu's harriers observed a minimum distance of on average (median) of 500 m from wind turbines. In seven farmland areas occupied by Montagu's harriers, where wind farms were constructed, only single broods occurred after the construction of wind turbines at a mean (median) distance of 500 m from the next wind turbine. No broods were found within wind farms. Avoidance of vertical landscape structures and the avoidance of power pylons and point vertical objects obvious from some of the logistic regressions provide evidence that wind turbines as a form of high vertical object may impair the habitat of Montagu's harriers.

The results of the present evaluations can be regarded as an indication that possible conflicts can be largely avoided by the present spatial separation of wind turbine locations and breeding areas of Montagu's harriers. It still cannot be excluded that suitable breeding areas for Montagu's harriers in the Hellweg Börde will be reduced to a size falling below the required minimum size of available open landscape especially for colonial breeding in a further development of wind energy. With the consumption of the remaining open landscape areas suitable for breeding Montagu's harriers located at a sufficiently large distance form settlements due to other land use (mining, expansion of settlements, industry, agricultural projects) a concentration of wind turbines in these areas due to planning guidelines may further reduce these areas. Especially the construction of further large wind farms will lead to direct area competition with Montagu's harrier breeding grounds.

Only indications of an avoidance of the close-up area of wind turbines or wind farms by nesting Montagu's harriers could be determined in the present study. If nesting Montagu's harriers do not actually avoid wind turbines, the risk that Montagu's harriers collide with wind turbines will be potentially higher, because they use the air space in the surrounding area of the nest more frequently. Although hunting flights are mostly performed low above the ground, many flight movements of Montagu's harriers in the surrounding area of the nest occur within the rotor swept area of wind turbines during the breeding season. These include courtship displays of breeding pairs, mobbing and attacks on avian predators, thermal soaring of male birds in arrival from or departure towards hunting grounds as well as practicing flights of young birds.

Acknowledgements We would like to thank the staff members of the districts Arnsberg, Paderborn, Soest and Unna, the municipalities Rüthen, Anröchte, Möhnesee and Ense as well as the German Wind Energy Association (BWE, regional association of Southern Westphalia) and the Enercon GmbH for the assistance in research and the data on the location and the date of construction of wind turbines. We are grateful to Dr. Leonid Rasran (NABU Michael-Otto-Institute) for statistical advise and to Birgit Beckers, Joachim Drüke and Claudia Pürckhauer for the review of the manuscript.

References

Arroyo BE (1995) Breeding ecology and nest dispersion of Montagu`s Harrier *Circus pygargus* in Central Spain. Dissertation, University of Oxford.

Arroyo BE, Garcia JT, Bretagnolle V (2002) Conservation of Montagu`s Harrier *Circus pygargus* in agricultural areas. Ornithologischer Anzeiger 41:119–134.

Arroyo BE, Garcia JT, Bretagnolle V (2004) Montagu's harrier. BWP update (The Journal of the Birds of the Western Palearctic) 6:41–55.

Arroyo B, Mougeot F, Bretagnolle V (2001) Colonial breeding and nest defence in Montagu's harrier (*Circus pygargus*). Behav Ecol Sociobiol 50:109–115.

Backhaus K, Erichson B, Plinke W, Weiber R (2006) Multivariate Analysemethoden. Springer, Berlin.

Bartelme N (2005) Geoinformatik. Modelle, Strukturen, Funktionen. 4th edn. Springer, Berlin.

Bauer H-G, Bezzel E, Fiedler W (2005) Das Kompendium der Vögel Mitteleuropas. Aula Verlag, Wiesbaden.

Barrios L, Rodrìguez A (2004) Behavioural and environmental correlates of soaring-bird mortality at on-shore wind turbines. J Appl Ecol 41:72–81.

Bergen F (2001) Windkraftanlagen und Frühjahrsdurchzug des Kiebitz (*Vanellus vanellus*): eine Vorher/Nachher-Studie an einem traditionellen Rastplatz in Nordrhein-Westfalen. Vogelkundliche Berichte aus Niedersachsen 33:88–96.

Boschert M (2005) Vorkommen und Bestandsentwicklung seltener Brutvogelarten in Deutschland 1997 bis 2003. Vogelwelt 126:1–51.

Braband D, Illner H, Salm P, Hegemann A, Sayer M (2006) Erhöhung der Biodiversität in einer intensiv genutzten Bördelandschaft Westfalens mit Hilfe von extensivierten Ackerstreifen. Abschlussbericht, Bad Sassendorf-Lohne.

Bunzel-Drüke M, Schulze-Schwefe KH (1994) Windkraftanlagen und Vogelschutz im Binnenland. Natur und Landschaft 69:100–103.

Bürgener M (1963a) Die naturräumlichen Einheiten auf Blatt 110 Arnsberg. Geographische Landesaufnahme 1:200.000. Naturräumliche Gliederung Deutschlands. Bundesanstalt für Landeskunde, Bad Godesberg.

Bürgener M (1963b) Die naturräumlichen Einheiten auf Blatt 111 Arolsen. Geographische Landesaufnahme 1:200.000. Naturräumliche Gliederung Deutschlands. Selbstverlag, Bundesanstalt für Landeskunde, Bad Godesberg.

Clarke R (1996) Montagu`s Harrier. Arlequin Press, Chelmsford.

De Cornulier T, Bretagnolle V (2006) Assessing the influence of environmental heterogeneity on bird spacing patterns: a case study with two raptors. Ecography 29:240–250.

Dickmann F, Zehner K (1999) Computerkartographie und GIS. Westermann, Braunschweig.

Glimm D, Hölker M, Prünte W (2001) Brutverbreitung und Bestandsentwicklung der Wiesenweihe in Westfalen. LÖBF Mitteilungen 2/01:57–67.

Grajetzky B, Nehls G (2013) Telemetrische Untersuchungen von Wiesenweihen in Schleswig-Holstein. In: Hötker H, Krone O, Nehls G (eds) Greifvögel und Windkraftanlagen: Problemanalyse und Lösungsvorschläge. Schlussbericht für das Bundesministerium für Umwelt, Naturschutz und Reaktorsicherheit. Michael-Otto-Institut im NABU,

Leibniz-Institut für Zoo- und Wildtierforschung, BioConsult SH, Bergenhusen, Berlin, Husum, http://www.nabu.de/themen/energie/erneuerbareenergien/windkraft/17417.html, Accessed 30 June 2014.
Griesenbrock B (2006) Habitat und Nistplatzwahl der Wiesenweihe *Circus pygargus* L. in der Hellwegbörde. Diploma thesis, Westfälische-Wilhelms-University of Münster.
Hölker M (1994–2006) Schutzprogramm für Wiesenweihen und Rohrweihen in Mittelwestfalen. Jahresberichte 1993 bis 2005. Arbeitsgemeinschaft Biologischer Umweltschutz, Bad Sassendorf Lohne.
Hötker H (2006) Auswirkungen des „Repowering“ von Windkraftanlagen auf Vögel und Fledermäuse. Michael-Otto-Institut im NABU, Bergenhusen, http://bergenhusen.nabu.de/imperia/md/images/bergenhusen/impact_of_repowering.pdf. Assessed 30 June 2013.
Hötker H (ed) (2009) Birds of Prey and Wind Farms: Analysis of Problems and Possible Solutions. Documentation of an international workshop in Berlin, 21st and 22nd October 2008. Extended abstracts of talks and brief summaries of discussions. Bergenhusen. http://bergenhusen.nabu.de/imperia/md/images/bergenhusen/bmuwindkraftundgreifwebsite/birds_of_prey_and_windfarms_documentation_2009.pdf. Assessed 30 June 2010.
Illner H (2007) Schutzprogramm für Wiesenweihen und Rohrweihen in Mittelwestfalen. Jahresbericht 2006. Arbeitsgemeinschaft Biologischer Umweltschutz, Bad Sassendorf-Lohne.
Illner H (2008) Schutzprogramm für Wiesenweihen und Rohrweihen in Mittelwestfalen. Jahresbericht 2007. Arbeitsgemeinschaft Biologischer Umweltschutz, Bad Sassendorf-Lohne.
Illner H, Steinwarz D (2003) Abschlussbericht der wissenschaftlichen Begleituntersuchungen zum E + E-Vorhaben des BfN „Biotopvernetzende Maßnahmen zur Reaktivierung des biologischen Potentials und zum Bodenschutz am Haarstrang bei Rüthen/Kreis Soest“. Teil I, Grundlagen. Arbeitsgemeinschaft Biologischer Umweltschutz, Bad Sassendorf-Lohne.
Jenness J (2005) Random Point Generator 1.3. http://www.jennessent.com/arcview/random_points.htm. Assessed 30 June 2010.
Kampik F (1995) Windkraft in Turbulenzen? ABUinfo19:32–34.
Kitowski I (2003) Inter-specific and temporal differences in brood defence of adult Montagu's harriers *Circus pygargus* and Marsh harrier *Circus aeruginosus* against foxes *Vulpes vulpes* and avian intruders during the post-feldging dependy period. Biota 4:57–66.
Koks B, Visser E (2000) Predatie door vossen *Vulpes vulpes* op broedende Grauwe Kiekendieven Circus pygargus. De Takkeling 8:211–217.
Koks BJ, Trierweiler C, Visser EG, Dijkstra C, Komdeur J (2007) Do voles make agricultural habitat attractive to Montagu's Harrier Circus pygargus? Ibis 149:575–586.
Krogulec J (1993) Status, population trends, breeding habitats and conservation problems of Montagu`s Harrier (Circus pygargus) in Poland. In: Clemens C (ed) International Montagu`s Harrier (Circus pygargus) Conference. Kiel, p 29–30.
Kruckenberg H, Borbach-Jaene J (2001) Auswirkung eines Windparks auf die Raumnutzung nahrungssuchender Blessgänse - Ergebnisse aus einem Monitoringprojekt mit Hinweisen auf ökoethologischen Forschungsbedarf. Vogelkundliche Berichte aus Niedersachsen 33:103–109.
LDS (Landesamt für Datenverarbeitung und Statistik Nordrhein-Westfalen) (2003) Statistische Rundschau für die Kreise Nordrhein-Westfalens. Kreis-Soest, Düsseldorf.
LVERMA NRW (Landesvermessungsamt Nordrhein-Westfalen) (1999, 2000) ATKIS-Geodaten. Geobasisdaten Landesvermessungsamt Nordrhein-Westfalen. Bad-Godesberg.
LVERMA NRW (Landesvermessungsamt Nordrhein-Westfalen) (2001): Das Digitale Geländemodell 5 (DGM5), Geobasisdaten. Landesvermessungsamt Nordrhein-Westfalen, Bad-Godesberg.
LVERMA NRW (Landesvermessungsamt Nordrhein-Westfalen) (2008) Landschaftsinformation – Höhenmodelle. Landesvermessungsamt Nordrhein-Westfalen, Bad-Godesberg.
LWK (Landwirtschaftskammer Westfalen Lippe) (2004) Zahlen zur Landwirtschaft in Westfalen-Lippe. Landwirtschaftskammer Westfalen Lippe, Münster.
LWL (Landschaftsverband Westfalen-Lippe) (2007) Kulturlandschaftlicher Fachbeitrag zur Landesplanung in Nordrhein-Westfalen. Landschaftsverband Westfalen-Lippe, Münster.

Larsen JK Madsen J (2000) Effects of wind turbines and other physical elements on field utilisation by pink-footed geese (*Anser brachyrhynchos*): a landscape perspective. Landscape Ecol 15:755–764.

Leddy KL, Higgins KF, Naugle DE (1999) Effects of wind turbines on upland nesting birds in Conservation Research Program grasslands. Wilson Bull 111:100–104.

Leukuona JM, Ursùla C (2007) Avian mortality in wind power plants of Navarra (northern Spain). In: de Lucas M, Janss G, Ferrer M (eds) Birds and Wind Power. Quercus, Madrid, p 177–192.

Mebs T, Schmidt D (2006) Die Greifvögel Europas, Nordafrikas und Vorderasiens. Biologie, Kennzeichen und Bestände. Kosmos, Stuttgart, pp 495.

Meisel S (1959) Die naturräumlichen Einheiten auf Blatt 98 Detmold. Geographische Landesaufnahme 1:200.000. Naturräumliche Gliederung Deutschlands. Bundesanstalt für Landeskunde, Remagen.

Meisel S (1960) Die naturräumlichen Einheiten auf Blatt 99 Münster. Geographische Landesaufnahme 1:200.000. Naturräumliche Gliederung Deutschlands. Bundesanstalt für Landeskunde und Raumforschung, Bad Godesberg.

Müller A, Illner H. (2001) Beeinflussen Windenergieanlagen die Verteilung rufender Wachtelkönige und Wachteln? Manuscript of a talk given at „Windenergie und Vögel - Ausmaß und Bewältigung eines Konfliktes" at the Technische Universität Berlin, 29/30 Nov 2001.

Newton I (1979) Population Ecology of Raptors. T and AD Poyser, Berkhamsted.

NWO (Nordrhein-Westfälische Ornithologengesellschaft (2002) Die Vögel Westfalens. Ein Atlas der Brutvögel von 1989 bis 1994. Beiträge zur Avifauna Nordrhein-Westfalens, vol 37, Bonn.

Pearce-Higgins JW, Leigh S, Langston RHW, Bainbridge IP, Bullman R (2009) The distribution of breeding birds around upland wind farms. J Appl Ecol 46:1323–1331.

Peitzmeier J (1969) Avifauna von Westfalen. Abhandlungen aus dem Landesmuseum für Naturkunde zu Münster in Westfalen 31, Münster.

Sayer M, Bittner HJ, Körner M, Schaefer M, Dornieden K, Illner H, Lücke K, Lutze A, Nickel H (2003) Straßenbedingte Auswirkungen auf die Pflanzen- und Tierwelt benachbarter Biotope. Forschung Straßenbau und Straßenverkehrstechnik Heft 865:1–136.

Smallwood KS, Thelander CG (2008) Bird mortality in the Altamont Pass Wind Resource Area, California. J Wildlife Manage 72:215–223.

Sudmann SR, Grüneberg C, Hegemann A, Herhaus F, Mölle J, Nottmeyer-Linden K, Schubert W, von Dewitz W, Jöbges M, Weiss J (2009) Rote Liste der gefährdeten Brutvogelarten Nordrhein-Westfalens. 5. Fassung, http://www.nw-ornithologen.de/index.php?cat=projects &subcat=2, Accessed 30 June 2010.

Südbeck P, Bauer H-G, Boschert M, Boye P, Knief W (2007) Rote Liste der Brutvögel Deutschlands, 4. Fassung, 30. November 2007. Berichte zum Vogelschutz 44:23–81.

Trierweiler C, Visser EG, Arisz J, Koks BJ (2008) Habitatgebruik van Grauwe Kiekendieven *Circus pygargus* in het agarisch landschap 2003-2006 onerzocht m.b.v radiotelemtrie. Bericht der Stichting Werkgroep Grauwe Kiekendief, pp 86, www.grauwekiekendief.nl. Assessed 30 June 2010.

Whitfield DP, Madders M (2006) A review of the impacts of wind farms on hen harriers *Circus cyaneus* and an estimation of collision avoidance rates. Natural Research Information Note 1 (revised), Natural Research, Banchory.

Chapter 6
Impacts of Wind Turbines on the Population and Nest Site Selection of the Montagu's Harrier in North Frisia

Leonid Rasran and Kai-Michael Thomsen

6.1 Introduction

This sub-study report is intended to provide information supplementing the comprehensive study carried out in the Hellweg Börde (NRW) on the nest site selection of Montagu's harriers and their behaviour towards wind turbines (see Chap. 5). Apart from the Hellweg Börde and the Mainfränkische Platten, North Frisia is one of the most important breeding areas of Montagu's harriers in Germany, strongly differing in landscape structure from the two other areas. At the same time, North Frisia is one of the leading areas in terms of wind energy development. Here, the first turbines were installed very early before the beginning of the 1990s. In contrast to the Hellweg Börde, the entire area is homogenous without being structured by higher elevations or connected forest and settlement areas with the exception of the Husum urban area. Aim of the present sub-study was to analyse, whether statements regarding the nest site selection and the impacts of wind turbines on Montagu's harrier breeding sites in the Hellweg Börde will also be true for a different area, or whether the geographical characteristics of North Frisia will have a particular influence on the nest site selection of Montagu's harriers.

L. Rasran (✉) · K.-M. Thomsen
NABU (Nature and Biodiversity Conservation Union), Michael-Otto-Institute, Goosstroot 1, 24861 Bergenhusen, Germany
e-mail: Leonid.Rasran@boku.ac.at

H. Hötker et al. (eds.), *Birds of Prey and Wind Farms*,
DOI 10.1007/978-3-319-53402-2_6

6.2 Methods

6.2.1 Study Area

The study area "North Frisian polders" (Fig. 6.1) comprises the coastal marshes of Schleswig-Holstein from the Danish border in the north up to the Eider estuary in the south. The total area covers 960 km^2, subdivided into 198 diked polders of varying sizes (between 1.2 and 100 km^2). Human settlement areas are wide-stretching but small-structured, consisting of small villages and single farmsteads. Forest areas make up a proportion of below 2%. Landscape areas suitable for breeding Montagu's harriers make up a proportion of more than 90%. These areas include arable cropland (mainly cereals), grassland, salt marshes, heathland and moorland. Cropland and permanent pastures prevail at a ratio of nearly 1:1. In North Frisia, similar to other distribution ranges, an increasing preference of Montagu's harriers for winter cereals as a site for nesting could be observed (Gahrau and Schmüser 2007). The coastal salt marshes on the seaward side of the dikes are an important foraging habitat (Glutz von Blotzheim et al. 1989; see also Chap. 4).

The North Frisia district is Germany's leading region in wind energy use. The wind energy power production was about 2 million MWh/year and covered more

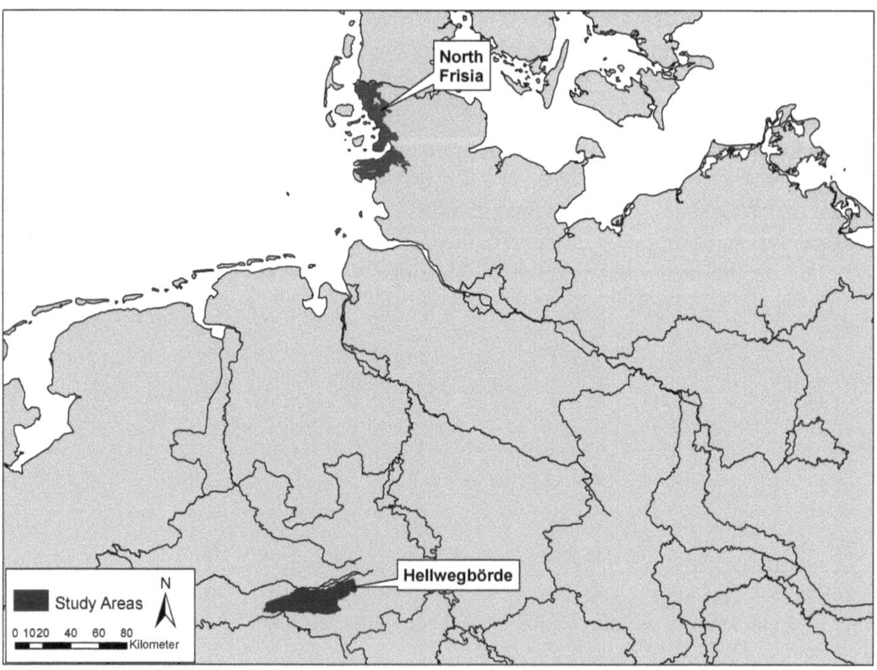

Fig. 6.1 Location of the two study areas in Germany (Hellweg Börde and North Frisia)

than 180% of the districts own needs (EnergyMap.info 2011). The number of wind turbines has risen from 44 in 1990 to 386 in 1995 (operator database (BDB), Keiler 2008) and stagnated at this level since (Fig. 6.2). Still, the total capacity of turbines increased, as old, smaller turbines were replaced by modern, larger and higher-capacity wind turbines in comprehensive repowering starting in 2002. The average nominal capacity per turbine has risen from 300 kWh in 1995 to over 1000 kWh in 2007. The sites and most important specifications of all wind turbines in the study area were researched and registered in a database and GIS-mapped.

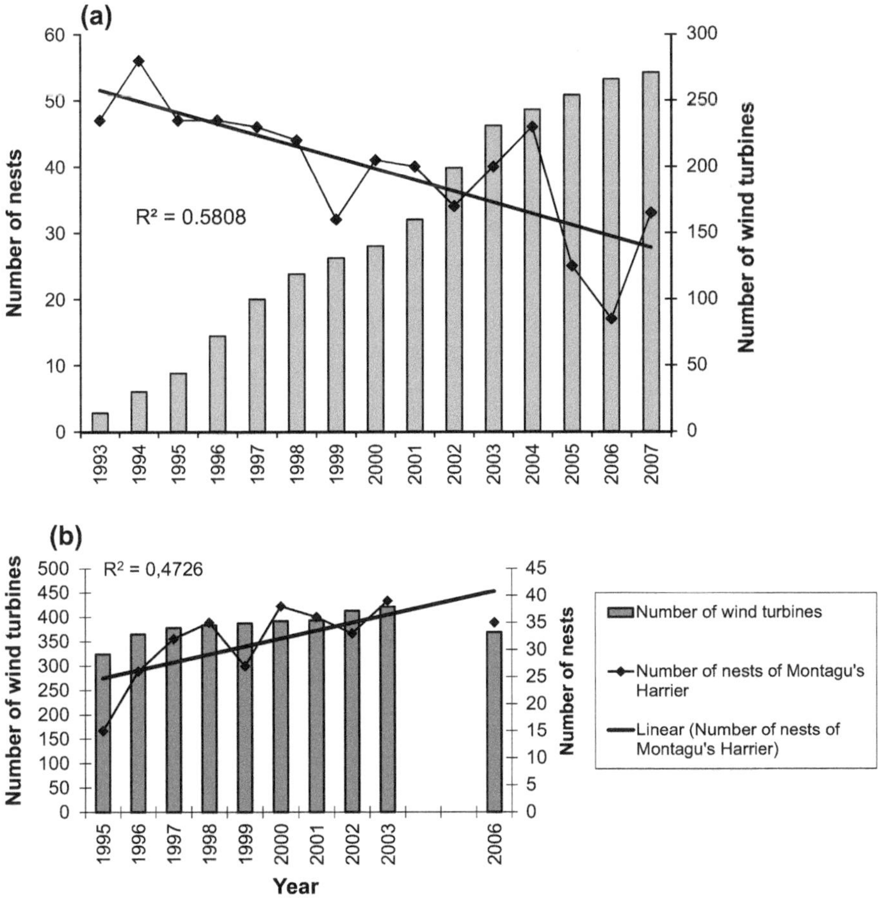

Fig. 6.2 Development of the Montagu's harrier population (*continuous line and dots* number of breeding pairs) in the Hellweg Börde (**a**) and North Frisia (**b**) in comparison to wind energy development (*grey bars* number of wind turbines)

6.2.2 *Survey of Montagu's Harriers*

Between 1995 and 2006, nest sites of Montagu's harriers in the North Frisian polders were surveyed with the help of numerous voluntary helpers and entered into GIS maps. Localisation errors depending on the observer did occur, but as a rule, locations could be determined exact to the crop or grassland plot. Where the nest could not be localised directly, the territorial behaviour of adult birds was assumed to be a proof for breeding.

6.2.3 *Data Analyses*

Structures in the Surrounding Area

To compare the influence of wind turbines and other landscape structures on the nest site selection of Montagu's harriers, the locations of nests were overlaid with the locations of wind turbines, including the date of construction or decommissioning as well as technical specifications. This way it was possible to provide information on the number and nominal capacity of wind turbines within a radius of one kilometre (3.14 km^2) around all known nests for every year of investigation. As Montagu's harriers establish a territory shortly after arrival from wintering grounds in the end of April, only those wind turbines were included in the evaluation, which went into operation until March of the respective year. The parameter "nominal capacity" was chosen, because this value is strongly correlated to parameters such as rotor diameter, footprint or sound level relevant for most ornithological issues. As a crosscheck for the nest sites, 50 reference points were randomly chosen within habitats of the North Frisian polders potentially suitable for Montagu's harriers, determining the same parameters (number and summarized nominal capacity of wind turbines). Linear mixed models were used to statistically analyse the influence of wind turbines on the nest site selection of Montagu's harriers. Using this method, a curve function serves to model temporal changes in number and capacity of wind turbines in the surrounding area of the nest. The development of wind turbines around the reference points, which are representative for the whole area, was modelled in the same way and both curves were compared in analyses of variance. Only if differences between the two curves are significant, rising above random fluctuations, an influence of wind turbines on the nest site selection can be assumed.

As most further parameters of a location such as height above sea level or slope gradient, which were tested for relevance in the Hellweg Börde, are not or only slightly relevant in North Frisia, only the distance between the nest site (or the random point of the reference data) and the next salt marsh were included in the further analyses. As data were not suitable for ANOVA in this case, differences in habitat parameters between nest sites and random points were described using medians and quartiles, which are robust towards deviations from the normal distribution, and tested for significance using the non-parametric Wilcoxon rank test.

Table 6.1 Distribution of grid cells in the two study areas with Montagu's harrier nests or wind turbines

	Regarded squares	Only nests	only wind turbines	Both
Hellweg Börde	127	63	48	16
North Frisia	157	66	55	37

The same method was used in the Hellweg Börde in the analysis of structures in the surrounding area.

Grid Analysis

To investigate the influence of wind energy development on the Montagu's harrier population in North Frisia, the study area was, like in the Hellweg Börde, overlaid with a grid of straight lines of two kilometres. This grid size was chosen to achieve a sufficiently large number of gird cells containing breeding Montagu's harriers as well as wind turbines. By using the same grid size (4 km^2) as in the Hellweg Börde, comparability of the results is ensured (see Table 6.1). For each of the grid cells, the number of nests per year (1995–2003 and 2006) and the number of wind turbines as well as the total capacity of the turbines were recorded. For grid cells with at least one brood or at least one wind turbine put into operation, the relation between the number of Montagu's harrier nests and the number of wind turbines was determined over time by means of an analysis of variance (ANOVA) with repeated measures. Using this method, the 0 hypothesis of the random distribution of nests within the grid cell is tested against the 1 hypothesis of a time-dependent distribution, which is in turn tested against the 2 or 3 hypotheses, including the number or the summarized capacity of wind turbines in the grid cell as an additional parameter. This serves to determine, whether the number of nests within a grid cell is influenced by the number of wind turbines within the same grid cell. Repeated measures take account of the fact that a time series within a grid cell is no independent sample.

All evaluations were done using software R (R development core team 2008).

6.3 Results

In contrast to the Hellweg Börde, where the number of Montagu's harrier nests showed a negative trend (linear regression $R^2 = 0.55$, $p < 0.001$, Fig. 6.2), the number of breeding pairs has slightly increased in North Frisia between 1995 and 2006 (linear regression, corr. $R^2 = 0.35$, $p < 0.05$). The population fluctuated between 15 pairs in 1995 and 39 pairs in 2003 (median = 34).

6.3.1 Structures in the Surrounding Area

Within the study area in North Frisia, no significant increase in the density of wind turbines occurred (Fig. 6.2). The number of individual turbines did not change significantly—a further distinctive feature in comparison to the Hellweg Börde. In the Hellweg Börde, the number of wind turbines increased considerably over the period of investigation. During the period of investigation in North Frisia, many of the old, smaller wind turbines were replaced by larger higher-capacity turbines in repowering, leading to a considerable increase in nominal capacity (Table 6.2). A significant difference between the density of turbines around the actual nest sites and the density around randomly chosen reference points could not be determined.

Number and capacity of wind turbines near the nest varied strongly between years and was found to be partly above the values determined for the reference areas (Fig. 6.3). Still, these fluctuations, resulting from the constantly changing nesting locations, showed no significant trend as compared to randomly chosen reference points (Table 6.2). No avoidance of or preference for wind turbines in the close-up area of one kilometre around the nest could be statistically proven.

Distances between nest site and random points from the next salt marsh differed only slightly (Fig. 6.4, nest: median = 4958 m, random point: median = 4239 m). A Wilcoxon rank sum test revealed no significant differences between the two samples ($W = 7592$, $p = 0.61$).

6.3.2 Grid Analysis

The number of grid cells with at least one wind turbine, which was also used at least once for nesting by Montagu's harriers, is considerably higher than determined in the Hellweg Börde for a similar total number of grid cells (Table 6.1). Time plays an equally important role in both study areas, as the distribution of nests within the grid differed from a random distribution independent of time (H0). For North Frisia, an additional correlation to the number of wind turbines could be proven. Different from the Hellweg Börde, more pairs bred in grid cells with wind turbines due to the constantly increasing population density (significant relation for H2, Table 6.3).

Table 6.2 Results of the analysis of variance between nest and reference locations over time (mixed linear model)

Factor			Number of wind turbines		Nominal capacity	
	Num DF	Den DF	F	*p*	F	*p*
Intercept	1	650	30.99	<0.0001	19.58	<0.0001
Time	1	650	1.36	n.s.	14.32	<0.001
Location (nest vs. reference point)	1	86	2.26	n.s.	1.37	n.s.
Time × location	1	650	0.2	n.s.	4.74	<0.05

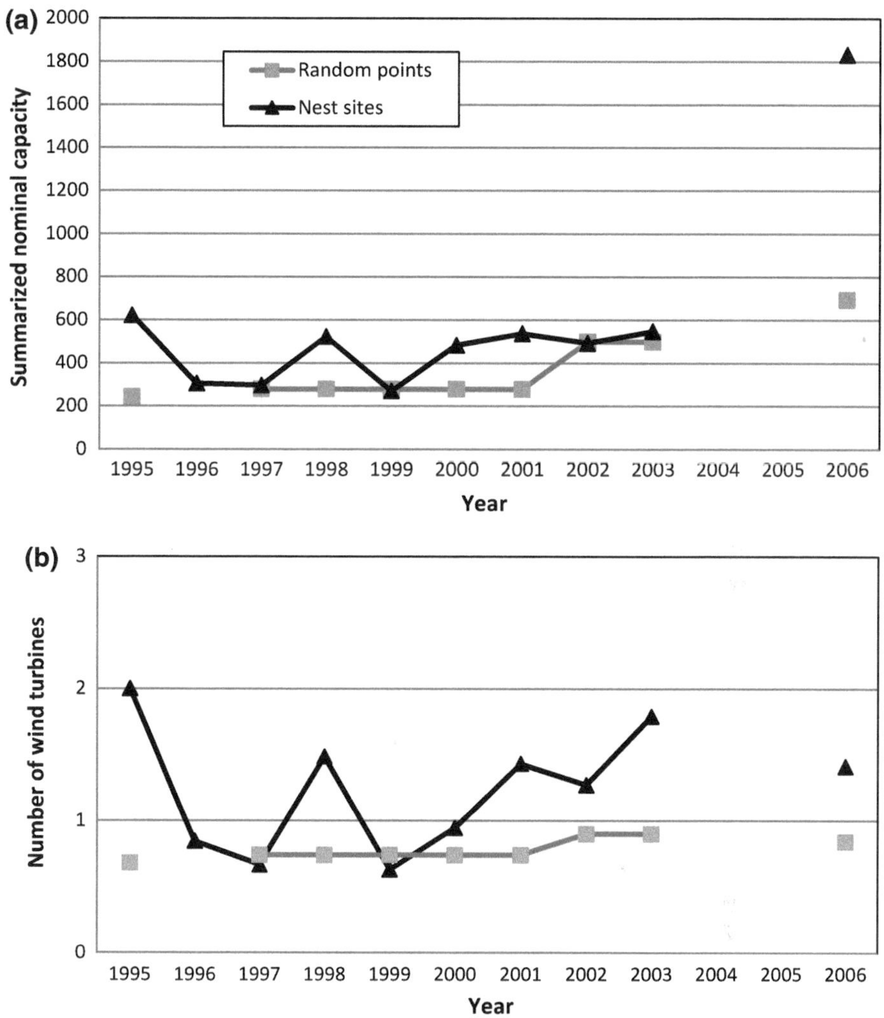

Fig. 6.3 Number (**a**) and summarized nominal capacity of wind turbines (**b**) within a radius of 1 km around nest sites and random points (mean values)

6.4 Discussion

The following facts need to be taken into consideration in the interpretation of the results: As opposed to Montagu's harrier nests, which are only used for one season, reference points are only chosen once for the whole period of investigation. Developments in the surrounding area of these points reflect the actual changes in number and capacity of wind turbines over time.

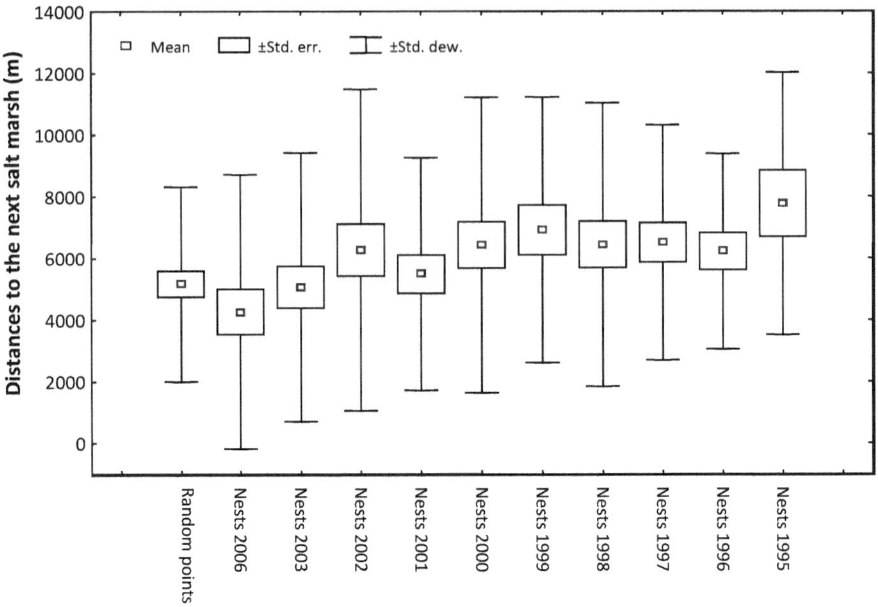

Fig. 6.4 Distances of random points and Montagu's harrier nests from the next salt marsh (seaward side of the dike) for different years

Table 6.3 Results of the analysis of variance with repeated measures on the influence of number and nominal capacity (Number_WEC, KW_WEC) of installed wind turbines on the number of Montagu's harrier nests (Number_nests) in 4 km^2 grid cells in the Hellweg Börde between 1993 and 2007 and in North Frisia between 1995 and 2006

Hellweg Börde						
Test	df	AIC	BIC	logLik	Lratio	p
H0	4	4111.209	4133.417	-2051.604		
H0 vs H1	**5**	**4108.646**	**4136.407**	**-2049.323**	**4.562969**	**0.0327**
H1 vs H2	6	4110.461	4143.774	-2049.23	0.184811	0.6673
H1 vs H3	6	4110.561	4143.874	-2049.28	0.085028	0.7706
North Frisia						
Test	df	AIC	BIC	logLik	Lratio	p
H0	4	2168.014	2189.449	-1080.007		
H0 vs H1	**5**	**2164.416**	**2191.211**	**-1077.208**	**5597.607**	**0.018**
H1 vs H2	**6**	**2162.02**	**2194.173**	**-1075.01**	**4396.167**	**0.036**
H1 vs H3	6	2163.758	2195.911	-1075.879	2658.391	0.103

Hypotheses: H0 = nests are randomly distributed, H1 = number of nests changes depending on time, H2 = distribution of nests over time depending on the number of wind turbines, H3 = Distribution of nests over time depending on the capacity of wind turbines (*df* degrees of freedom, *AIC* Akaike Information Criterion, *BIC* Bayesian Information Criterion, *logLik* Likelihood of the logarithmic function, *L.Ratio* values of the likelihood ratio tests, *p* significance level). Parameters with significant influence are highlighted in *bold* font

Distribution and density of Montagu's harrier nests however varies strongly between years. Under these circumstances, the probability of a β error is increasing, i.e. the probability to find a non-existing correlation. The results of the model calculations (Table 6.2) show nevertheless that wind turbines within a radius of one kilometre do not have an influence on the selection of a nest site on this rather large-scale level. Avoidance behaviour in the close-up area of wind turbines can still not be excluded.

With core areas of about one kilometre, the activity ranges of birds breeding in the North Frisian polders were predominantly small (see also Chap. 4). Longer hunting flights were rare and the birds were mainly hunting in the vicinity of the nest. Contrary to observations in coastal areas of the Netherlands (e.g. Trierweiler et al. 2008), foraging flights to salt marshes at a distance of several kilometres were rare and this behaviour proved to be insignificant for the birds. The proximity of these potential foraging grounds did not play an important role in the selection of a nesting habitat.

Whereas in the Hellweg Börde a clear natural separation between the lowland areas along the Lippe river (main area of distribution of the Montagu's harrier) and the higher elevations of the Haarkamm (main area of distribution of wind turbines) lead to spatial separation of nest sites and wind turbines. These naturally separating conditions cannot be found in North Frisia. Thus, the probability of an overlap of nest sites and wind turbines sites was larger in North Frisia than in the Hellweg Börde. Furthermore, only few larger settlement areas or forests exist in the North Frisian polders, so that almost the entire study area is potentially suitable for Montagu's harriers. The same applies to designated wind farm areas. The fact that birds and wind farm operators use the same areas, may explain the statistical correlation between the number of nests and the number of wind turbines within the grid. It is however unlikely that this is a causal correlation.

On the whole, Montagu's harriers in North Frisia showed no clear avoidance behaviour towards wind turbines in the selection of a nest site. Rather on the contrary, mostly the same areas were used. Under these circumstances, the collision risk should not be underestimated. This situation was further investigated in Chaps. 4 and 14. The present results cannot be compared to results for other breeding areas of the species without taking account of the natural conditions in these areas.

References

EnergyMap.info (2011) Deutsche Gesellschaft für Sonnenenergie e.V., http://www.energymap.info. Assessed 30 June 2013.

Gahrau C, Schmüser H (2007) Wiesenweihe. In: Landesjagdverband Schleswig-Holstein (ed) Wildzustandsbericht 2007. Landesjagdverband Schleswig-Holstein, Flintbek.

Glutz von Blotzheim UN, Bauer KM, Bezzel E (1989) Handbuch der Vögel Mitteleuropas. Vol 4, 2nd edn, AULA-Verlag, Wiesbaden.

Keiler J (2008) Windkraft Betreiber-Datenbasis: Register/Errichtungsdaten deutscher Windanlagen. http://www.btrdb.de/. Assessed 30 June 2013.

R development core team (2008): http://www.r-project.org/.

Trierweiler C, Visser EG, Arisz J, Koks BJ (2008) Habitatgebruik van Grauwe Kiekendieven *Circus pygargus* in het agarisch landschap 2003-2006 onerzocht m.b.v radiotelemtrie. Bericht der Stichting Werkgroep Grauwe Kiekendief, pp 86, www.grauwekiekendief.nl. Assessed 30 June 2013.

Chapter 7
Satellite Tracking of White-Tailed Sea Eagles in Mecklenburg-Western Pomerania and Brandenburg

Oliver Krone, Gabriele Treu and Thomas Grünkorn

7.1 Introduction

In autumn 2010, the guidance document "Wind Energy Developments and Natura 2000" has been issued by the European Commission listing the bird species which are considered to be particularly vulnerable to wind farms in Annex II. Potential risk or impact has been classified into five classes from non-significant to substantial. For the white-tailed sea eagle the collision risk has been assessed to be substantial.

Collisions of white-tailed sea eagles with wind turbines have been reported in 75 cases in Germany between 2002 and 2012 (Dürr 2012). As no systematic carcass search has been carried out in German wind farms so far, white-tailed sea eagle fatalities are mainly random findings. The actual collision rate can be substantially higher as carcasses are usually quickly eaten or removed by scavengers. A study of Smallwood et al. (2010) where 63 large, raptor-sized carcasses have been placed on the site of a wind farm covering 250 ha (one to five carcasses at a time) showed that the majority of the carcasses (42%) has been removed by scavengers within 15 days. An estimate of the actual rate of raptors colliding with wind turbines calculated on the basis of this data resulted in a mortality rate which was 68% higher than in conventional, less natural setups where a large number of carcasses (10–20) were usually laid out simultaneously.

Furthermore, there is no general agreement on the size of the area to be searched. Hull and Murir (2010) apply a Monte-Carlo approach to ballistics theory to describe

O. Krone (✉) · G. Treu
Leibniz Institute for Zoo and Wildlife Research (IZW),
Alfred-Kowalke-Str. 17, 10315 Berlin, Germany
e-mail: Krone@izw-berlin.de

T. Grünkorn
BioConsult SH, Schobüller Str. 36, 25813 Husum, Germany

H. Hötker et al. (eds.), *Birds of Prey and Wind Farms*,
DOI 10.1007/978-3-319-53402-2_7

the fall zone of birds of different sizes after colliding with turbines of different sizes to ensure that an adequate area is being searched. For large birds a search radius between 70 and 100 m has been determined depending on turbine height. Grünkorn et al. (2009) have defined the search radius for small birds as height of the turbine.

In Germany, the white-tailed sea eagle is not an endangered species any more, which was recognised in the Red List but is still listed in Annex I of the EC Birds Directive (2010). Chapter 1, Article 1 of the Federal Nature Conservation Act—BNatSchG states: "In order to permanently safeguard biological diversity, the following actions are to be taken, in accordance with the relevant threat levels: (1) viable populations of wild plants and animals, including their living sites, shall be protected, and exchange between relevant populations, and migrations and resettlement, shall be facilitated, (2) threats to naturally occurring ecosystems, biotopes and species shall be countered."

The white-tailed sea eagle is one of the long-lived raptor species and adult mortality has a stronger influence on the population density than for short-lived species (Newton 1998). The example of the Egyptian vulture (*Neophron percnopterus*) showed how long-lived raptors are affected or even endangered by the impacts of wind power due to lower abundances and reproduction rates (Carrete et al. 2009).

The aim of the white-tailed sea eagle partial project was to analyse how often and under which circumstances white-tailed sea eagles approach wind turbines in order to assess and minimize the collision risk in the future by means of satellite telemetry and behavioural observations. For this purpose adult and immature birds were to be fitted with satellite transmitters. Nestlings were tagged to investigate whether wind turbines will have an influence on the birds ranging behaviour shortly after leaving the nest and whether a breeding location in close vicinity of a wind farm is associated with an increased collision risk for the young birds.

7.2 Data and Method

7.2.1 *Choice of the Study Animals*

Preferably adult territorial birds occupying territories within windfarms should be tagged to evaluate the influence of wind turbines on habitat and spatial use. The survey has been carried out within the German core breeding area of the white tailed sea eagle in Brandenburg and Mecklenburg-Western Pomerania. Analysis of all known breeding sites resulted in five suitable breeding pairs in Mecklenburg-Western Pomerania and three in Brandenburg. The eagles were equipped with satellite transmitters between 2008 and 2009. Due to better prospects of success, the catching was done mainly in winter. The birds were captured near the bait using different net throwing techniques (Bloom et al. 2007) and subsequently equipped with GPS satellite transmitters.

Table 7.1 Project-external white-tailed sea eagles included in the analysis of the data

Eagle ID	Age	Federal state	Gender	First GPS location	Last GPS location	Number of locations
472	Adult	MV	f	24.07.2003	14.01.2004	448
655	Adult	MV	f	29.07.2004	22.06.2005	590
964	Adult	MV	f	17.7.2004	17.5.2005	260
956	Juvenile	MV	m	12.06.2004	13.05.2005	271
958	Juvenile	MV	f	05.06.2004	28.07.2005	333
959a	Juvenile	MV	m	13.06.2004	29.1.2005	175
960	Juvenile	NI	f	03.06.2004	01.07.2005	337
962a	Juvenile	MV	f	29.07.2004	01.12.2004	107
963	Immature	MV	m	26.07.2004	20.06.2005	294
966	Immature	MV	f	12.09.2005	03.05.2006	190
990	Juvenile	MV	f	22.06.2005	26.04.2006	168
1527	Juvenile	MV	m	09.10.2005	12.10.2006	75
1528	Juvenile	BB	m	18.06.2006	23.01.2007	117
1531	Juvenile	NI	f	17.06.2006	21.01.2007	99
5848	Juvenile	MV	m	27.06.2009	21.02.2010	27
6018	Subadult	MV	f	02.02.2009	10.10.2012	1352

Age: *adult* >5 J, *sub-adult* 4 J, *immature* 2–3 J, *juvenile* 1 J, f = female, m = male, Federal States MV = Mecklenburg-Western Pomerania, NI = Lower Saxony, BB = Brandenburg

Additionally the telemetry data of three adult birds and another 13 young eagles, which have been tagged for other projects between 2003 and 2010, has been evaluated for the first time in the present research project. Characteristics of these birds are given in Table 7.1.

Three additional adult white-tailed sea eagles tagged prior to the start of this study were included in the dataset to analyse the ranging behaviour of adult territorial birds in relation to distance from the nest.

7.2.2 *Satellite Transmitters*

The VECTRONIC Aerospace satellite transmitters contained GPS receiver, GSM bi-directional communication, two acceleration senders and VHF beacon. At regular intervals locations were automatically transmitted per SMS to a central computer. Satellite transmitters on young birds were set to deliver between two and four locations per day. Transmitters of adult birds were activated as soon as the bird's axis moved to an angle of below 30° to record flight locations. The average accuracy of the recorded locations (coordinates) is 5.9 m (SD = 3.69) at a height of 5 m (SD = 3.96) (unpublished data). With a weight of about 150 g transmitters made up fewer than 3% of the body weight for eagles of over 5 kg weight.

All transmitters were attached as backpack harness using Teflon ribbons which were sewn at one point over the sternum as a predetermined breaking point resolving itself after 2–3 years depending on strain (Kenward 1983).

7.2.3 *Spatial Analysis of Telemetry Data*

In order to analyse the flight behaviour of the tagged eagles the determined GPS locations have been overlaid with topographical maps using the geographical information system GIS ArcView 3.3 (ESRI, USA). Specific attention has been payed to the analysis of structures of the GPS locations in the vicinity of wind turbines to identify possible reasons why eagles approach.

Currently newly installed wind turbines in Germany are on average 105 m high with a rotor diameter of 83 m (Windenergiereport Deutschland 2011). It is generally expected that turbine height as well as rotor diameter will increase over the next years, as sufficiently good wind conditions can only be reached at a considerable height (>130 m) for inland sites. Consequently, a zone of 150 m around a wind turbine has been defined as risk area in the analysis of the approaching behaviour. All locations recorded for this area have been analysed subsequently in terms of topographical structure. In doing this the actual location has been used instead of interpolating a possible location between two points. Home ranges have been calculated using the GIS extensions "spatial analyst" and "animal movement" as minimum convex polygon (MCP) and as kernel density estimate (kernel analysis).

7.2.4 *Statistics*

The statistical data analysis was performed using SPSS 18 (SPSS Inc., USA). After testing the distribution of all observations for normality, results of univariate tests were expressed as mean ± standard deviation (SD), degree of freedom (DF), range or number and percentage of observed GPS locations. In descriptive statistics the name of the test, the sample size as well as the confidence level were indicated. The significance level was set to 5%. Avoidance of wind turbines has been tested using a Wilcoxon signed-rank test comparing the mean distance (distance of the locations) from the wind turbine with the mean distance from randomly created point within the 100% MCP home range. Comparison of the average number of locations per square kilometer for habitat type "cropland" and the reference area has been calculated using the binominal test.

7.3 Results

One territorial adult, two sub-adults and one immature bird have been fitted with satellite transmitters between 2007 and 2010. Three juvenile white-tailed sea eagles were already tagged before fledging. A total of seven white-tailed sea eagles were tagged with transmitters (Table 7.2).

Evaluation of the telemetry data shows that at least three of the investigated seven project internal white-tailed sea eagles moved as close as at least 150 m to wind turbines. Among these an adult bird, a sub-adult and a juvenile white-tailed sea eagle (Table 7.3). As all turbines within the home range of the birds were located on cropland, locations of the comparable habitat "cropland without wind turbine" have been included in the analysis of the data. The number of locations and the mean distances of approach are given in Table 7.3.

7.3.1 Telemetry of Adult Birds

The transmitter of the adult bird (5900) delivered a total of 4803 GPS locations on 143 days between February and June 2009. This means an average of 34 locations

Table 7.2 White-tailed sea eagles tracked during the project

Eagle ID	Age	Federal state	Gender	First GPS location	Last GPS location	Number of locations
5900	Adult	BB	f	31.01.2009	22.06.2009	4803
2356	Subadult	MV	m	19.08.2008	13.02.2009	171
5846	Subadult	BB	m	22.01.2009	23.11.2009	439
5847	Immature	BB	f	19.12.2009	03.07.2011	510
5841	Juvenile	MV	m	06.06.2008	07.07.2010	1892
5901	Juvenile	MV	f	09.06.2008	12.06.2009	1463
5902	Juvenile	MV	m	11.06.2008	25.06.2009	1466

Age: *adult* >5 J, *sub-adult* 4 J, *immature* 2–3 J, *juvenile* 1 J, f = female, m = male, Federal States MV = Mecklenburg-Western Pomerania, BB = Brandenburg

Table 7.3 Number and distance of GPS locations in the vicinity of wind turbines for three white-tailed sea eagles

Eagle ID	Age	Total number of GPS locations (n)	GPS locations in cropland (n)	GPS locations close to wind turbines (up to 150 m)	Proportion of GPS locations close to wind turbines in cropland (%)	Mean distance of GPS locations close to wind turbines (m)
5900	Adult	4803	1123	38	3.38	89
2356	Subadult	171	86	2	2.33	122
5902	Juvenile	1466	250	1	0.40	80

Table 7.4 Size of the different types of home ranges calculated for the adult bird 5900

	100% MCP	95% MCP	95% fixed kernel home range
Size (km^2)	526.17	159.82	141.14
Number of wind turbines	93	69	42
Number of GPS locations	4803	4563	4630
GPS locations per km^2	9.12	28.55	32.8

per day with a minimum of four and a maximum of 84 locations. Initially, spatial and temporal autocorrelation of the locations was eliminated using the ArcView extension animal movement, calculating the index of autocorrelation according to Swihart and Slade (1985). Subsequently the sizes of the home ranges have been calculated. The 100% MCP home range, obtained by connecting the boundary locations, equalled 526 km^2 including four windfarms with 93 wind turbines. The 95% MPC home range excluding 5% of the outliers was considerably smaller covering an area of 160 km^2 and including 69 wind turbines. The area with a 95% probability of use (95% fixed kernel home range) comprised 141 km^2 with 42 wind turbines (Table 7.4).

On ten of a total of 121 days of the recorded data, the adult bird stayed in the risk area of a wind turbine with a total of 38 GPS locations transmitted due to the activity profile. In general, with 3.4% of the locations the adult bird approached wind farms more often than young birds, traversing repeatedly the risk area and flying through the rows of wind turbines (Fig. 7.1).

In the evaluation no avoidance behaviour of the adult bird towards wind turbines could be identified for the different calculated home ranges by comparing average distances to random distances (Fig. 7.2).

The number of GPS locations per km^2 of the adult eagle (5900) is significantly higher within the wind farm area than in the reference area (cropland without wind turbine) (Fig. 7.3). This result is remarkable, as no wind turbines were located on the direct flight path from the nest to the main foraging habitat (food rich lakes).

7.3.2 Telemetry of Young Birds

One of the three tagged eaglets (bird 5929) has been recorded within a wind farm during the period of transmission (Table 7.2). GPS data as well as visual observations have shown that the young bird repeatedly traversed the wind farm in the vicinity of its nest and showed no fear of the turbines. Although all three tagged young eagles (immature and sub-adult) had very large home ranges and might have crossed a number of wind farms, only one individual approached the risk area of a wind turbine. Sub-adult white-tailed sea eagle (2356) landed in two different wind farms in the proximity of a Feldsoll (kettle hole) or a lake respectively (Fig. 7.4). Only one of the 13 young eagles visited a wind farm with no apparent reason for approach.

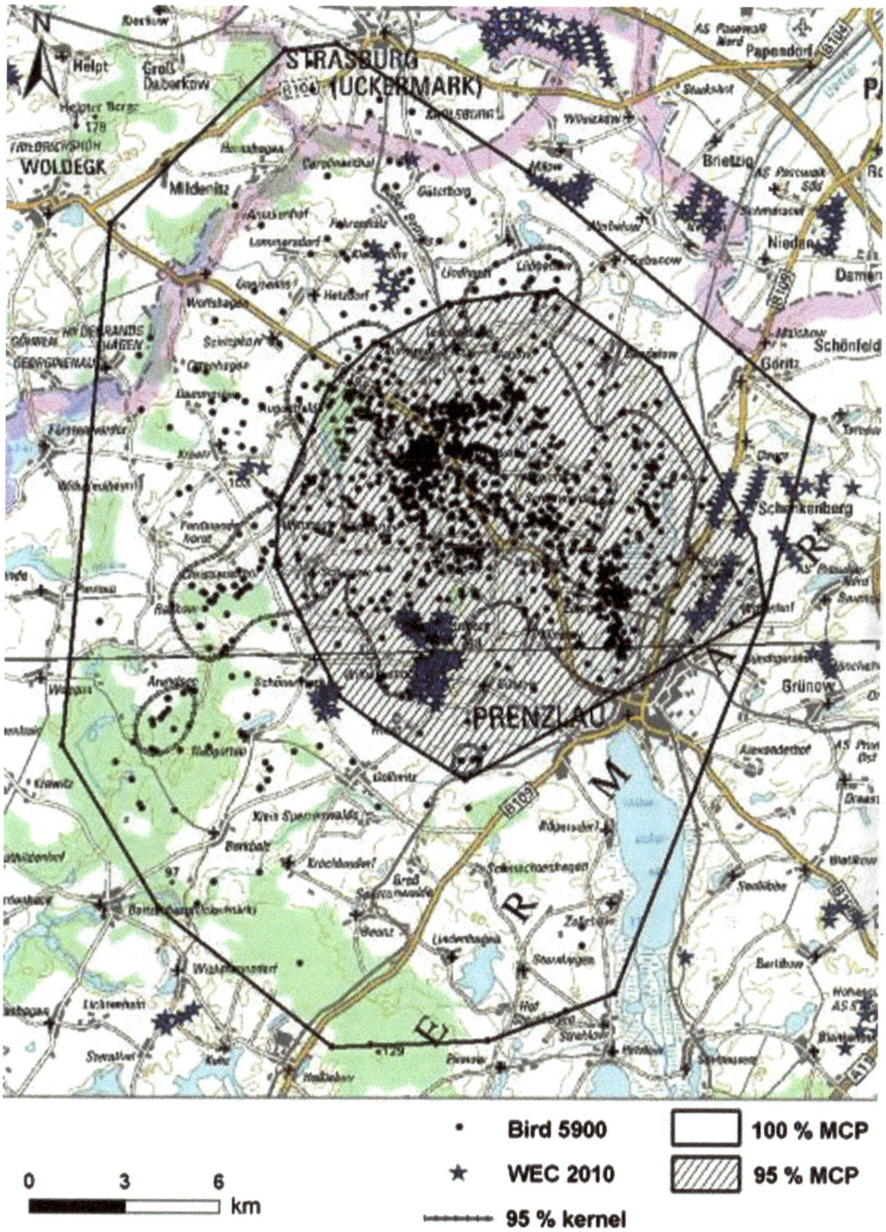

Fig. 7.1 Visualisation of the home ranges (100% MCP, 95% MCP and 95% Fixed kernel) recorded by GPS locations of white-tailed sea eagle 5900 near Prenzlau, Brandenburg. WEC = wind power plant

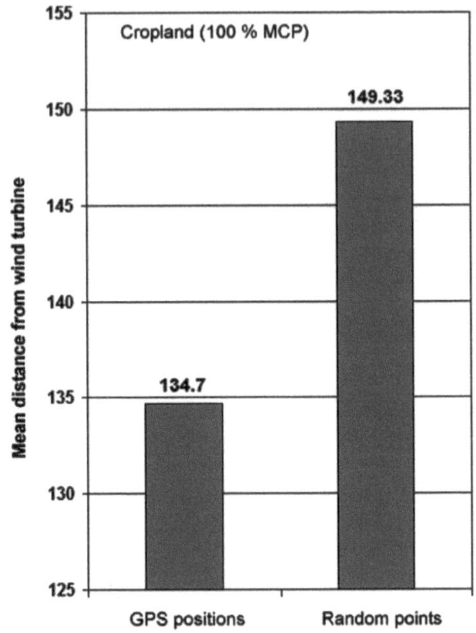

Fig. 7.2 Testing avoidance behaviour: comparison of mean distance from wind turbine with random distances, 100% MCP (Wilcoxon signed-rank test: n = 200, p = 1.44)

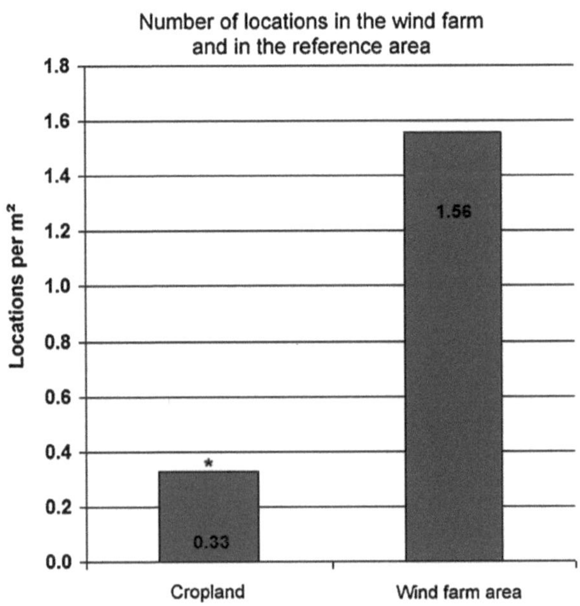

Fig. 7.3 Comparison of the average number of GPS locations of the adult eagle 5900 per km^2 for the wind farm and reference area (*asterisk* binomial test: n = 121, p = 0.002)

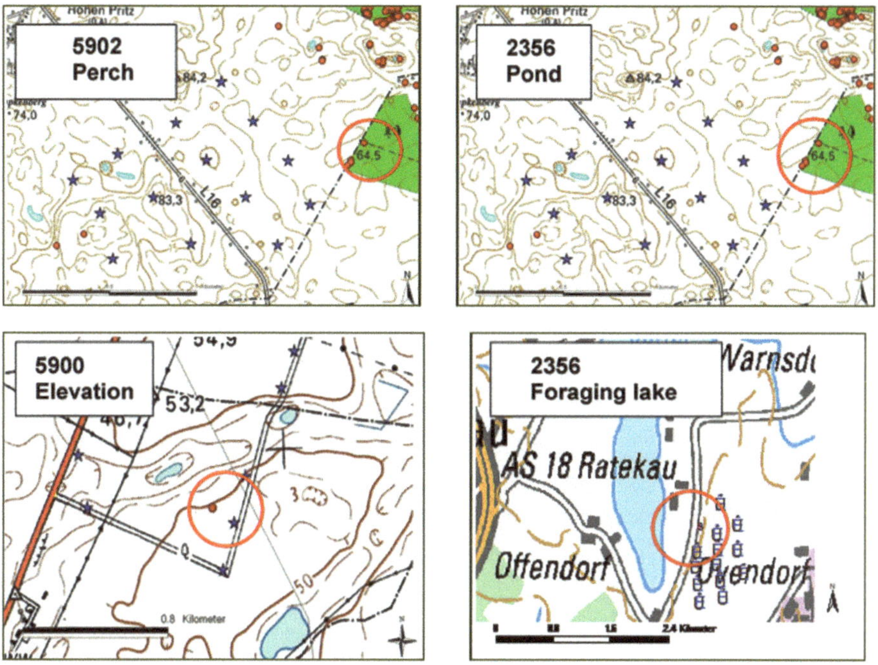

Fig. 7.4 Examples for attractive structures used by eagles in wind farms (*small circle* GPS location, *large circle* area of the attractive structure)

7.3.3 Structural Analysis of GPS Locations of Eagles in the Surroundings of Wind Turbines

The terrain structure around the GPS locations was analysed using topographic maps. It was possible to identify structures which were repeatedly visited by young as well as adult birds and could therefore be rated as attractive. Based on the terrain, it was possible to identify all GPS locations of the young eagles within a radius of 300 m around a wind turbine as ground locations (30 m accuracy) and to analyse their structure. Among the attractive structures were: kettle holes, ponds, large bodies of water, perching trees on forest edges and local elevations (Table 7.5; Fig. 7.4).

Analysis of the near wind turbine locations (up to 150 m) of the adult white-tailed sea eagle resulted in 38 locations on 10 days. Seven locations were identified as flight locations and the majority (n = 13) as ground locations (Table 7.6). A stay of 3 h 40 in a wind farm with 17 locations was recorded on 18.04.2009. 12 locations were recorded on 19.04.2009 over a period of 2 h 10 within the same wind farm (Fig. 7.5).

In another wind farm near the nest, several cases were observed where buzzards (*Buteo buteo*) and red kite (*Milvus milvus*) circled together with adult white-tailed

Table 7.5 Structure of the GPS locations approached by all young birds within 300 m around a wind turbine

Young bird	Date	Habitat type	Structural analysis within 300 m around a wind turbine
2356	08.01.2009	Cropland	Lake
	13.01.2009	Cropland	Temporary structure?
	21.12.2008	Cropland	Drainage ditch
			Kettle hole
	17.09.2008	Cropland	Pond
5902	19.08.2008	Cropland	Forest edge, perching point
	06.09.2008	Cropland	Pond
	19.12.2008	Cropland	Temporary structure?
	20.12.2008	Cropland	Temporary structure?
6018	13.04.2010	Cropland	Close to a motorway (searching for a roadkill?)

Table 7.6 GPS flight hights of the adult eagle 5900 at 18.04.2009 inside the wind farm

Date	Time	Flight height (m above sea level)	Terrain height (m)	Flight height corrected[a]	Habitat type
18.04.2009	16:27	233	78	155	Cropland
18.04.2009	17:27	111	65	46	Cropland
18.04.2009	17:37	99	64	35[b]	Cropland
18.04.2009	17:47	104	73	31[b]	Cropland
18.04.2009	17:57	105	74	31[b]	Cropland
18.04.2009	18:07	108	74	34[b]	Cropland
18.04.2009	18:17	113	74	39	Cropland
18.04.2009	18:27	108	74	34[b]	Cropland
18.04.2009	18:47	114	74	40	Cropland
18.04.2009	18:57	115	74	41	Cropland
18.04.2009	19:07	116	74	42	Cropland
18.04.2009	19:17	109	64	45	Cropland
18.04.2009	19:37	114	69	45	Cropland
18.04.2009	19:47	112	66	46	Cropland
18.04.2009	19:57	114	66	48	Cropland
18.04.2009	20:07	100	68	32[b]	Cropland
18.04.2009	20:17	100	68	32[b]	Cropland

[a]Flight height after correcting for height of terrain and the inaccuracy of approximately +/-5 m
[b]Ground location assumed

sea eagles over the manure spread in the wind farm (an example for a temporary structure) searching for food (e.g. dead rats). In another case during a catching campaign near Kladrum (MV), an adult sea eagle pair was observed landing near a wind turbine on the manure of a chicken farm searching for food (dead chicks).

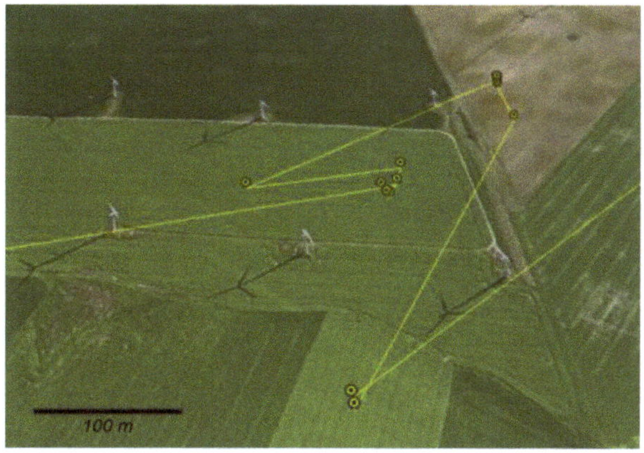

Fig. 7.5 Locations of an adult white-tailed sea eagle in a wind farm near Prenzlau of 18.04.2009

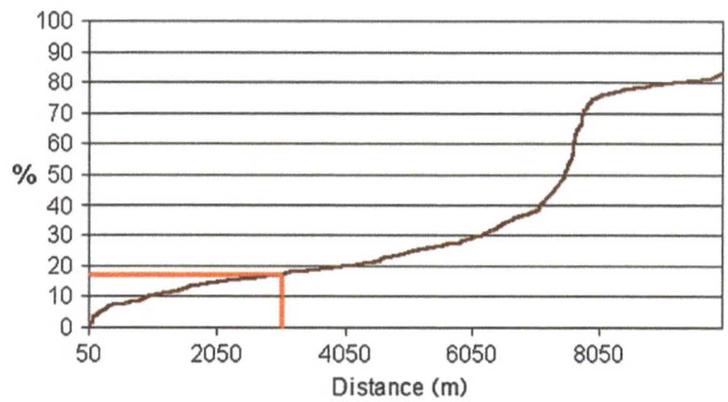

Fig. 7.6 Cumulative distances (%) of GPS locations of the territorial adult bird (5900) at more than 50 m from the nest (n = 2587 from a total of n = 4803). Only 17.3% of the GPS locations are recorded at a distance of up to 3000 m around the nest (*red mark*), the majority further away

7.3.4 Movements Patterns of Territorial Adults in Relation to the Distance from the Nest

The tagged bird 5900 mainly stayed near the nest over the period of the study. Of 4803 GPS locations 50% were recorded at a distance of 0–359 m from the nest. After nesting failure, the eagle mainly stayed in a foraging ground north-west of the nest site (Fig. 7.1). This is clearly reflected by an accumulation of GPS locations at distances between 7000 and 8000 m from the nest (Fig. 7.6). To estimate a possible collision risk with a wind turbine in the vicinity of the nest, only locations at a distance of at least 50 m from the nest were included in the analysis of the tagged adult. In order not to arbitrarily define a distance from the nest as risk area, the proportion of locations at a distance of 3000 m was looked at, a distance which is discussed in the recommendations for distances of wind turbines to important areas for birds by the Working Group of German State Bird Conservancies (Länderarbeitsgemeinschaft der Vogelschutzwarten LAG VSW 2007) (see also Sect. 7.4.4). Only 17.3% of the locations at distances of at least 50 m from the nest were recorded in an area of 3000 m around the nest. Half (50%) of the localizations occur within a distance of 7486 m (Fig. 7.6).

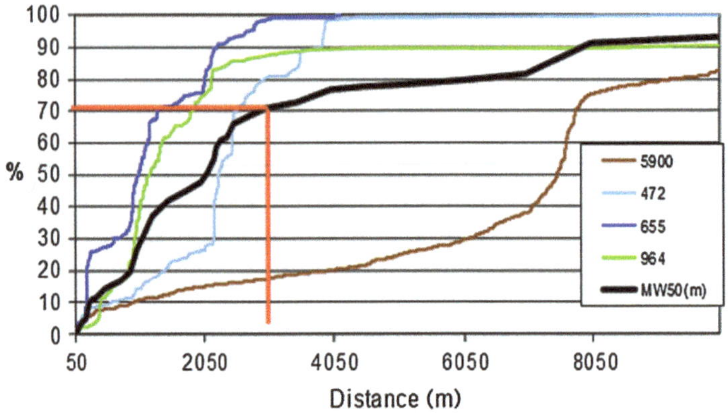

Fig. 7.7 Distances (%) of four territorial adult eagles above 50 m around the nest. On average (*black line*), 71% of the locations were recorded at a distance of up to 3000 m around the nest (*red mark*)

The analysis of GPS locations more than 50 m from the nest of another three territorial white-tailed sea eagles, which were satellite tagged in other projects, resulted in 80% locations in an area of 50–3000 m around the nest (eagle 472, n = 463, total n = 475), 87% locations up to 3000 m (eagle 964, n = 260, total n = 261) and in one case even 98% of the recorded locations at distances up to 3000 m around the nest (eagle 655, n = 463, total n = 573). The mean of all locations at distances from 50 to 3000 m around the nest is 71% for all four white-tailed sea eagles (Fig. 7.7).

7.4 Discussion

Juvenile (one year old), immature (two to three years old) and sub-adult (four years old) white-tailed sea eagles roam extensively. Only adult birds (over five years old) occupy a territory permanently and usually stay there all year round. For this reason, several territorial adult birds with wind turbines in the vicinity of the nest (<3 km distance) should have been captured and tracked. This proved to be extremely difficult. White-tailed sea eagles are shy birds and often have an escape distance of more than 500 m rendering capture difficult. Additionally, the capturing technique failed several times due to the extreme cold of winters 2008/2009 and 2009/2010. Blizzards and snowdrift complicated the work in the field obstructing access to some of the capturing sites. The deep and loose powder snow enabled two adult white-tailed sea eagles to escape the net. The bait was less attractive to the eagles as many animals starved to death (hoofed game, water birds) providing plenty of food for scavengers. Despite these difficulties, seven young and one territorial adult bird were fitted with satellite transmitters for the project. To enlarge the sample size, data on white-tailed sea eagles collected for other studies was included in the evaluation.

7.4.1 Telemetry of Adult Birds

In comparison to adult white-tailed sea eagles satellite tagged in other projects (Krone et al. 2009; Krone, unpublished data) the home range size of the investigated territorial adult eagle (5900) was rather large, 160 km^2 (95% MCP) respectively 141 km^2 (95% fixed Kernel). This may be due to the feeding conditions of the habitat used, as 70% of the area consisted of agricultural land and the eagle had to range over large distances to forage. After nesting failure, the tagged white-tailed sea eagle mainly stayed 7–8 km away from the nest, probably due to good food conditions there. The reason why the eagle flew more often into a wind farm than the young birds might be due to the fact that the adult bird occupies a territory with many wind turbines near by. Studies by Grünkorn et al. (2009) and Krijgsveld et al. (2009) on the collision risk at modern wind turbines showed that mainly diurnal birds from local populations are at risk. A before and after study in the USA carried out by Garvin et al. (2011) showed that most bird of prey species avoid wind turbines (no approach below 100 m or above the rotor area). Flight behaviour in a wind farm strongly depends on species, e.g. turkey vulture (*Cathartes aura*) and red-tailed hawk (*Buteo jamaicensis*) showed no avoidance behaviour at all but performed dangerous flight manoeuvres in the risk area.

The observation that the adult white-tailed sea eagle stayed significantly more frequent in a wind farm than randomly to be expected and stayed in one location within the wind farm for several hours suggests that the bird was attracted by a special habitat structure or food. Studies on the behaviour of other bird species by Orloff and Flannery (1992), Barrios and Rodriguez (2004) and Smallwood and Thelander (2004) showed that many birds enter the risk area of wind turbines frequently, using turbine towers (mostly lattice towers) as perching points representing the main reason for approaching the wind turbines.

A new approach previously not taken into account to clarify the high bird fatality rates on wind turbines, are the studies by Martin (2010, 2011) on visual perception. Complex correlations could be proven between the visual abilities of a bird, the interpretation of sensory information and the behaviour in free flight. To look downwards, birds are able to turn their heads so that they either use the binocular field of vision or the lateral visual fields to look at the area below. In flight, they may temporarily be blind or nearly blind to objects ahead of them in the direction of flight. This was proved for three species (kori bustard, blue crane, white stork). It is assumed that visual perception of birds is rather tuned to detect movement than spatial detail. This ability is particularly pronounced in birds of prey while hunting, to better perceive the movement of prey animals and thus enhancing the success of capture.

To our knowledge there are hardly any investigations on the size of the field of vision and the visual perception of birds of prey. Martin and Katzir (1999) determined a relatively small field of vision (20° width and 81° height) for the short-toed snake eagle (*Circaetus gallicus*). A slight turn of the head downwards severely impairs the frontal vision and may even lead to temporal blindness. The temporary blindness of birds ahead of them may explain why white-tailed sea eagles collide with wind turbine blades even under excellent visibility conditions (Krone and Scharnwerber 2003). Investigations of sensory perception and visual field of the white-tailed sea eagle could provide important clues to minimize the collision risk with wind turbines and should urgently be carried out.

An example of a very large density of wind turbines (68 wind turbines per 274 km^2) with a simultaneously very high density of white-tailed sea eagles (approx. 60 eagle territories) in a relatively small area is the situation on the Norwegian archipel of Smøla. Studies on the breeding success concluded that eagles breeding on the wind farm site experienced significantly lower breeding success than pairs breeding outside the wind farm (Follestad et al. 2007; Bevanger et al. 2008). Visual observations however indicated that some of the adult and young birds on Smøla show no fear of wind turbines, as they have been observed on the site of the wind farm and in close proximity to the turbines (Bevanger et al. 2009).

To better understand the behaviour of territorial white-tailed sea eagles towards wind turbines, before-after-control-impact (BACI) studies are necessary. Farfán et al. (2009) showed that some raptor species (especially common kestrels) avoided the site of a newly erected wind farm in Spain. Garvin et al. (2011) found that the number of all previously recorded bird of prey species was reduced by 47% after the construction of a wind farm. Although BACI studies are an effective way to study the impacts of wind turbines on the behavioural ecology of a species they are still not very often implemented in Europe (Langston and Pullan 2003; Madders and Whitfield 2006). Additionally, it will be important to examine, whether sensitive species will be displaced from their breeding sites by the construction of new wind turbines. White-tailed sea eagles for instance are known to abandon the nest after repeated disturbances (Altenkamp et al. 2007).

Similar results were obtained in a study on the influence of wind turbines on the distribution of breeding birds in English upland by Pearce-Higgins et al. (2009). Golden eagles (*Aquila chrysaetos*) and hen harriers (*Circus cyaneus*) left the breeding habitat when wind turbines were erected and operated.

7.4.2 Telemetry of Young Birds

The aim of tagging the nestlings was to investigate whether wind turbines will have an influence on the bird's behaviour shortly after leaving the nest, and whether a breeding site close to a wind farms increases the collision risk for young birds. Our results show that young birds fledged close to a wind farm are at risk, as they show no fear of wind turbines and sometimes fly in the height of the rotor blades.

A young bird repeatedly traversed the risk area of 150 m around a wind turbine. Evaluation of data collected on earlier tagged white-tailed sea eagles revealed that three of the 19 tracked young eagles approached the immediate risk area of a wind turbine.

7.4.3 *Structural Analysis in the Surroundings of Wind Turbines*

Depending on region, terrain, wind speeds, thermals and the expected number of windy days, wind turbines are erected in areas with the highest potential for wind. These are often characterised by topographic structures like hilly landscapes, mountain ridges or fjords. Several studies could prove that such topographies increase the collision risk for birds of prey because the birds take advantage of favourable conditions like thermals or upwind for energetically cheaper flight (Barrios and Rodriguez 2004; Smallwood and Thelander 2004; Follestad et al. 2007). A study of Katzner et al. (2012) showed that at least golden eagles fly especially low in areas prone to wind like hill crests and slopes (on average 150 m) in comparison to open areas (on average 300 m) resulting in an increased collision risk. Smallwood et al. (2009) suggested to leave sufficiently large gaps between groups of turbines to allow birds to travel and forage without having to necessarily fly close to wind turbines.

In Germany, 17% of the total onshore wind capacity is installed in coastal areas, 55% in the North German Plain and 28% in the low mountain ranges (Windenergiereport Deutschland 2011). Consequently, many wind turbines are erected in windy lowland areas and are not concentrated in particular regions so that collisions may occur in almost all types of landscape (Krone et al. 2008). The majority of wind turbines were constructed on agricultural land. Forests, wetland and bird reserves were avoided so far (Illner 2011, 2012). More recent considerations include forest areas as potential wind turbine sites (Landesbund für Vogelschutz study 2010), as turbulences above the canopy layer have a relatively little impact on high wind turbines (Fallen et al. 2006). There are almost no studies on conflicts between birds and wind turbines erected in forests, but direct loss of habitat of sensitive species requiring large territories (e.g. white-tailed sea eagles) should be expected (Landesbund für Vogelschutz 2010). In Germany, white-tailed sea eagles mostly nest in trees and might be significantly impaired by the development of wind turbines in the vicinity of the nest. Hötker et al. (2004) and Albrecht et al. (2006) recommend the exclusion of wind turbines near wetlands, on mountain ridges and slopes. Apart from the size of home range of individuals of an endangered species, the intensity of use of different areas and structures is of particular interest. The Länderarbeitsgemeinschaft der Vogelschutzwarten (2007) therefore recommends establishing ranges of verification around a wind turbine to check for feeding habitats of particular bird species.

7.4.4 Movements of Territorial Adults in Relation to the Distance from the Nest and Recommendations

Due to the conservation status of protected raptor species, minimizing fatalities at wind turbines is a major concern. This becomes even more important in the context of a significantly increasing number of wind turbines to fulfill the German political goals of renewable energy production. The existing regional distance recommendations for the construction of wind turbines near white-tailed sea eagle nests are given in Table 7.7.

Distance recommendations serve as basis for assessment in the approval of wind farm projects (LAG-VSW 2007).

As a large proportion of the recorded GPS locations of tagged adult birds can be found at a distance of 3000 m around the nest (average 71%, minimum 17%, maximum 98%), the probability that the eagles enter the area of a wind turbine decreases above a certain distance. If a preferred flight path towards a foraging habitat will be occupied by wind turbine, the probability of contact with wind turbines is increasing. How this leads to an increase in collision risk will be discussed in the following sections.

Table 7.7 Distance recommendations of governmental institutions for wind turbines to nests of white-tailed sea eagles in Germany

Federal state	Minimum distance from the nest (m)	Recommendations of different institutions
Brandenburg	3000	Recommendations MUGV: Keep flight corridor to feeding waters clear Radius: 6000 m around breeding site, width: 1000 m
Mecklenburg-Western Pomerania	2000	Recommendations LANU (BRB): Keep potential feeding grounds clear, ensure accessibility Radius: 6000 m around the nest
Lower Saxony	3000	Recommendations NLT: Keep a 6000 m radius of feeding habitats clear as well as associated flight paths
Schleswig-Holstein	3000	Recommendations LANU (SH): Keep feeding habitats and flight paths clear
Saxony	3000	In line with LAG-VSG recommendations: range of verification to check for feeding habitats Radius: 6000 m around the nest Flight paths should be kept clear
Saxony-Anhalt	3000	In line with LAG-VSG recommendations: range of verification to check for feeding habitats Radius: 6000 m around the nest Flight paths should be kept clear

Especially in harsh winters when access to prey is difficult or even impossible, white-tailed sea eagles may enlarge their home range considerably. Estimating the average home range size of white-tailed sea eagles by visual observation alone is difficult, as sizes may vary significantly depending on the quality of the habitat. Only few studies have been carried out on this subject so far.

Acknowledgements We are grateful to the Nossentiner/Schwinzer Heide National Park Administration and the Müritz National Park Administration for the logistic support. We feel obliged to the State Forest Administrations of Brandenburg, Mecklenburg-Western Pomerania and Schleswig-Holstein and to all forestry offices providing us with game. Furthermore we would like to thank all farmers and wind farm operators for granting access to the wind farms. Special thanks are due to all helpers and members of staff supporting us in the project, especially to Mirjam Gippert and Heiner Schumann. Finally we are grateful to Hubertus Illner for helpful comments improving the quality of the manuscript.

References

Albrecht R, Knief W, Mertens I, Göttsche M, Götsche M (2006) Empfehlungen zur Berücksichtigung tierökologischer Belange bei Windenergieplanungen in Schleswig-Holstein. Landesamtes für Umwelt und Natur des Landes Schleswig-Holstein, Flintbek

Altenkamp R, Stoewe D, Krone O (2007) Verlauf und Scheitern einer Brut des Seeadlers (*Haliaeetus albicilla*) in Berlin und Konsequenzen für den Schutz der Brutplätze. Berliner ornithologische Berichte 17:31–41

Barrios L, Rodriguez A (2004) Behavioural and environmental correlates of soaring-bird mortality at on-shore turbines. J Appl Ecol 41:72–81

Bevanger K, Follestad A, Gjershaug JO, Halley D, Hansen F, Johnson L, May R, Nygård T, Pedersen HC, Reitan O, Steinheim Y (2008) Pre- and post-construction studies of conflicts between birds and wind turbines in coastal Norway. Status report, NINA Report, Trondheim, pp 355

Bevanger K, Berntsen F, Clausen S, Dahl EL, Flagstad Ø, Follestad A, Halley D, Hanssen F, Hoel PL, Johnsen L, Kvaløy P, May R, Nygård T, Pedersen HC, Reitan O, Steinheim Y, Vang R (2009) Pre- and post-construction studies of conflicts between birds and wind turbines in coastal Norway (BirdWind). Progress Report 2009. NINA Report 505, Trondheim, pp 70

Bloom P, Clark WS, Kidd JW (2007) Capture techniques. In: Bird DM, Bildstein KL (eds) Raptor research and management techniques. Hancock house publisher, Surrey, Blaine

Carrete M, Sánchez-Zapata JA, Benítez JR, Lobón M, Donázaret A (2009) Large scale risk-assessment of wind-farms on population viability of a globally endangered long-lived raptor. Biol Conserv, doi:10.1016/j.biocon.2009.07.027

Dürr T (2012) Vogelverluste an WKA in Deutschland und Fledermausverluste weltweit. Summe der Funde seit 1989. Stand Dezember 2012. Daten aus dem Archiv der Staatlichen Vogelschutzwarte im Landesamt für Umwelt, Gesundheit und Verbraucherschutz Brandenburg. Status: December 2012. http://www.lugv.brandenburg.de/cms/media.php/lbm1.a.3310.de/wka_vogel_de.xls, http://www.lugv.brandenburg.de/cms/media.php/lbm1.a.3310.de/wka_voegel_eu.xls

EG-Vogelschutzrichtlinie (2010): Richtlinie 2009/147/EG des Europäischen Parlaments und des Rates vom 30. November 2009 über die Erhaltung der wildlebenden Vogelarten. Anhang 1

Fallen M, Puga JF, Tobias K, Jung C (2006) Hinweise zur Planung von Windenergieanlagen auf Waldstandorten. AG Windenergie, University of Kaiserslautern

Farfán MA, Vargas JM, Duarte J (2009) What is the impact of wind farms on birds. A case study in southern Spain. Biodivers Conserv 18:3743–3758

Follestad A, Flagstad Ø, Nygård T, Reitan O, Schulze J (2007) Vindkraft og fugl på Smøla 2003–2006. NINA Rapport 248, Trondheim, pp 78

Fraunhofer Institut für Windenergie und Energiesystemtechnik (IWES) (2011) Windenergiereport Deutschland 2011. Fraunhofer Institut für Windenergie und Energiesystemtechnik, Bereich Energiewirtschaft & Netzbetrieb, Kassel. http://www.fraunhofer.de/content/dam/zv/de/fors. Assessed 30 June 2013

Garvin JC, Jennelle CS, Drake D, Grodsky SM (2011) Response of raptors to a windfarm. J Appl Ecol 48:199–209

Grünkorn T, Diederichs A, Poszig D, Diederichs B, Nehls G (2009) Wie viele Vögel kollidieren mit Windenergieanlagen? Natur und Landschaft 84:309–314

Hötker H, Thomsen K-M, Köster H (2004) Auswirkungen regenerativer Energiegewinnung auf die biologische Vielfalt am Beispiel der Vögel und der Fledermäuse – Fakten, Wissenslücken, Anforderungen an die Forschung, ornithologische Kriterien zum Ausbau von regenerativen Energiegewinnungsformen. Endbericht für das Bundesamt für Naturschutz, Förd. Nr. Z1.3-684 11-5/03. Michael-Otto-Institut im NABU, Bergenhusen, pp 80

Hull CS, Murir S (2010) Search Areas for Monitoring Bird and Bat Carcasses at Wind Farms Using a Monte-Carlo Model. Aust J Envir Manage 17:77–87

Illner H (2011) Comments on the report "Wind Energy Developments and Natura 2000", edited by the European Commission in October 2010. http://www.abu-naturschutz.de/images/H_Illner_15Febr2011_comments_EU-Guidance_wind_turbines_NATURA_2000.pdf

Illner H (2012) Kritik an den EU-Leitlinien „Windenergie-Entwicklung und NATURA 2000", Herleitung vogelartspezifischer Kollisionsrisiken an Windenergieanlagen und Besprechung neuer Forschungsarbeiten. Eulen-Rundblick Nr. 62:83–100

Katzner TE, Brandes D, Miller T, Lanzone M, Maisonneuve C, Tremblay JA, Mulvihill R, Merovich GT (2012) Topography drives migratory flight altitude of golden eagles: implications for on-shore wind energy development. J Appl Ecol 49:1178–1186

Kenward RE (1983) A Manual for Wildlife Radio Tagging. Academic Press, London

Krijgsveld KL, Akershoek K, Schenk F, Dijk F, Dirksen S (2009) Collision risk of birds with modern large wind turbines. Ardea 97:357–366

Krone O, Scharnweber C (2003) Two white-tailed sea eagles (*Haliaeetus albicilla*) collide with wind generators in northern Germany. J Raptor Res 37:174–176

Krone O, Grünkorn T, Gippert M, Dürr T (2008) White-tailed Sea Eagles and wind power plants in Germany - preliminary results. In: Hötker H (ed) Birds of Prey and Wind Farms: Analysis of Problems and Possible Solutions. Documentation of an international workshop in Berlin, 21st and 22nd October 2008. http://bergenhusen.nabu.de/imperia/md/images/bergenhusen/bmuwindkraftundgreifwebsite/birds_of_prey_and_windfarms_documentation_2009.pdf, Assessed 30 Jun 2013

Krone O, Berger A, Schulte R (2009) Recording movement and activity pattern of a White-tailed Sea Eagle (*Haliaeetus albicilla*) by a GPS datalogger. J Ornithol 150:273–280

Landesbund für Vogelschutz in Bayern (LBV) (2010) Literaturstudie über die Auswirkungen von WKA auf die Avifauna. Landesbund für Vogelschutz in Bayern, Hilpoltstein

Länder-Arbeitsgemeinschaft der Vogelschutzwarten (LAG-VSW) (2007) Abstandsregelungen für Windenergieanlagen zu bedeutsamen Vogellebensräumen sowie Brutplätzen ausgewählter Vogelarten. Berichte zum Vogelschutz 44:151–153

Langston RWH, Pullan JD (2003) Wind farms and birds: an analysis of the effects of wind farms on birds, and guidance on environmental assessment criteria and site selection issues. Report written by BirdLife International on behalf of the Bern Convention, Sandy

Madders M, Whitfield DP (2006) Upland raptors and the assessment of wind farm impacts. Ibis 148:43–56

Martin GR (2011) Understanding bird collisions with man-made objects: a sensory ecology approach. Ibis 153:239–254

Martin GR, Katzir G (1999) Visual field in short-toed eagles *Circaetus gallicus* and the function of binocularity in birds. Brain Behav Evolut 53:55–66

Martin GR, Shaw JM (2010) Bird collision with power lines: Failing to see the way ahead? Biol Conserv 143:2695–2702

Newton I (1998) Population limitation in birds. Academic Press, London

Orloff S, Flannery A (1992) Wind turbine effects on avian activity, habitat use and mortality in Altamont Pass and Solano County wind resource areas, 1989-1991. Prepared by BioSystems Analysis, Inc. for the California Energy Commission, Sacramento, California

Pearce-Higgins JW, Leigh S, Langston RHW, Bainbridge IP, Bullman R (2009) The distribution of breeding birds around upland wind farms. J Appl Ecol 46:1323–1331

Smallwood SK, Thelander CG (2004) Developing methods to reduce bird mortality in the Altamont Pass Wind Resource Area. Final Report by BioResource Consultants to the California Energy Commission, Public Interest Energy Research-Environmental Area, Contract No. 500-01-019, pp 363

Smallwood SK, Rugge L, Morrison ML (2009) Influence of Behavior on Bird Mortality in Wind Energy Developments. J Wildlife Manage 73:1082–1098

Smallwood KS, Bell DA, Snyder SA, Didonato JE (2010) Novel Scavenger Removal Trials Increase Wind Turbine-Caused Avian Fatality Estimates. J Wildlife Manage 74: 1089–1097

Swihart R, Slade N (1985a) Testing for independence of observations in animal movements. Ecology 66:1176–1184

Chapter 8
Behavioural Observations of White-Tailed Sea Eagles in the Vicinity of Wind Turbines

Thomas Grünkorn, Mirjam Gippert, Gabriele Treu and Georg Nehls

8.1 Introduction

Aim of this part of the study was to investigate the flight behaviour of white-tailed sea eagles in the vicinity of existing wind farms by means of visual observations to determine especially the collision risk. To date, white-tailed sea eagles found dead under wind turbines usually are chance discoveries (Dürr 2011) providing no further information on the circumstances of the collision. For this reason direct behavioural observations seemed adequate to determine the immediate small-scale flight behaviour. Direct monitoring was aimed to provide hints on the circumstances and possible reasons for collisions and to identify avoidance measures.

8.2 Methods

Inside wind farms in Mecklenburg-Western Pomerania and Schleswig-Holstein flight paths and heights were recorded complemented by direct observations of reactions and behaviour of birds of prey. A binocular with integrated rangefinder (Leica Geovid 10×42) mounted on a tripod was used. The inclination angle of the

T. Grünkorn (✉) · G. Nehls
BioConsult SH, Schobüller Str. 36, 25813 Husum, Germany
e-mail: t.gruenkorn@bioconsult-sh.de

M. Gippert · G. Treu
Leibniz Institute for Zoo and Wildlife Research (IZW), Alfred-Kowalke-Str. 17, 10315 Berlin, Germany

H. Hötker et al. (eds.), *Birds of Prey and Wind Farms*,
DOI 10.1007/978-3-319-53402-2_8

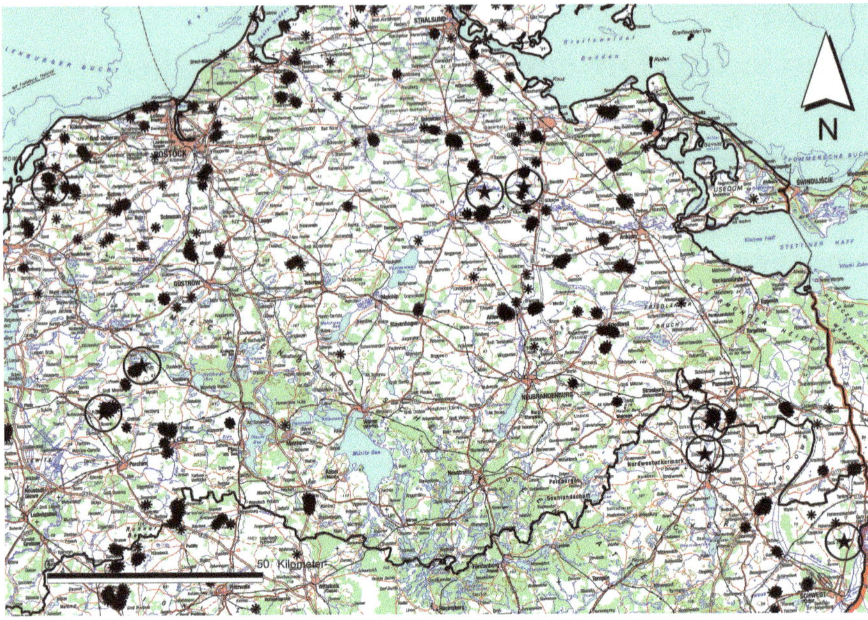

Fig. 8.1 Locations of selected breeding sites in close vicinity of wind turbines (*circle around star*) and breeding sites chosen for visual observations in Mecklenburg-Western Pomerania and Brandenburg (1 to 4)

tripod head could be read from a scale. Using distance and angle, the flight height of white-tailed sea eagles can be calculated. In the close-up area of wind turbines, the flight height was estimated using the known height of the tower and rotor diameter.

Breeding sites with wind turbines within the home range of the breeding birds suitable for visual observations (and attempts to capture birds) were identified using GIS (Figs. 8.1 and 8.2).

Series of vantage point watches (SNH 2005) of at least four hours duration were mainly carried out from elevated terrain. Flight activity of white-tailed sea eagles was ideally watched from a point where the forest (forest edge) with the mostly concealed eagle nest as well as near wind turbines were visible. In choosing further vantage points at different sides of the forest attempts were made to determine further directions of departure from the nest.

In addition eyewitness reports on collisions of white-tailed sea eagles with wind turbines were collected.

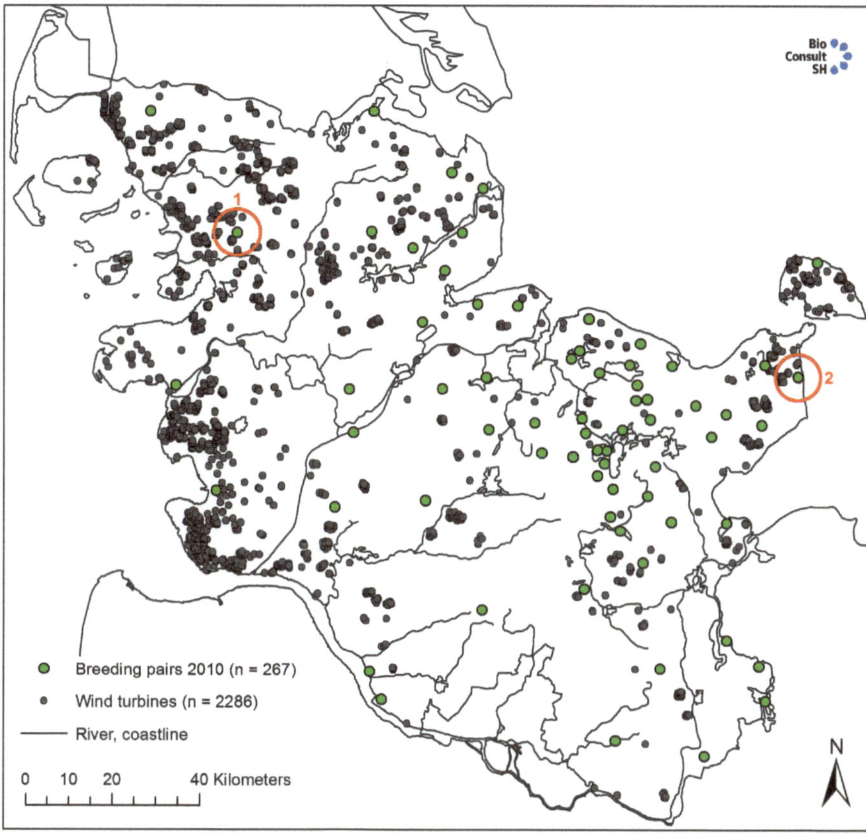

Fig. 8.2 Locations of breeding sites with wind turbines for visual observations of adult birds in Schleswig-Holstein (*1* and *2*)

8.3 Results

8.3.1 Visual Observations in Mecklenburg-Western Pomerania and Brandenburg

8.3.1.1 Untagged Adult Birds

In summer 2008 visual observations were carried out at two breeding sites (1/MV and 2/MV) and in June 2010 at two further breeding sites (3/BB und 4/BB). Attempts were made to observe adult birds flying near wind turbines and to determine flight paths. Despite the long watching series no results were achieved. Visual observations of untagged birds proved to be unsuitable, as no result were achieved despite the high personnel effort. During the observation period, no white-tailed sea eagles were spotted in the wind farm although they were visiting the wind farm on a regular basis according to local residents.

8.3.1.2 Tagged Adult Bird (VHF and GPS/GSM)

Visual observations of tagged bird 5900 near Schönwerder 22 to 25 June 2009.

In addition to automatically recording GPS coordinates via VHF transmitter, the adult bird tagged on 30 January 2009 near Schönwerder/Uckermark in Brandenburg was located on 22 June and 25 June 2009 by means of a portable directional antenna. Over the whole period the bird was tracked 7 km from the nest site in a floodwater area near Holzdorf/Uckermark (Fig. 8.3). Only small-scale

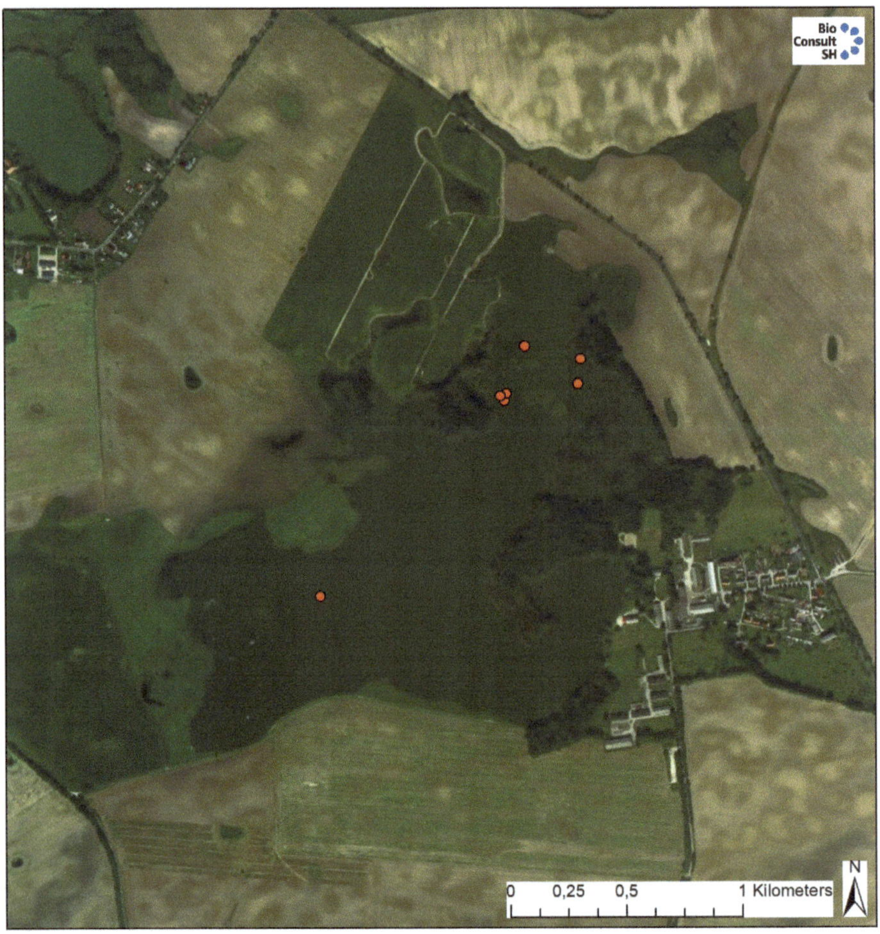

Fig. 8.3 Locations recorded via VHF telemetry (*red dots*), *black outline* floodwater area

changes in location over a few 100 m and short hunting flights over the floodplain occurred. The battery of the transmitter unit for GPS coordinates and VHF telemetry has run out of life on 25 June after transmitting more than 4800 locations. No wind turbine was located in nearby areas. Further recorded locations and the results of satellite telemetry are described in Sect. 7.3.1.

8.3.1.3 Tagged Young Birds (VHF and GPS/GSM)

In June 2008, three nestlings were fitted with GPS and VHF transmitters in Mecklenburg Western-Pomerania (two young birds near Wismar (2/MV) and one young bird near Parchim (1/MV)). From leaving the nest minute-by-minute time-use surveys and behavioural observations were carried out for these birds being unfamiliar with wind farms.

The young birds of 2/MV left the breeding site with the adjacent wind turbines shortly after fledging. On 12 August 2008 several of this year's eagles could already be VHF tracked using a beam aerial at a distance of 26 km from the breeding site hunting at a lake.

The tagged young eagle 5902 (1/MV) was tracked and observed over a total of 113 h staying 97% of the time out of sight in the forest mainly near the nest perching in a tree. The rest of the time this bird was observed flying outside the forest where the flight height was recorded. In this period, the young bird was entering the wind farm three times, flying significantly above the turbines in one case but moving at the height of the dangerous rotor swept area and below as well (Fig. 8.4).

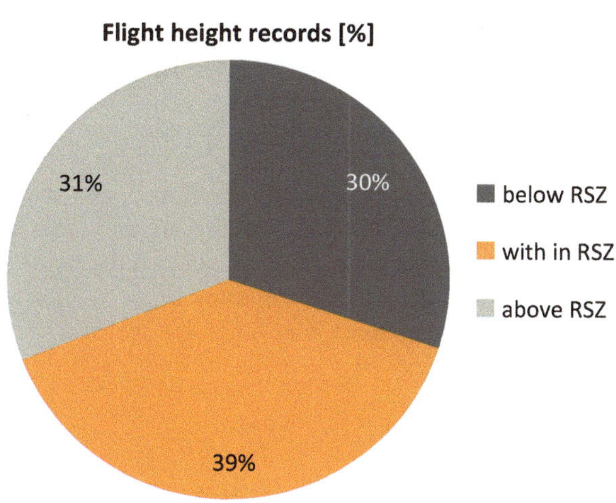

Fig. 8.4 Relative proportion of flight heights of young eagle 5902 (breeding site Kukuk/Parchim) in relation to the rotor swept area of present wind turbines (n = 203 min)

8.3.2 *Visual Observations in Schleswig-Holstein*

In Schleswig-Holstein, the Projektgruppe Seeadlerschutz Schleswig-Holstein e.V. had refused to tag young eagles so that only observations of breeding birds were carried out.

In summer 2007, flight movements of breeding birds took place at two sites in Schleswig-Holstein (1/SH in North Frisia and 2/SH in Ostholstein). Many wind turbines are located in the vicinity of these breeding sites.

8.3.2.1 Breeding Site 1/SH in North Frisia

The breeding birds at 1/SH did not use the wind farm sites for hunting but traversed them on a regular basis on the way between breeding site to foraging grounds (coastal areas). The numerous wind farms are located across the direct flight line. No large-scale avoidance behaviour towards wind farms was observed. In three recorded approaches, an adult bird crossed the western wind farms in Breklumer Koog at the height of the rotors (approximately 90 m) but outside the risk area (Fig. 8.5). Two flyovers were recorded in the Eastern wind farm near Bohmstedt/NF clearly above the rotor swept area at a height of approximately 200 m (Fig. 8.5).

8.3.2.2 Breeding Site 2/SH in Ostholstein

At the breeding site 2/SH exclusively departure paths which were not leading into the local wind farm were found in spring and summer 2007. In spring 1998, members of the Projektgruppe Seeadlerschutz e.V. still recorded flight activity in the direction of the site where wind turbines were constructed later (Figs. 8.6 and 8.7).

Standardised observations in the wind farms near both eagle breeding sites partially lasting for days resulted in only few behavioural observations. Due to the low flight activity and the long daylight hours with possible hunting or roosting flights, the exclusive visual observations were stopped in 2008 and 2009. Subsequently, there were plans to increase the number of observed foraging flights by individually tracking the breeding birds by means of VHF transmitters. Due to the missing success to capture birds this was not successful in Schleswig-Holstein.

8.3.2.3 Breeding Season 2010

In breeding season 2010, visual observations were repeated at breeding site 1/SH in North Frisia.

Fig. 8.5 Flight paths of adult white-tailed sea eagles at breeding site 1/SH in North Frisia (*green dot*) in summer 2007 in relation to the local wind farm (*blue dots* wind turbines), *arrow* width = 1, 2 and 3 observations, (in the western part of the map extract: wind farms in the Breklumer Koog, in the eastern part: one wind farm near Bohmstedt)

On four days in Mai and early June 2010, attempts were made to record flight sequences. However, no foraging flights to the nest by white-tailed sea eagles were observed. Therefore, no observations near the different wind turbines within the flight corridor were possible either. Observations were discontinued in June upon request from the contracting authority.

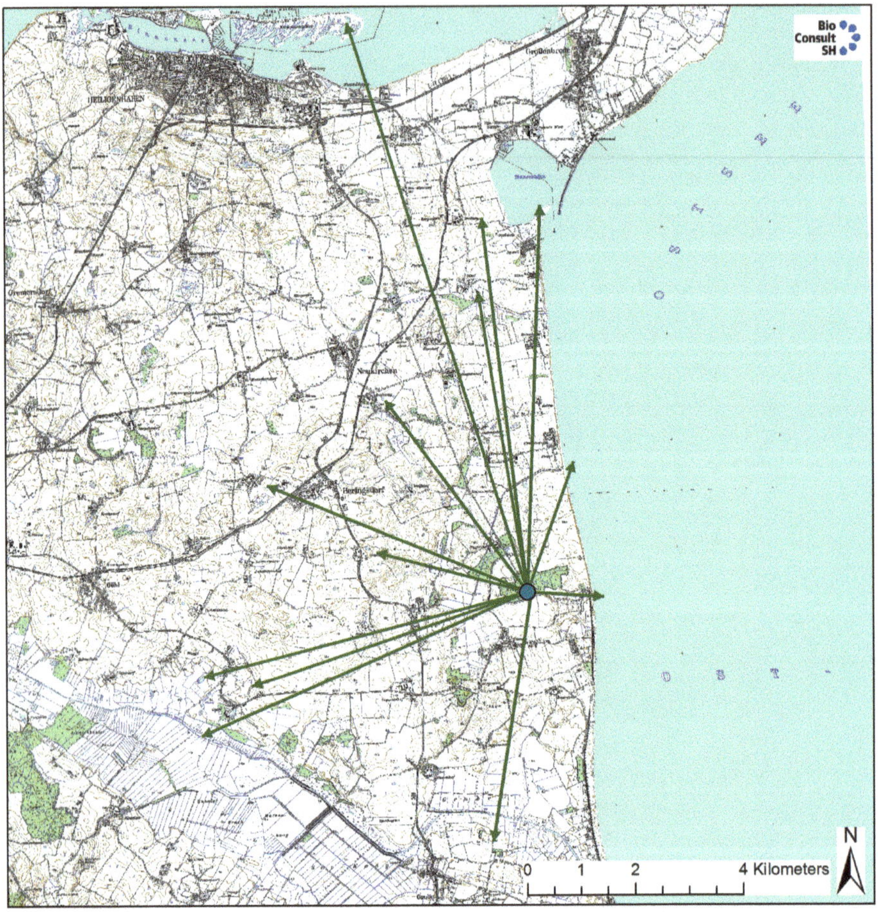

Fig. 8.6 Directions of departure (and expected aims) of adult white-tailed sea eagles for breeding site 2/SH (D. + P. Göhrke pers. comm., n = 25) in May 1998 (no wind farms yet)

8.4 Previously Documented Observations of Collisions

Due to the low number of direct behavioural observations (and especially possible avoidance behaviour) in the close-up area of wind turbines, the three known cases of directly observed collisions of white-tailed sea eagles with wind turbines in Mecklenburg Western-Pomerania (Krone and Scharnweber 2003) and Schleswig-Holstein (Projektgruppe Seeadlerschutz Schleswig-Holstein e.V.) are described.

Case 1 (Krone and Scharnweber 2003)
On 01 April 2002 (9:20 am), a walker heard a muffled bang near Anklam and thereupon saw an eagle dropping to the ground at a wind turbine. The flight

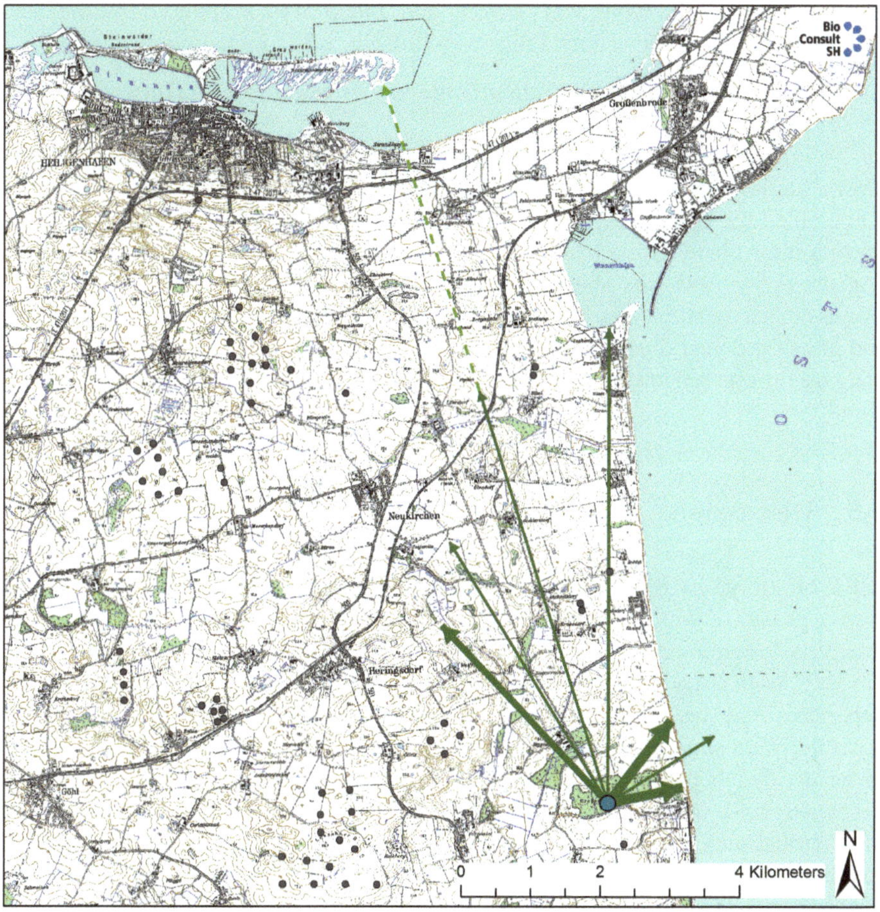

Fig. 8.7 June and July 2007: Directions of departure from breeding site 2/SH (*arrow* width = 1, 2 and 3 observations by T. Grünkorn and D. + P. Göhrke postal comm., n = 13). Existing wind turbines are represented as *dots*

behaviour of the eagle prior to the collision was not observed. It was a sunny day. No information on wind speeds is available. The bird was a sub-adult male in the 5th year, which was ringed at a distance of 25 km on Usedom as a young bird. Because of the severe injuries the bird was euthanized.

Case 2 (Projektgruppe Seeadlerschutz Schleswig-Holstein e.V., annual report 2007) On 12 September 2007, a hunter observed a flying sub-adult white-tailed sea eagle being attacked by two buzzards near Sprengerhof/RD. The buzzards repeatedly flew brief attacks targeted on the white-tailed sea eagle. In an evasive manoeuvre the white-tailed sea eagle was caught by a strong gust of wind and unexpectedly got into the rotor swept zone of one of the three wind turbine on a crest. The bird got hit by a rotor blade and fell to the ground. On the harvested field the dead bird was quickly found. One wing was broken and blood was seeping from the beak (Projektgruppe Seeadlerschutz Schleswig-Holstein e.V., annual report 2007).

Case 3: (Projektgruppe Seeadlerschutz Schleswig-Holstein e.V., via Internet OAGSH on 26.06.2011)
On 24 June 2011, a walker observed a white-tailed sea eagle being attacked by two crows above the wind turbines near Rethwisch/PLÖ east of Preetz. As the white-tailed sea eagle moved downwards in an evasive manoeuvre, it got hit by the tip of a rotor blade with a loud bang and was propelled away. With the help of his dog the eyewitness was able to retrieve the bird in a near wheat field, where it would otherwise surely not have been found in this season. The bird was not ringed and in its third year. Left wing and left foot were cut off. The right foot was broken, hanging on one tendon.

8.5 Discussion

All in all, only few flight movements of young and adult birds could be observed. Active phases of white-tailed sea eagles are generally short. While rearing young, a relatively low number of approaches to the nest can be expected due to the size of the prey. With a daylight period of 14–17 h during the breeding season, arrivals and departures may be distributed over the whole day and only few observations are possible. With several possible directions of departure, the observation effort needs to be divided between these directions. Furthermore the detection probability is decreasing with increasing distance from the breeding site. In densely wooded and hilly terrain, only a small part of the air space will be visible and the observed flight sequences may be short. A limited amount of flight movements could be followed over a longer distance in open and flat terrain (especially near 1/SH in North Frisia). In its presumably first approach, the young bird did not avoid the risk area of the rotor swept zones of the wind farm.

In case 1 of the so far described direct observations of collisions, no external causes could be detected. Cases 2 and 3 though indicate similar circumstances which may lead to collisions with wind turbines. Apparently white-tailed sea eagles may be distracted by conflicts with crows or birds of prey and probably intraspecific competition preventing proper perception of the danger arising from the rotor. In inter- and intraspecific conflicts, the restricted view of birds described by Martin (2011) might well lead to an increased collision risk. May et al. (2010) report a high collision risk of white-tailed sea eagles on the island of Smøla/Norway due to a high population density leading to a large amount of social and/or territorial flight activity. No targeted visual observations were carried out on Smøla.

References

Dürr T (2011) Zentrale Fundkartei der Kollisionsopfer (Vögel). mugv.brandenburg.de/cmc/detail.php/bb2.c.451792.de. Assessed 30 June 2013

Krone O, Scharnweber C (2003) Two White-tailed Sea-eagles (*Haliaetus albicilla*) collide with wind-generators in northern Germany. J Raptor Res 37:174–176

Martin GR (2011) Understanding bird collisions with man-made objects: a sensory ecology approach. Ibis 153:239–254

May R, Hoel PL, Langston R, Dahl EL, Bevanger K, Reitan O, Nygård T, Pedersen HC, Røskaft E, Stokke BG (2010) Collision risk in white-tailed eagles. Modelling collision risk using vantage point observations in Smøla wind-power plant. NINA Report 639, Trondheim, pp 25

Projektgruppe Seeadlerschutz Schleswig-Holstein (2007) Seeadler kollidiert mit Windernergieanlage im Dänischen Wohld. Großvogelschutz im Wald. Jahresbericht 2007, pp 20

Scottish Natural Heritage (SNH) (2005) Survey methods for use in assessing the impacts of onshore wind farms on bird communities. www.snh.gov.uk/docs/C278917.pdf. Accessed 30 June 2013

Chapter 9
Effects of Individuals Lost Due to Wind Turbines on the Time of Occupying a Territory, the Survival of Breeding Territories and the Breeding Success of White-Tailed Sea Eagles (Case Studies)

Thomas Grünkorn and Georg Nehls

9.1 Introduction

By means of case studies, nature and extent of the effects of individuals lost due to wind turbines on establishment and survival of breeding territories and the breeding success of the white-tailed sea eagle will be presented. Main focus is on collision victims, which are due to their age sexually mature and therefore potential breeding birds. Collision victims may have all ages but will be of different value for the population. Further assessment of the effects of additional mortality of white-tailed sea eagles due to wind turbines on the size of the population taking into account all age groups is carried out by Prof. Dr. O. Krüger (University of Bielefeld) using matrix models within the framework of the PROGRESS study supported by the Federal Ministry for the Environment, Nature Conservation, Building and Nuclear Safety (BMUB).

9.2 Methods

Collisions of potentially sexually mature white-tailed sea eagles were identified using the central database of collision victims (Dürr 2011). For birds ringed in the nest, the year of birth is absolutely certain. Estimating the age of birds, which were not ringed, by the plumage is not as precise. In the fourth year, first time breeding is possible (physiological maturity). Single attempts to breed in the third year were recorded (Struwe-Juhl and Grünkorn 2007). As a rule, white-tailed sea eagles breeding for the first time are older (ecological maturity, depending on population

T. Grünkorn (✉) · G. Nehls
BioConsult SH, Schobüller Str. 36, 25813 Husum, Germany
e-mail: t.gruenkorn@bioconsult-sh.de

H. Hötker et al. (eds.), *Birds of Prey and Wind Farms*,
DOI 10.1007/978-3-319-53402-2_9

density). Through identification of individual colour rings (Struwe and Grünkorn 2007) and collection of moulted feathers (Struwe and Schmidt 2002), the individual breeding birds and the breeding success (number of fledged young birds) is known for many breeding birds in Schleswig-Holstein.

9.3 Results

9.3.1 Schleswig-Holstein

Between 1993 and 2011, 26 collisions of white-tailed sea eagles were recorded in Schleswig-Holstein. 19 of these birds were ringed (18 thereof in Schleswig-Holstein) (Table 9.1). It was therefore possible to determine the exact age of the collision victims in most cases. Fifteen birds were at least four years old but only three of them could be connected to a known breeding territory. This could only be attributed to a known breeding territory in three cases. The rest of the fatalities were non-breeding or migrating birds. Territorial coherence can be excluded for the nearest breeding pair but not for breeding pairs further away. In three cases (2 × Seegalendorf/OH, Rodenäs-Dreieckskoog/NF), collisions of this year's ringed young birds could be connected to a breeding site.

It is remarkable that one of the collisions occurred at a single small wind turbine (40 m total height only) near Krieseby/RD. Further fatalities concerned wind farms with higher and larger numbers of turbines.

In the following, the three collision victims where a connection with the respective breeding pair or site could be made are described. None of the collisions occurred during the breeding season.

Case 1: White-tailed sea eagle in the fourth year found in November 2003 under a wind turbine near Damlos, Ostholstein. In spring 2003, the establishment of a not-breeding new territorial pair has been found in this area. In 2004 the territory was vacant. In 2005 breeding could be proved at a distance of 3 km from the site of the find near Manhagen. It is not known, if one of the breeding birds originates from the 2003 pair. Circumstances suggest that the collision fell into the time of establishing a territory, which was as a result delayed by one year.

Case 2: Discovery of a white-tailed sea eagle in the eighth year in December 2004 in the Desmerciereskoog, North Frisia. In spring 2004, the establishment of a not-breeding new territorial pair has been found in this area. In the next year (2005), the territory was vacant again. Evidence of nesting was found in 2006 at a distance of 8 km. Circumstances suggest that the collision fell into the time of establishing a territory, which was as a result delayed by one year.

Case 3: Discovery of a white-tailed sea eagle in the fourth year in September 2007 near Sprengerhof, district Rendsburg-Eckernförde. In 2007, no evidence of nesting was found in the surrounding area. In the following year, evidence of nesting was found at a distance of approximately 1 km from the location of

Table 9.1 Fatalities of white-tailed sea eagles at wind turbines in Schleswig-Holstein (*red* potential breeding birds >= 4th year, *blue* collisions of this year's birds, which cannot be breeding)

Turbine/farm	Location	Record	Age [calendar]	Breeding?
wind farm	Heringsdorf/Göhl	11.03.2003	2	no
wind farm	Zarpen	20.03.2003	4	no
wind farm	Kronprinzenkoog	30.03.2003	5	no
wind farm	Damlos	11.11.2003	4	yes, see case 1
wind farm	Timmaspe	15.03.2004	5	no
wind farm	Koselau-Riepsdorf	06.04.2004	3	no
2 turbines	Desmercierеskoog	06.12.2004	8	yes, see case 2
wind farm	Seegalendorf	31.08.2005	1	no
1 turbine	Krieseby	23.01.2006	7	no
wind farm	Heringsdorf	08.10.2006	2	no
wind farm	Sprengerhof	12.09.2007	4	yes, see case 3
wind farm	Rellin	05.01.2008	4	no
wind farm	Rodenäs-Dreieckskoog	16.10.2008	1	no
wind farm	Rehhorst	28.11.2008	3	no
wind farm	Hardebek	16.12.2008	5	no
wind farm	Eichede	06.04.2009	4	no
wind farm	Wasbek	18.05.2009	2	no
wind farm	Barkau	27.02.2010	adult	no
3 turbines	Steinberghaff	16.03.2010	5	no
wind farm	Puls	22.03.2010	4	no
wind farm	Tralau	11.08.2010	3	no
wind farm	Schwochel	06.09.2010	5	no
wind farm	Seegalendorf	21.10.2010	1	no
wind farm	Brokstedt	01.03.2011	5	no
wind farm	Grömitz	26.03.2011	2	no
wind farm	Rethwisch	24.06.2011	3	no

discovery. The dead eagle was replaced ten day after discovery. Circumstances suggest a loss during establishment of a territory, but no evidence of a resulting delay was found.

9.3.2 *Lower Saxony*

One of the three collision fatalities registered in Lower Saxony so far was an adult bird.

Case 4: Found dead in January 2008 near Nöpke, district Hanover, 12th year, evidently the death of the female breeding bird resulted in the abandonment of the nest site at a distance of 5 km at the Steinhuder Meer. The male left the territory, and no evidence of nesting was found since 2008.

9.3.3 *Brandenburg*

The individual mapping of birds found dead to specific territories is in most cases impossible in Brandenburg (and Mecklenburg Western-Pomerania) due to the partly high population density of white-tailed sea eagles and the limited knowledge of individual birds (ringing, identification of individual rings or moulted feathers).

In Brandenburg, 18 collisions of white-tailed sea eagles with wind turbines became known (Table 9.2). Eight of the birds had reached at least the fourth year. In one case it was impossible to determine the age of the bird. In four cases the collisions occurring during the breeding season and connection to a known breeding site could be made due to missing adult birds and missing breeding

Table 9.2 White-tailed sea eagles found dead at wind turbines in Brandenburg (*red* potential breeding birds, adult >= 4th year, *blue* collisions of this year's young birds)

Turbine/farm	Location	Record	Age [calendar]	Breeding?
wind farm	Beerfelde	12.04.2003	adult	yes
wind farm	Duben-Schiebsdorf	15.01.2004	adult	no
wind farm	Wilhelmsfelde	30.07.2006	5 or 6	no
wind farm	Buckow-Nord (Klein Rietz)	10.04.2006	1	no
wind farm	Kerkow	03.12.2007	adult	no
wind farm	Wilmersdorf	07.03.2008	4	no
1 turbine	Bomsdorf	04.05.2008	adult	no
wind farm	Sembten	12.11.2008	8	no
3 turbines	Markee-Süd	24.07.2009	adult	yes
wind farm	Lüdersdorf-Haselberg (Wriezen)	16.08.2009	2	no
wind farm	Ketzin-I-b	28.07.2009	3	no
wind farm	Barenthin	07.10.2009	2	no
wind farm	Klettwitz-III (Salgast-Poley)	13.06.2009	?	no
wind farm	Zernitz.Holzhausen	08.03.2010	3	no
wind farm	Prötzel	21.05.2010	adult	yes
wind farm	Jahnkeshof	07.11.2003	2	no
wind farm	Drehnow - Turnow	A.07.2009	1	no
wind farm	Gr. Schacksdorf	26.02.2011	adult	yes

success (distance of the wind turbine from breeding site 2.6, 2.9, 5,6 and 20 km) (Dürr and Langgemach 2011). Two of the breeding sites were occupied again in the following year. No information on the other two sites is available.

It was possible to allocate two further collisions of two this year's birds to a breeding site (found near Drehnow/district Spree-Neiße and Buckow-Nord/district Oder-Spree).

9.4 Discussion

In Schleswig-Holstein, breeding birds are known individually, and established relations are conclusive. In Brandenburg the breeding birds are not known individually and cannot be clearly allocated to known breeding sites.

Apart from adult birds, this year's young birds were found dead under wind turbines. Allocation of young birds, which are not ringed, to known white-tailed sea eagle territories is not always possible with certainty. If the collision occurs shortly after leaving the nest in the vicinity of a known breeding site, a connection can be made with a certain probability depending on the population density. In Schleswig-Holstein, young collision victims which were ringed in the nest were found additionally.

Three of the 26 collision victims in Schleswig-Holstein could be allocated to local breeding pairs even though collisions occurred outside the breeding season. One of the collisions took place within a radius of 3 km around the breeding site.

Death of sexually immature eagles (1st to 3rd year) will surely have no influence on nesting in specific not yet existing territories. Eagles in the 4th and 5th year are physiologically mature, but quite often form a pool of non-breeders. This kind of reserve is typical for vital populations. It serves to replace dead breeding birds in breeding territories existing for many years and to naturally extend the breeding range as shown for Denmark and The Netherlands or to increase densities.

Impacts of individuals lost due to collisions with wind turbines vary in severity between none at all (case 3), one year of delay in establishing a breeding territory (cases 1 and 2) or permanent abandonment of the territory for the time being (case 4). Case studies from Schleswig-Holstein indicate that white-tailed sea eagles are especially prone to collisions in the time of establishing a breeding territory. Experienced birds familiar with the home range possibly do not collide as often. This preliminary assumption needs to be validated with additional data.

Of the total of 13 collisions, which could be attributed to a breeding site (8 breeding birds, 5 this year's birds), seven occurred within the protection zone (radius 3 km) and 10 within the range of verification for flight paths (radius 6 km) recommended by the Working Group of German State Bird Conservancies (Länderarbeitsgemeinschaft der Vogelschutzwarten LAG VSW 2007). No wind turbines within these distances around the nest site might have prevented seven out of 13 collisions.

References

Dürr T (2011) Zentrale Fundkartei der Kollisionsopfer (Vögel). mugv.brandenburg.de/cmc/detail.php/bb2.c.451792.de. Assessed 30 June 2013

Dürr T, Langgemach T (2011) Verlustmonitoring von Großvogelarten im Land Brandenburg. Auszug vom 26. Juni 2011, LUGV BB

Länder-Arbeitsgemeinschaft der Vogelschutzwarten (2007) Abstandsregelungen für Windenergieanlagen zu bedeutsamen Vogellebensräumen sowie Brutplätzen ausgewählter Vogelarten. Berichte zum Vogelschutz 44:151–153

Struwe-Juhl B, Grünkorn T (2007) Ergebnisse der Farbberingung von Seeadlern *Haliaeetus albicilla* in Schleswig-Holstein mit Angaben zu Ortstreue, Umsiedlung, Dispersion, Geschlechtssreife, Altersstruktur und Geschwisterverpaarung. Vogelwelt 128:117–129

Struwe-Juhl B, Schmidt R (2002) Möglichkeiten und Grenzen der Individualerkennung von Seeadlern anhand von Mauserfederfunden in Schleswig-Holstein (1955–2000). Corax 19, Sonderheft 1:37–50

Chapter 10
Distances Between Breeding Sites and White-Tailed Sea Eagles Found Dead Near Wind Turbines in Schleswig-Holstein

Thomas Grünkorn and Georg Nehls

10.1 Introduction

The development of the breeding population and breeding success of white-tailed sea eagles is particularly well documented in Schleswig-Holstein (data of the Projektgruppe Seeadlerschutz Schleswig-Holstein e.V.).

In the last ten years, the breeding population in Schleswig-Holstein experienced an exponential growth (Projektgruppe Seeadlerschutz Schleswig-Holstein e.V. 2007). This highly significant growth, which to date still is exponential, will according to model calculations by Krüger et al. (2010) change into a sigmoid curve limited by the carrying capacity in the decade ahead (Fig. 10.1).

Furthermore, the range has extended to the west coast of Schleswig-Holstein over the last ten years. White-tailed sea eagles now breed throughout Schleswig-Holstein (four breeding sites on the west coast in 2010, Fig. 10.2, data: Projektgruppe Seeadlerschutz Schleswig-Holstein e.V.). The main areas of distribution are the districts Plön and Ostholstein. The number of fledged eagles is still increasing exponentially with no density effect so far (Fig. 10.3).

In the following, distance from breeding site, spatial distribution of birds found dead and location of foraging waters used will be put into context with existing wind turbines.

T. Grünkorn (✉) · G. Nehls
BioConsult SH, Schobüller Str. 36, 25813 Husum, Germany
e-mail: t.gruenkorn@bioconsult-sh.de

H. Hötker et al. (eds.), *Birds of Prey and Wind Farms*,
DOI 10.1007/978-3-319-53402-2_10

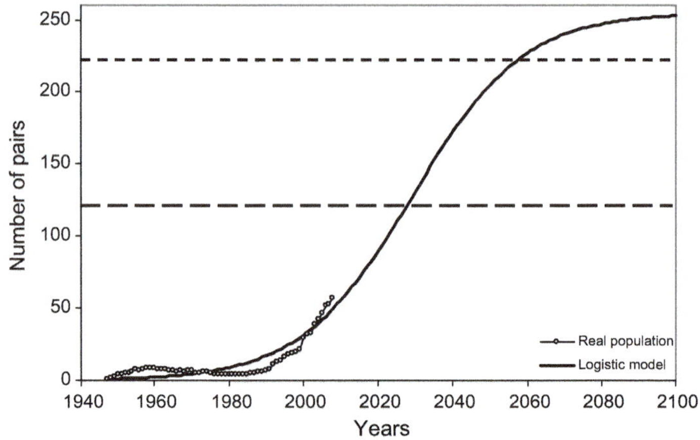

Fig. 10.1 Carrying capacity model based on the growth of the white-tailed sea eagle population in Schleswig-Holstein. The *column* indicates the final size of the population after 100 years after applying 1000 matrix model simulations of the years 1975–2006 with an initial population as of 2006 (95% confidence interval). The *dashed line* indicates the estimate of a maximum population for Schleswig-Holstein of 120 breeding pairs taking into account the maximum densities recorded so far as well as the different natural habitat conditions (Krüger et al. 2010)

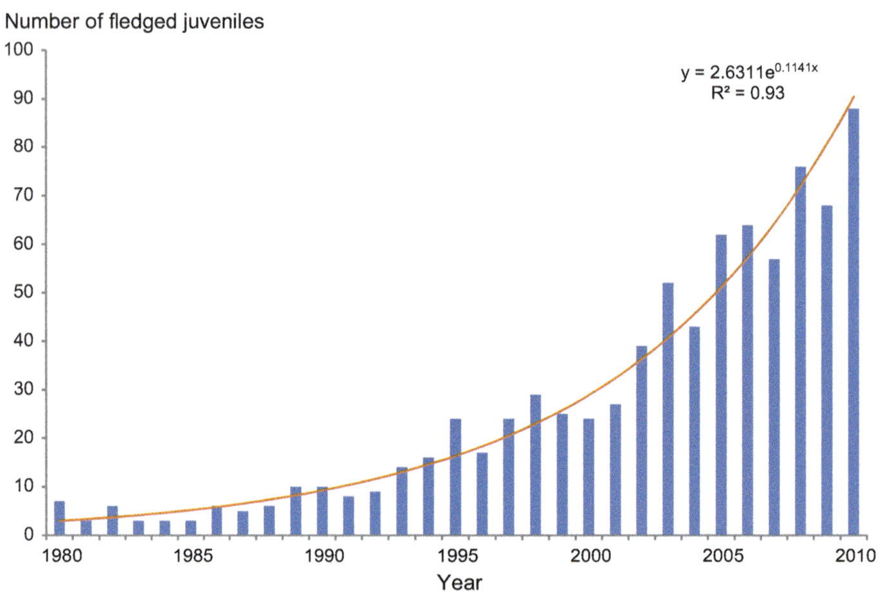

Fig. 10.2 Number of juvenile eagles fledged in Schleswig-Holstein between 1980 and 2010

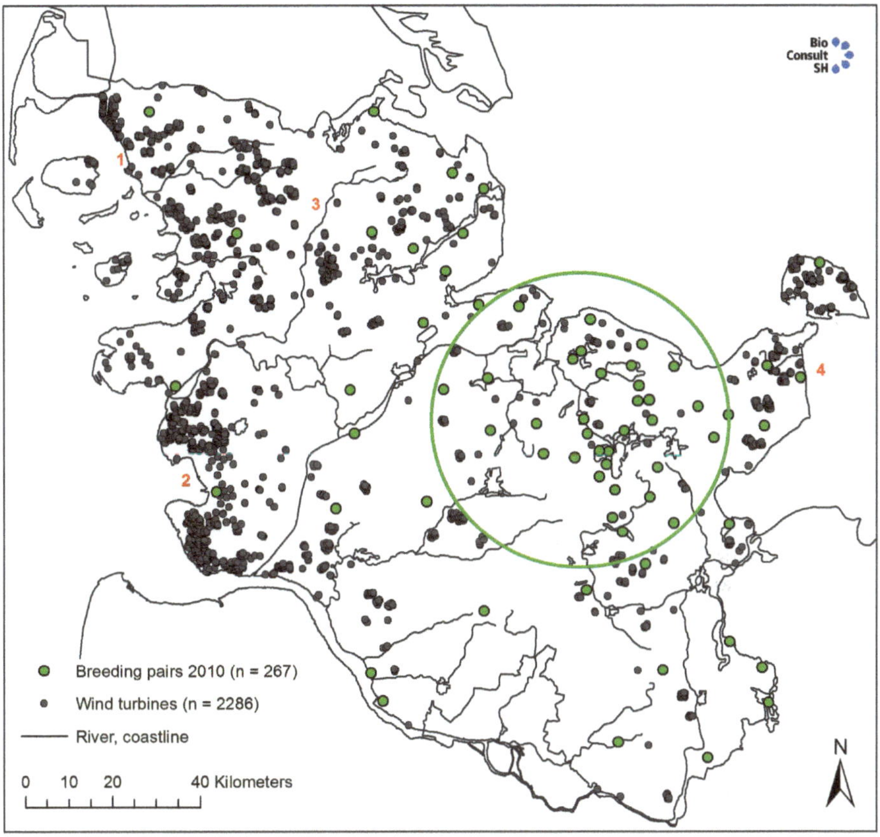

Fig. 10.3 Distribution of the breeding population of white-tailed sea eagles (*green dots*, n = 67 incl. main distribution area) and installed wind turbines (*grey dots*, n = 2286) main distribution in North Frisian (*1*) and Dithmarschen (*2*) marshland, the northern Geest Ridge (*3*) and the northern Ostholstein including Fehmarn (*4*)

10.2 Methods

Reference will be made to data collected by the Projektgruppe Seeadlerschutz Schleswig-Holstein e.V. on the breeding population, which was published in the annual reports on hunting and species protection by the Federal Agency for Nature Conservation (MELUR). The distance between breeding site and wind turbines was determined using the geographic information system (GIS) ArcView 3.2. Coordinates of the exact locations of breeding sites were made available by the Projektgruppe Seeadlerschutz Schleswig-Holstein e.V.

Identification of foraging flights at selected breeding sites involved significant participation of B. Struwe-Juhl (Projektgruppe) including earlier studies (Struwe-Juhl 1996, 2000) as well as more recent observations of nest supervisors.

In each case, the departure direction of the eagles was recorded. Using GIS, the next potentially suitable foraging water in flight direction and its distance were determined. Direction and length of the predicted foraging flights is represented in topographic maps including the location of wind turbines.

10.3 Results

According to the maps of the land surveying office of Schleswig-Holstein (ATKIS_Shape_Folie103_ObjArt_2327.dbf), 2286 wind turbines were installed in this federal state. They were mainly located in the North Frisian (1) and Dithmarschen marshland (2), the northern Geest Ridge (3) and the northern Ostholstein including Fehmarn (4). A general spatial separation of breeding sites of the white-tailed sea eagle and currently installed wind turbines is obvious (Fig. 10.3).

Within a distance of 3 km around the 67 nest sites of the year 2011, no wind turbines can be found in 51 cases (76%) and up to one in 59 (88%) (Fig. 10.4). Within a distance of 6 km around the nest site, no wind turbine can be found in 22 cases (33%) and up to one in 37 (55%) (Fig. 10.5). Near single white-tailed sea eagle breeding sites however, several wind turbines may be present.

In the eight breeding sites with the highest number of wind turbines within 6 km around the nest, observed foraging flights and those assumed to be foraging flights

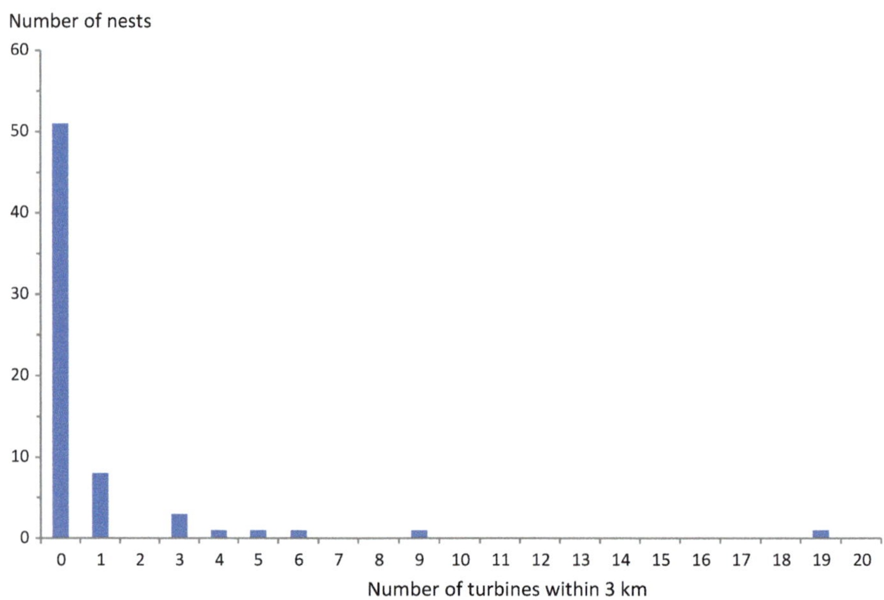

Fig. 10.4 Wind turbines at a distance of 3 km around white-tailed sea eagle nests

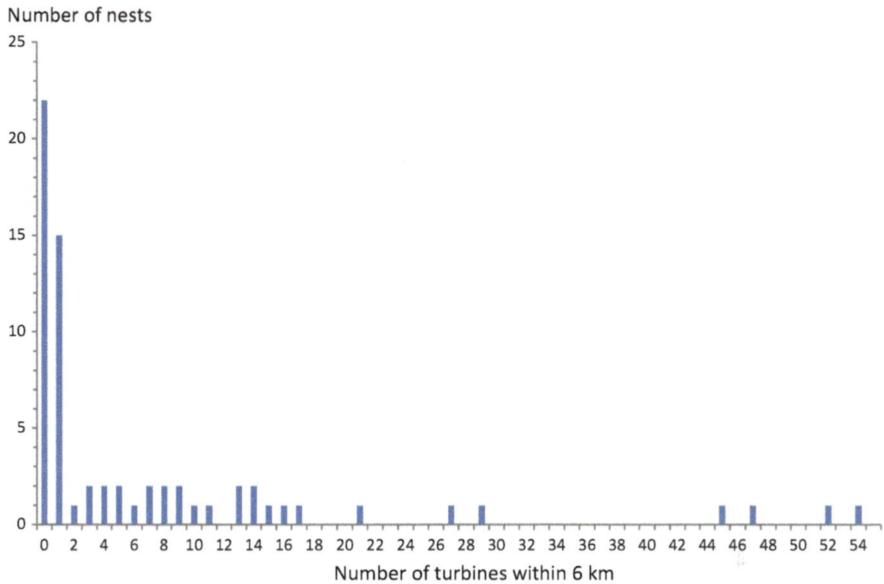

Fig. 10.5 Wind turbines at a distance of 6 km around white-tailed sea eagle nests

due to known foraging waters were evaluated in cooperation with the Projektgruppe Seeadlerschutz Schleswig-Holstein e.V. over the last years. The mean distance of the thereby determined 40 flight paths equals 5.4 km (Fig. 10.6). No collisions of adult birds, which could be linked to the breeding pair, became known for any of the above territories. The polar grid with all eight nest sites in the centre shows no preferred direction but an accumulation of foraging flights at a distance of up to 3 km and a maximum distance of 12 km (Fig. 10.7).

The cartographical representation of the distribution of birds found dead at wind turbines indicates a connection between the breeding distribution of the white-tailed sea eagle and the density of wind turbines in Schleswig-Holstein (Fig. 10.8). On the west coast with the relatively low number of breeding pairs, only three birds were found dead. In the districts Plön (PLÖ) and Ostholstein (OH) the results are clearly different. The highest number of collisions was found in the northern and southern part of Ostholstein (with a high density of wind turbines). This also coincides with the highest densities of white-tailed sea eagle breeding pairs. In the Plön district (with less wind turbines) only one collision victim became known (Fig. 10.8).

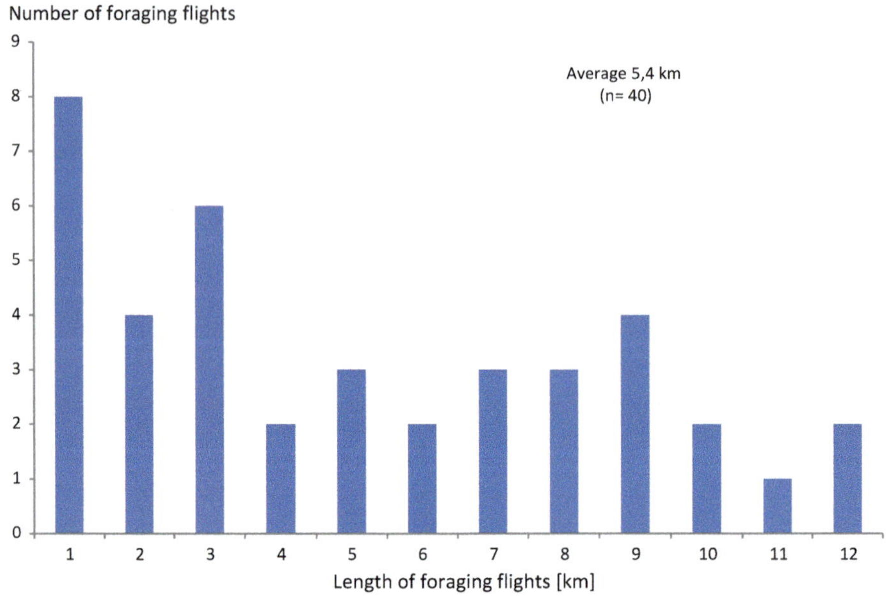

Fig. 10.6 Length of foraging flights (n = 40) for eight white-tailed sea eagle breeding territories (with the highest number of wind turbines within the home range of the breeding birds)

Fig. 10.7 Summarised presentation of distance and direction of foraging flights for eight white-tailed sea eagle breeding sites (n = 46)

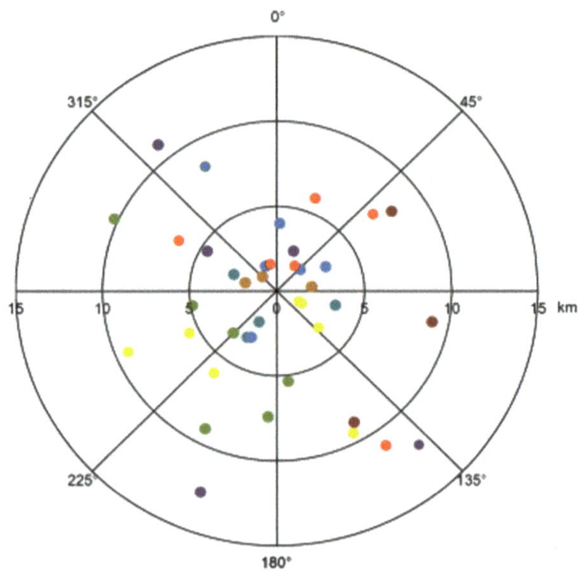

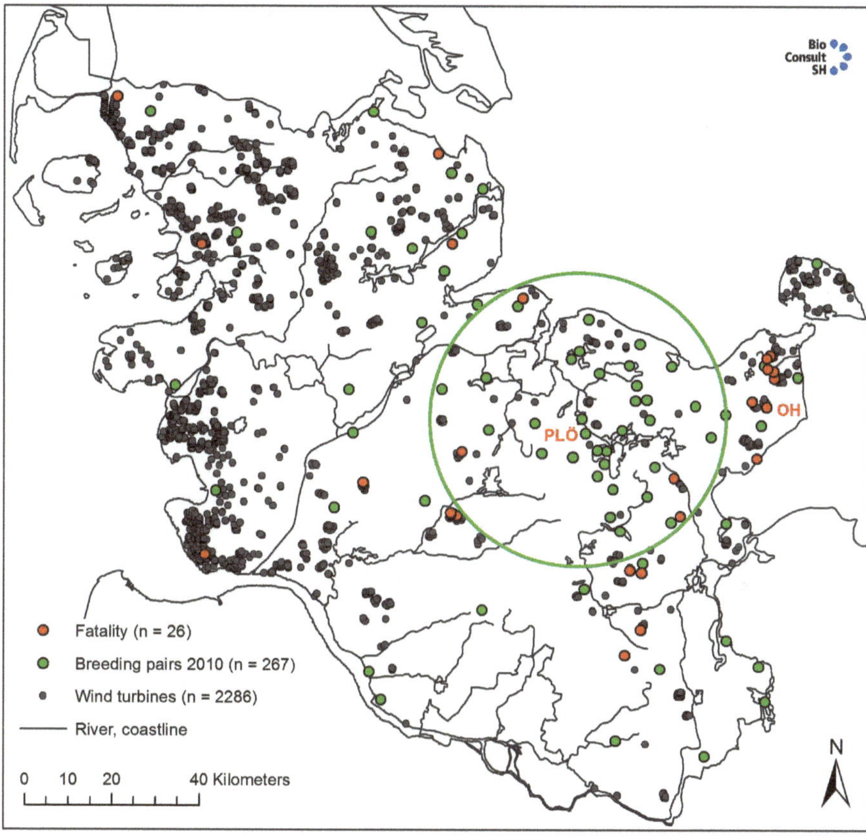

Fig. 10.8 Locations of white-tailed sea eagles found dead at wind turbines (*red*, n = 26), wind turbine sites (*grey*, n = 2.286) and breeding sites of the white-tailed sea eagle (*green*, n = 67 with concentration range, PLÖ = district of Plön, OH = district of Ostholstein)

10.4 Discussion

Natural distribution of the breeding population, home range and habitat use

Due to the natural structure of the landscape, the highest density of white-tailed sea eagle breeding pairs concentrates on the forest and lake landscape of the districts Plön and Ostholstein in Schleswig-Holstein. Documented by ring recoveries and individual ring identification, colonisation of western parts of the country (as well as Denmark and the Netherlands) started from these areas. The extension of the range of the white-tailed sea eagle population inevitably results in the occupation of suboptimal and pessimal habitats after the best habitats are taken. An increasing number of nests was found in very small woodland areas or even rows of trees, and the distribution of white-tailed sea eagles is increasingly less depending on the distribution of woodland and lakes in the country. With the extension of the breeding range and larger distances to lakes, the distance from foraging grounds has mostly increased. Increasingly new potential foraging grounds and sources of food like the small waters, ditches and fish ponds of agricultural areas and possibly site of wind farms as well are used increasingly (cf. Hauff 2010).

In the choice of a nest site, the number and distance from wind turbines is not a decisive factor. Several examples of establishing a territory despite the vicinity of wind turbines are known in Schleswig-Holstein (near Seegalendorf/OH a territory was established at a distance of 750 m from a large wind turbine in 2009).

According to studies carried out by Struwe-Juhl (1996) the home range of white-tailed sea eagles covers an area of 62 ± 35 km^2 in the breeding season. This equals a radius of 5.4 km around the nest. The longest hunting flight ranged up to 13 km. The mean size of a water body used for foraging by a pair is 812 ± 466 ha (Struwe-Juhl 1996). The identified flight distances show that the most important foraging grounds can be expected within an area of 5–6 km around the breeding site. This corresponds to the determined value of 5.4 km. For the evaluation of the impacts of wind farms on white-tailed sea eagles, the state authority for nature and the environment (LANU 2008) recommends examining a range of verification of 6 km around the nest site. In less favourable territories longer distances may be necessary to find food (see also Sect. 15.3.1, white-tailed sea eagle tagged near Schönwerder for this project).

Number and location of birds found dead

On the west coast of Schleswig-Holstein (North Frisia and Dithmarschen), less breeding pairs as well as non-breeding white-tailed sea eagles are present than in the eastern part of the state (Projektgruppe Seeadlerschutz SH, postal comm.). Due to the low density of white-tailed sea eagle breeding pairs, only three birds were found dead despite the large number of wind turbines. Although young birds and non-breeders are equally abundant in the districts Plön and Ostholstein and quite often accumulations of up to 20 (30) non-breeding eagles were observed, the number of birds found dead differs considerably between the two districts. Most collision victims were found in the northern and southern part of Ostholstein with a high density of wind turbines. In the Plön district a considerably lower number of wind turbines is installed and only one collision victim was found.

The areas of Ostholstein, where most of the collisions occur, lie within the migratory flyway between Scandinavia and Central Europe. White-tailed sea eagles migrating from Scandinavia will probably add to the fatalities of the breeding and non-breeding population in these areas.

Presumably for energetic reasons, foraging flights of white-tailed sea eagles concentrate on foraging waters within a distance of 3 km, although distances of up to 12 km were found (mean distance 5.4 km). There is no indication of an influence of wind turbines on the direction of the flight path. The birds appropriately use the randomly distributed foraging waters within a distance of 12 km around the breeding site. Apparently, wind farms do not cause large scale evasive movements like flying round or over wind turbines. A reduction of the collision risk can therefore only be achieved by excluding the construction of new wind turbines in the vicinity of known nest sites.

Barrier and displacement effects of wind turbines

During the breeding season, the white-tailed sea eagle is a species generally considered to be shy and sensitive to human activities (e.g. Glutz von Blotzheim and Bauer 1971; Looft and Busche 1981). In selecting a breeding site, white-tailed sea eagles usually avoid the proximity of human settlements. In Sweden, the distance between eagle nests and human settlements or industrial estates was not less than 1 km (Helander and Mizera 1997). According to Koivusaari et al. (1988) disturbance caused by human activities is one of the two main limiting factors for reproduction success in Finland. Folkestad (2003) determined a minimum distance of 500 m between established eagle nests and public roads and buildings. Breeding pairs nesting in the proximity of human activities apparently have a lower reproduction rate than those in remote, interference-free areas. This is especially apparent on the Island of Smøla/Norway hosting 65–70 white-eagle pairs, where the reproduction rate of breeding pairs within a radius of 1 km around human settlements/activities compared to those breeding further away is considerably lower. Considering these results, Folkestad (2004) concluded that white-tailed sea eagles have a limited ability to adapt to human disturbances near the breeding site.

Wind turbines as such may cause direct displacement effects and furthermore are associated with increased human activity in the construction and the operation phase, representing potential additional disturbance of breeding white-tailed sea eagles (e.g. Bright et al. 2006). Recent results for Smøla show that population density as well as reproduction rate of white-tailed sea eagles is reduced on the site of the wind farm as compared to sites outside the wind farm (Bevanger et al. 2009). This can also be attributed to disturbances caused by the construction of roads and maintenance works, impairing especially white-tailed sea eagles breeding on the ground. Due to these investigations, Folkestad (2004) estimates that wind turbines will cause a reduced reproduction rate of white-tailed sea eagles up to a distance of approximately 1 km around a turbine. Therefore, the distance of a wind turbine to the white-tailed sea eagle breeding site is of particular importance in the approval of new turbines. At present a range of verification of 6 km is taken into account.

Acknowledgements Bernd Struwe-Juhl (Projektgruppe Seeadlerschutz) made the individual-based evaluation in Schleswig-Holstein possible.

References

Bevanger K, Berntsen F, Clausen S, Dahl EL, Flagstad Ø, Follestad A, Halley D, Hanssen F, Hoel PL, Johnsen L, Kvaløy P, May R, Nygård T, Pedersen HC, Reitan O, Steinheim Y, Vang R (2009) Pre- and post-construction studies of conflicts between birds and wind turbines in coastal Norway (BirdWind). Progress Report 2009, NINA Report 505, Trondheim, pp 70

Bright JA, Langston RHW, Bullman R, Evans RJ, Gardner S, Pearce-Higgins J, Wilson E (2006) Bird Sensitivity Map to provide locational guidance for onshore wind farms in Scotland. RSPB Research Report No 20, Sandy, pp 136

Folkestad AO (2003) Nest site selection and reproduction in the white-tailed sea eagle in M¢re & Romsdal county, western Norway in relation to human activity. In: Helander B, Marquiss M, Bowerman W (eds) Sea Eagle 2000. Proceedings from an international conference at Björkö, Sweden, 13–17 September 2000. Swedish Society fori Nature Conservation, Stockholm

Folkestad AO (2004) Project Sea Eagle: Smøla Wind Park Phases 1 and 2: Population and reproduction monitoring 2004. Annual Report 2004 Nr. 248, NINA Norway, Trondheim

Glutz von Blotzheim UN, Bauer K, Bezzel E (1971) Handbuch der Vögel Mitteleuropas, vol 4, Falconiformes. Akademische Verlagsgesellschaft, Frankfurt/M, p 169–203

Hauff P (2010) Brutplätze von Seeadlern *Haliaeetus albicilla* in Deutschland auf Pappeln *Populus spec.* und Weiden *Salix spec.* Vogelwelt 130:67–76

Helander B, Mizera T (1997) *Haliaeetus albicilla* White-tailed Sea Eagle. In: Hagemeijer WJM, Blair MJ (eds) The EBCC Atlas of European Breeding Birds: Their Distribution and Abundance. T & AD Poyser, London

Koivusaari J, Nuuja I, Palokangas R (1988) White-tailed eagle nesting sites in jeopardy (summary in English). Suomen Luonto 47:13–17

Krüger O, Grünkorn T, Struwe-Juhl B (2010): The return of the white-tailed eagle (*Haliaeetus albicilla*) to northern Germany: modelling the past to predict the future. Biological Conservation 143 (2010) 710–721

LANU (Landesamt für Natur und Umwelt des Landes Schleswig-Holstein, 2008): Empfehlungen zur Berücksichtigung tierökologischer Belange bei Windenergieplanungen in Schleswig-Holstein. Flintbek.

Looft V, Busche G (1981) Vogelwelt Schleswig-Holsteins. Vol 2 Greifvögel. Ornithologische Arbeitsgemeinschaft für Schleswig-Holstein und Hamburg, Wachholtz-Verlag, Neumünster

Projektgruppe Seeadlerschutz Schleswig-Holstein (2007) Seeadler kollidiert mit Windernergieanlage im Dänischen Wohld. Großvogelschutz im Wald. Jahresbericht 2007: 20

Struwe-Juhl B (1996) Untersuchungen zur Habitatausstattung von Seeadler-Lebensräumen in Schleswig-Holstein. Abschlussbericht im Auftrag des Ministeriums für Ernährung, Landwirtschaft, Forsten und Fischerei Schleswig-Holstein, Kiel

Struwe-Juhl B (2000) Funkgestützte Synchronbeobachtung – eine geeignete Methode zur Bestimmung der Aktionsräume von Großvogelarten (*Ciconiidae, Haliaeetus*) in der Brutzeit. In: Stubbe M, Stubbe A (eds) Populationsökologie von Greifvogel- und Eulenarten. Bd. 4. Materialien des 4. Internationalen Symposium in Meisdorf/Harz 05.11.-08.11.1998. Martin-Luther-University of Halle/Wittenberg, p 101–110

Chapter 11
White-Tailed Sea Eagle—Summary and Conclusions

Oliver Krone, Gabriele Treu, Thomas Grünkorn and Georg Nehls

The aim of the project was to investigate to what extent white-tailed sea eagles fly near wind turbines, to estimate the collision risks and to make proposals on how to reduce collisions. To verify a spatial link between breeding population and collisions, adult and young birds were fitted with satellite transmitters, behavioural observations were carried out and available data was evaluated.

The telemetry data of an adult bird was used to determine the size of the home range, to analyse flight movements and to identify preferred structures. The size of the home range of this territorial bird comprises 160 km^2 (95% MCP) or 141 km^2 (95% fixed kernel). Compared to reported home ranges this relatively large home range size can be explained by unfavourable habitat conditions with only few water bodies and mainly cropland. Avoidance behaviour towards wind turbines could not be proven, as the number of locations within the wind farm was significantly higher than in areas of the same habitat type (cropland). This may be due to potential foraging flights and perching points. After nesting failure, the tagged adult bird used a foraging ground at a distance of 7 km, which was approached deliberately. There were no wind turbines present in the flight corridor. Using additional data from other projects it was possible to clearly show that the majority of locations (on average 71%) of tagged adult white-eagles are located within a distance of up to 3000 m around the nest in high quality habitas. In suboptimal habitats eagle fly much further as shown in our study (this book) and only 17% of the positions fall within the range of 3000 m.

The results of the telemetric monitoring of young birds after fledging from the nest showed that young eagles partly entered the risk area of wind turbines. Evaluation of the recorded data of further birds resulted in only few locations of

O. Krone (✉) · G. Treu
Leibniz Institute for Zoo and Wildlife Research (IZW),
Alfred-Kowalke-Str. 17, 10315 Berlin, Germany
e-mail: Krone@izw-berlin.de

T. Grünkorn · G. Nehls
BioConsult SH, Schobüller Str. 36, 25813 Husum, Germany

H. Hötker et al. (eds.), *Birds of Prey and Wind Farms*,
DOI 10.1007/978-3-319-53402-2_11

tagged young eagles in the vicinity of wind turbines. While searching for prey young eagles ranged widely in the first years, approaching wind turbines randomly as well as on purpose if these were located in the flight path or in potential habitats of prey animals.

Despite the effort involved, behavioural observations resulted in only few indications on the reasons why white-tailed sea eagles approach wind turbines (e.g. foraging flights). Observations did not show any clear avoidance behaviour of white-tailed sea eagles towards wind farms.

Of the 75 known cases of white-tailed sea eagle collisions with wind turbines (LUGV 2012) 38, where age and origin of the fatalities could be determined because the birds were ringed, were examined more closely in Schleswig-Holstein. Seven cases were identified as collisions of adults of a breeding pair from the vicinity. The evaluation of known collisions did not allow drawing conclusions on especially risky areas or distances from breeding sites. The majority of collisions of white-tailed sea eagles in Schleswig-Holstein accounted for juvenile and sub-adult white-tailed sea eagles with no direct association to a breeding site or foraging ground.

White-tailed sea eagles found dead under wind turbines got hit by the rotor blade. It can therefore be concluded that white-tailed sea eagles perceive the movement of the rotor too late and underestimate the speed at the circumference. They are at risk, if they do not show avoidance behaviour towards the structure. Recent investigations on the visual perception of birds of prey show that they are blind ahead when turning the head down in flight. This partial blindness in the direction of flight while focussing on the prey results in the white-tailed sea eagle not or too late noticing the danger represented by the rotor of a wind turbine.

Due to missing avoidance behaviour a reduction of the collision risk can therefore especially be achieved by excluding the construction of new wind turbines in the vicinity of known nest sites. To minimize the collision risk for white-tailed sea eagles it needs to be taken into consideration that certain structures on the site of a wind farm may attract the birds (perching points for foraging flights, grazing grounds for geese, kettle holes and ponds as habitat for waterbirds, etc.) and may lead to an increased risk. Furthermore, specific flight corridors of white-tailed sea eagles to foraging grounds further away need to be taken into account in the planning of wind turbines.

The results of the investigations lead to the following assessments and recommendations:

1. For wind farms within the distribution range of the white-tailed sea eagle there will always be a risk of collision with wind turbines.
2. The collision risk of a breeding bird is decreasing with increasing distance of the wind turbine from the nest site.
3. Depending on habitat conditions and quality white-tailed sea eagles may travel long distances on their way to foraging grounds.
4. In wind farm planning, sites at a radius of 6 km around known nest sites should as rule be checked for the presence of eagles (flight movements).

5. Location and structure of potential wind farm sites should be checked with respect to preferred flight paths of white-tailed sea eagles between breeding site and foraging grounds.
6. Areas with a high density of white-tailed sea eagles have to be considered adverse for the construction of wind farms in terms of collision risk. This should be taken into account in the choice of potential areas for wind energy use.

Reference

LUGV (2012) Informationen über Einflüsse der Windenergienutzung auf Vögel. http://www.lugv.brandenburg.de/media_fast/4055/vsw_dokwind_voegel.pdf

Chapter 12
Collisions of Birds of Prey with Wind Turbines—Analysis of the Circumstances

Leonid Rasran and Tobias Dürr

12.1 Introduction

Collisions with wind turbines are the most obvious negative effect of wind power development on birds of prey. From a nature conservation point of view it is therefore essential to investigate the circumstances leading to such events as accurately as possible, to develop strategies to minimize the risks.

Already the estimates of collision mortality rates per wind turbine are widely differing however (Percival 2005; Everaert 2003). One of the reasons is the difficulty to obtain reliable data for wind farm areas. Complete recording of collision fatalities requires regular and personnel intensive searches of the surrounding area of wind turbines, as well as possible additional surveys to account for carcasses removed by scavengers (Barrios and Rodriguez 2004; see also Strickland et al. 2007). This is only rarely possible (Möckel and Wiesner 2007) and cannot be met in practice on a large scale.

In the present sub-study, results of regular surveys of some of the German wind farms, carried out with differing periodicity and intensity over the past years, as well as available carcass survey data will be evaluated.

Most authors agree that local topographical features of the surrounding area of wind farms are decisive for the degree of impairment of birds (Madders and Whitfield 2006). Determination of the location of a collision as well as recording of

L. Rasran (✉)
NABU (Nature and Biodiversity Conservation Union) Michael-Otto-Institute, Goosstroot 1, 24861 Bergenhusen, Germany
e-mail: Leonid.Rasran@boku.ac.at

T. Dürr
Staatliche Vogelschutzwarte (State Bird Observatory and Field Centre) of Landesamt Für Umwelt, Gesundheit Und Verbraucherschutz Brandenburg (Brandenburg Environmental Agency), Buckower Dorfstr. 34, 14715 Nennhausen-Buckow, Germany

H. Hötker et al. (eds.), *Birds of Prey and Wind Farms*,
DOI 10.1007/978-3-319-53402-2_12

characteristics of the surrounding area (proximity to forests or water bodies, etc.) are important factors in collision risk assessment.

Potential correlations between collision risks and technical specifications of wind turbines are especially important in estimates of bird of prey mortality in the context of a replacement of many smaller wind turbines by a few high performance turbines in repowering, already proposed or approved for many sites (Smallwood and Neher 2004; Hötker et al. 2006).

Weather conditions, especially the wind speed, as a further factor influencing the risk of collision of birds of prey, have been insufficiently investigated so far (but see Barrios and Rodriguez 2004). Wind speed was assessed and tested for significance in the present study.

Based on the findings of this study, some conclusions will be presented, which may be relevant to minimise the risk in the planning and construction phase of wind turbines.

12.2 Methods

12.2.1 Material

Since 2002, reported avian collisions with wind turbines as well as birds found dead near wind farms, which may have collided, are recorded in a central German database by the State Bird Observatory and Field Centre of the Brandenburg Environmental Agency (T. Dürr). A total of 730 fatalities reported between 1989 and end of 2008 were included in the present study. Reports originate from systematic studies of differing duration with regular monitoring and observations in selected wind farms, as well as from attentive citizens and service personnel of wind farms (random finds), ornithologists and bat experts (targeted search) but also press releases and professional publications. Data are therefore rather heterogeneous and of differing quality levels. Reports became more frequent towards the end of the study period due to an increasing number of wind turbines, but also because the central database became more widely known, receiving an increasing number of reports. The actual number of reports included in the analysis was considerably smaller, as only birds of prey and a few further large bird species were in the focus and only data with additional information required for the respective analysis could be used.

Information on monitoring activities at different wind farms going beyond own investigations was compiled by the State Bird Observatory Brandenburg from different sources of literature, diploma theses and expert reports. These data are equally heterogeneous due to the differing duration of surveys, different methods and various observers. For this reason and because some reports of collision victims are based on systematic searches whereas others are not, wind farms as a whole and single collision events will be looked at separately.

12.2.2 Evaluation

12.2.2.1 Preliminary remarks

In a first step, attempts were made to supplement information on wind turbines, where birds were found dead, on the surrounding area and on the circumstances of the discovery. Information on the wind turbine model, technical specifications (height, nominal capacity) and the size of the wind farm are based on the reports of fatalities and on the operator database (Keiler 2008). The circumstances of collisions (for events with known time, meteorological data of the Deutscher Wetterdienst were included), as well as data on the exact location of the collision (if map or sketch was available to allocate the collision victim to a specific turbine) were also added.

For statistical analyses, different reference groups of wind turbines were used depending on the analysis. Data used for the reference group was likewise based on the operator database. By randomly selecting turbines from the database, a representative sample of technical specifications was created and used in comparisons of the surrounding area of wind turbines.

Land use in the surrounding area (within a radius of 5 or 1 km) of wind turbines was compiled using the CORINE Land Cover data (CLC2000). Land use within a radius of 5 km around the wind turbines was grouped into the following categories: residential areas, roads and infrastructure, dumps and mining, farmland, forest, grassland, water bodies and other. For the smaller radius of 1 km around wind turbines, the categories farmland, grassland and forest were further investigated only. Distances of wind turbines from ecologically relevant structures and objects (water bodies, forests, smaller woodland areas, buildings and roads) were determined using the Google Earth Teleatlas (Geocontent 2009).

12.2.2.2 Wind farms

In the statistical analysis of the number of birds found dead per wind farm, only wind farms with a least ten searches were included. The total number of collisions victims of relevant species found in all searches served as a dependent variable. Influences of single parameters on the number of collision victims was investigated in stepwise backward multiple regressions. Apart from the number of searches and the size of the wind farm, information on the surrounding area, i.e. the distances from forests, roads, residential areas and water bodies (measured from the centre of the respective wind farm) and the proportions of different types of land use (arable land, grassland, forest, etc.) within a radius of 5 km were included in the extended models. Due to land use correlation, only one type of land use was included per model.

12.2.2.3 Collision events

Differences between locations of collisions (exact to the wind turbine) and randomly selected wind turbines was analysed in analyses of variance (MANOVA, one-way ANOVA). Technical specifications (tower height, rotor diameter, nominal capacity), the position of the wind turbine within the wind farm (for categories see results, positions of turbines within the wind farm, wind farm size) and characteristics of the surrounding area (distances from roads, buildings or other structures, proportion of land uses within a radius of 5 or 1 km) were looked at. As land uses are correlated, only one type of land use was included per model.

To be able to investigate the relative importance of single parameters, wind turbines with reported collisions of relevant bird species were compared to a reference group of wind turbines. This reference group consisted of a random selection of all turbines with no reported fatality so far. This approach needs to be seen critically, but is necessary for evaluation in regression analyses. Subsequently, a stepwise backward multiple regression was applied.

The statistical analyses (analyses of variance, multiple regressions) were carried out using software R 2.8.1 (R development core team 2008). The principal component analysis war carried out using software PAST (Hammer and Harper 2005).

12.3 Results

12.3.1 Number of Collision Fatalities

By far the most frequently killed bird of prey species were red kite and common buzzard (Fig. 12.1). Further bird of prey species such as white-tailed sea eagle, common kestrel and black kite were disproportionately often reported as collision victims as well.

In addition to the above mentioned species, further bird of prey species (Eurasian hobby, Eurasian sparrowhawk, etc.) as well as a number of large bird species relevant for nature conservation issues, such as white stork, crane and the Eurasian eagle-owl, were included, which, together with birds of prey, made up for about 40% of all collision victims recorded during the study period between 1989 and 2008.

Table 12.1 gives an overview of the investigated species and the respective number of fatalities as well as the conservation status in Germany.

Most reports registered in the data base are birds found dead. Only about 10% of the birds survived the collision. Most of these had to be euthanized due to the serious injuries; some were set free after intensive care. In almost all cases the injuries due to collisions with wind turbines were clearly visible and involved massive bone fractures. Quite often, single body parts were completely detached. Birds of prey without external injuries after a collision were very rare. This

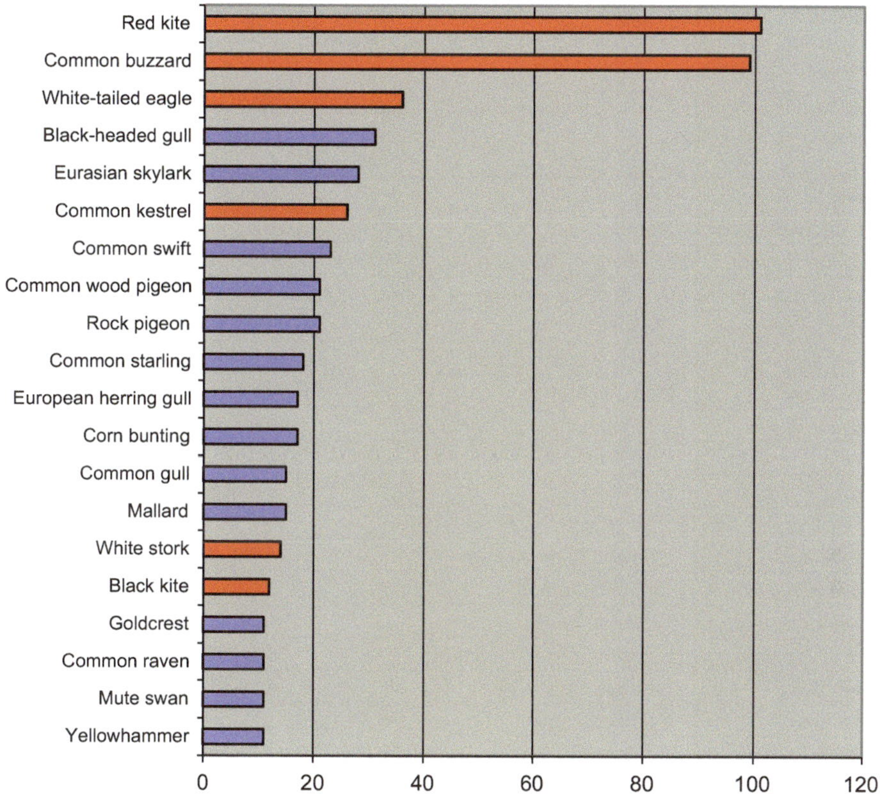

Fig. 12.1 Number of collision victims of the most frequently reported species in the central database of the State Bird Observatory Brandenburg (as from end of 2008). Species investigated in this study are highlighted

indicates that large birds are above all killed by direct contact with rotor blades, whereas mortality due to barotrauma (lung injuries due to differences in pressure caused by moving rotor blades; Baerwald et al. 2008), shock (triggered by turbulences in the rotor zone) or collisions with the tower is most common in bats and small birds.

12.3.2 Age Structure of Collision Victims

Most collision victims are adult birds. 70% of the victims with known age of the three most frequent species (red kite, white-tailed sea eagle and common buzzard) are mature birds (Fig. 12.2). In late summer and autumn, some juvenile birds were found among the collision victims at a too small proportion to carry out age-specific evaluations.

Table 12.1 Number of collision victims (central database of the State Bird Observatory Brandenburg, as from 2008) in Germany, breeding population and Red List status of species covered in this study

Species		Number of collision victims	Breeding population	RL status
Falco subbuteo	Eurasian hobby	4	2600–3400	3
Pandion haliaetus	Osprey	1	500	3
Accipiter gentilis	Northern goshawk	3	11,000–13,000	*
Buteo buteo	Common buzzard	99	77,000–110,000	*
Falco columbarius	Merlin	2	nb	–
Buteo lagopus	Rough-legged buzzard	1	nb	–
Circus aeruginosus	Marsh harrier	4	5900–7900	*
Milvus milvus	Red kite	101	10,000–14,000	*
Aquila pomarina	Lesser spotted eagle	1	100	1
Milvus migrans	Black kite	12	5000–7500	*
Haliaeetus albicilla	White-tailed sea eagle	36	500	*
Accipiter nisus	Eurasian sparrowhawk	6	15,000–21,000	*
Falco tinnunculus	Common kestrel	26	43,000–65,000	*
Falco peregrinus	Peregrine falcon	1	810–840	*
Pernis apivorus	European honey buzzard	1	3800–5000	V
Bubo bubo	Eurasian eagle-owl	8	1400–1500	*
Grus grus	Common crane	2	5200–5400	*
Ciconia ciconia	White stork	14	4200–4300	3
Ciconia nigra	Black stork	1	500–530	*
Larus ridibundus	Black-headed gull	31	140,000–150,000	*
Alauda arvensis	Eurasian skylark	28	2,100,000–3,200,000	3

Information of rarity and Red List status following Südbeck et al. (2007) (*nb* species not breeding in Germany). The two most common collision victim species other than birds of prey (black-headed gull and skylark) are given for comparison with target species of the present study
* species not listed on the RL

12.3.3 Seasonal Distribution of Collision Fatality Records

Most collision victims were recorded in spring (March to May) and late summer (August and September). Only few fatalities were recorded in winter (Fig. 12.3), when red kite are usually wintering in areas outside Germany. Additionally, searches for collision victims were limited in comparison to the rest of the year. Between May and July high growing and dense vegetation will have reduced the probability to find collision victims.

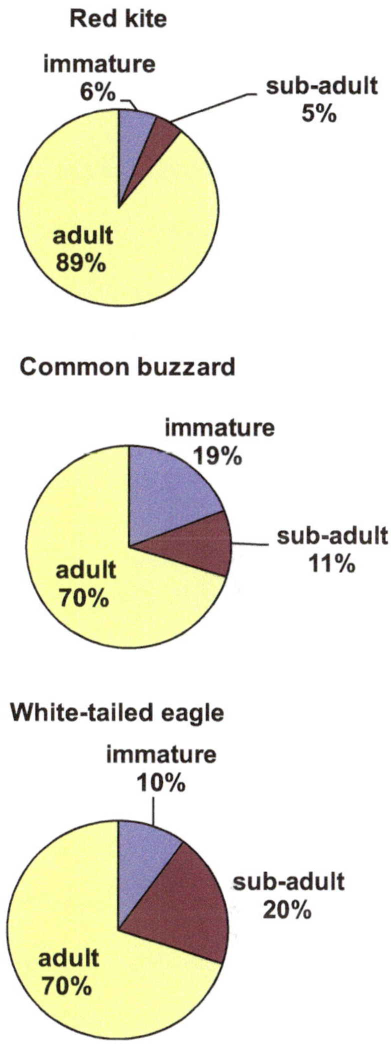

Fig. 12.2 Age structure of collision victims of the three most frequent species. Immature birds are fully fledged young birds of the respective calendar year, sub-adult birds are birds older than a year but without full adult plumage

12.3.4 Influence of Wind Speed on Collisions

Weather conditions, and especially the wind speed, have an influence on the flight behaviour of birds of prey as well as on the operation of wind turbines. For collisions, where the time of collision was known exact to the day, the maximum wind speeds (m/s) were taken from the website of the Deutscher Wetterdienst (www.dwd.de). The nearest meteorological station from the wind turbine site (maximum distance: 100 km) was used. The cut-in wind speed of most wind turbines is between 2 and 3 m/s, required to produce a measurable power output. Full power is reached at wind speeds between 10 and 15 m/s. Typically, wind turbines

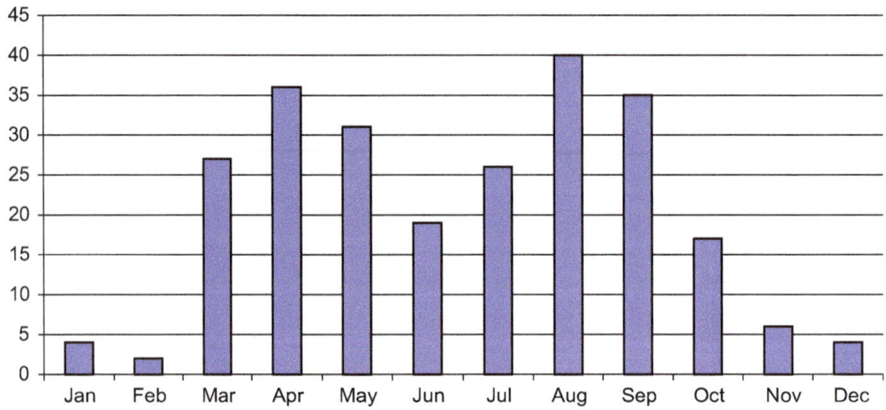

Fig. 12.3 Seasonal distribution of recorded collisions of birds of prey and other large bird species with wind turbines (n = 247)

Fig. 12.4 Relative frequency of different daily wind speed maxima for all days of a year (meteorological station Potsdam, n = 3311; *blue columns*) and days with known collisions (n = 100; *red columns*)

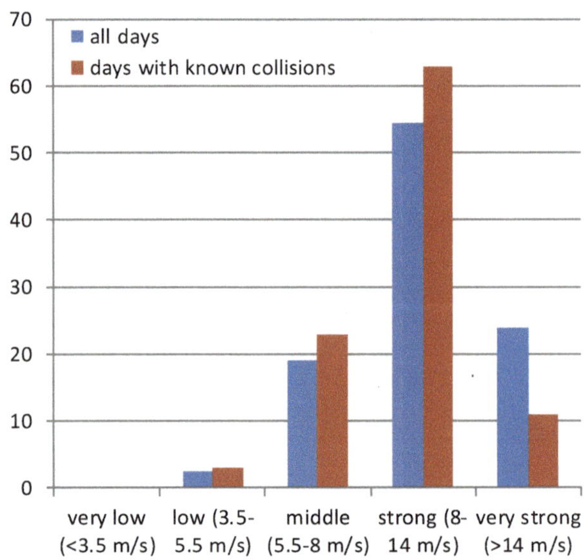

shut down at a wind speeds of more than 20 m/s. All but one of the analysed collision events with known date (n = 100) occurred at wind speeds between 3.5 and 20 m/s. For a typical inland weather station in Germany (e.g. Potsdam, used a reference station) most of the daily maximum values of the past ten years fall within this range.

A number of days with higher wind speeds could be found in regular weather records, which did not occur on days with collision events (Fig. 12.4). On average, wind speeds during collision events were slightly lower than under general conditions in the area ($F = 9.67$; $p < 0.01$). Classification of wind speeds (3.5–5.5 m/s—light, 5.5–8 m/s—moderate, 8–14 m/s—strong and >14 m/s—very strong; see also Barrios and Rodriguez 2004) shows that 0.035 of the collisions per day occur at wind speeds of the first three categories when looking at the number of collisions in relation to the total number of days with a particular wind speed. For higher wind speeds the number of collisions drops to 0.014/day. Therefore wind speeds play a minor role for collision risks under normal operating conditions. As expected, the collision risk declines in very low or high wind speeds, as rotors are shut down under these conditions and flight activity of the potentially affected species is considerably lower (in strong winds).

12.3.5 *Influences on the Probability of Collisions in Regularly Searched Wind Farms*

Only few long-term systematic studies on the number of collision victims accounting for carcasses removed by scavengers as well as search efficiency were carried out in Germany so far (e.g. see Sect. 12.2). Still, the database of the State Bird Observatory Brandenburg (Dürr, unpublished) contained information on about 12,000 partly regular searches of 454 wind turbines in Brandenburg at the time of the present analyses. This number and the number of birds found dead resulted in an average search success rate of 0.018 birds per turbine and search (an average of 18 birds after 1000 searches). The relevant bird of prey and other large bird species made up a proportion of about 40% of all victims. The actual mortality rate will presumably be two to three times higher, the removal of carcasses by scavengers being a major factor. In addition observers faced difficulties to detect collision victims in summer in areas with high growing vegetation (set-aside areas, crops; Möckel and Wiesner 2007).

It can be assumed that a large proportion of collision victims of large and intensively monitored species as white stork and white-tailed sea eagle is found and reported, but for smaller and/or more abundant species such as Eurasian kestrel, Eurasian hobby, common buzzard and Montagu's harrier, the number of collision victims which is not found will be many times higher than the number of recorded victims.

40 searched wind farms with at least ten searches each (24 with and 16 without recorded collision victims of the target species) were included in further analyses. A stepwise backward multiple regression returning a high explanatory value (corr. $R^2 = 0.7336$; $F = 15.46^{***}$) showed that the number of collision victims depends on the following parameter (in order of significance):

- Number of searches (more searches—more birds found dead): $t = 5.965$, $p < 0.0001$
- Distance between wind turbine and next residential area (larger distance—more birds found dead): $t = 3.817$, $p < 0.01$

- Distance between wind turbine and next forest (larger distance—more birds found dead): t = 2.695, $p < 0.05$
- Proportion of residential area within a radius of 5 km (larger area—more birds found dead): t = 1.879, $p = 0.078$

Of all types of land use in the surrounding area of wind farms, models including residential areas returned the highest explanatory value, as this parameter had an almost significant influence, in contrast to models including farmland or grassland. As different types of land use made up a varying proportion of the total area and were therefore correlated, use of all types of land use in a single regression model was not possible and the different types of land use had to be tested in separate models.

Influences of further tested parameters (distance from roads, distance from water bodies and total wind farm area) were negligible.

12.3.6 Influence of Wind Turbine Specifications

The vast majority of wind turbines currently in operation in Germany are horizontal turbines with three-bladed rotors. Most of the manufacturers use massive steel or reinforced concrete towers. Only a few turbines (Fuhrländer) have lattice towers. Further differences between manufacturers consist of minor external design features such as shape of the nacelle or the coloured design of tower or rotor blades. This kind of difference seems to be negligible in the determination of the collision risk for birds of prey or large bird species. The distribution of collision victims per manufacturer largely corresponds to the overall German market share of the respective manufacturer (correlation: R = 0.97). A particular risk arising from a certain wind turbine model can therefore be excluded (see Fig. 12.5).

Nominal capacity and rotor diameter are strongly correlated and are decisive for the space requirement and the effect distance of the wind turbine. The wind turbine operator database includes turbines from 1 kW or a rotor diameter of 3 m. The largest turbines so far are 6 MW turbines with a rotor diameter of 126 m. The majority of wind turbines installed in Germany varied between 0.5 and 2 MW (rotor diameter: 40–70 m). In the comparison of the locations of collisions to a random sample of similar size, a significantly larger proportion of birds was found dead at larger wind turbines (one-way ANOVA, F = 8.96; $p < 0.01$ for nominal capacity, Fig. 12.6; F = 9.87; $p < 0.01$ for rotor diameter, Fig. 12.7).

In Germany, no collisions of birds of prey were reported for smaller turbines (below 200 kW) so far. For medium-sized wind turbines, reported fatality rates of birds of prey and other large bird species vary between 0.01 and 0.02 per turbine and year for all size classes.

Looking at regularly searched wind farms while relating the number of reported collisions to a power unit (MW), larger wind turbines returned better results than smaller turbines (Fig. 12.8).

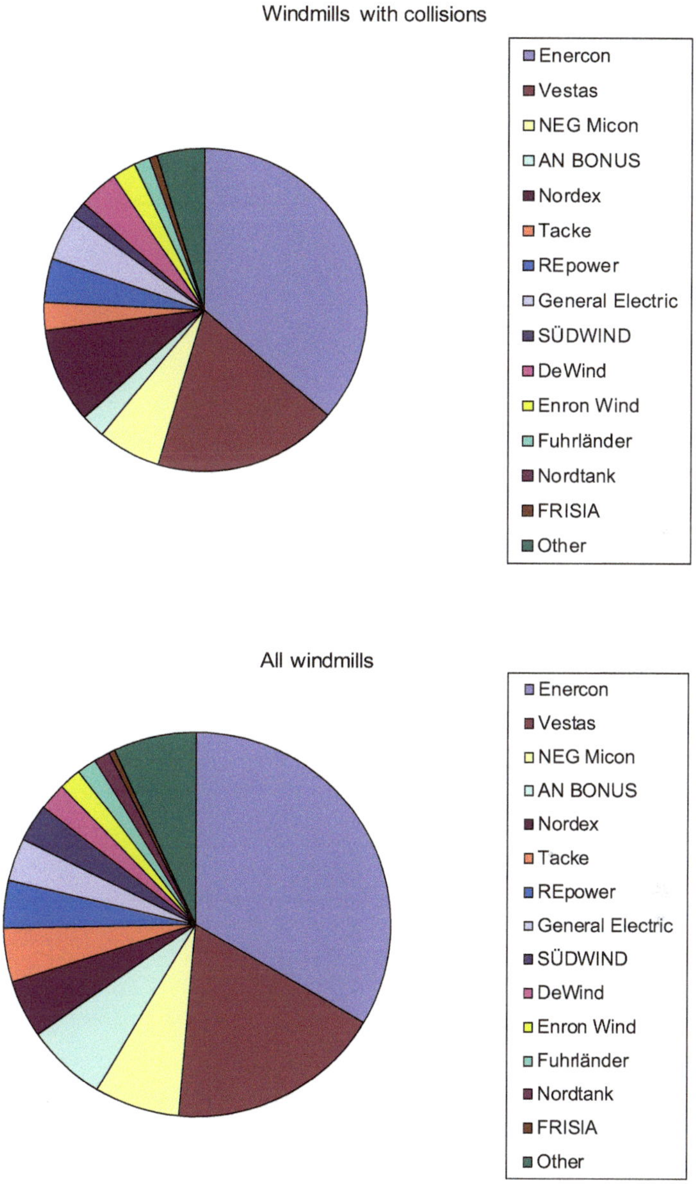

Fig. 12.5 Share of different models of the total number of wind turbines in Germany (n = 19,733) in comparison to wind turbine models with known collisions of birds of prey and further large bird species (n = 251)

The parameter "tower height" (hub height) is less influenced by the capacity of the wind turbine than by the rotor diameter, although it is correlated to the power output. A comparison of hub heights within a random sample of all wind turbines

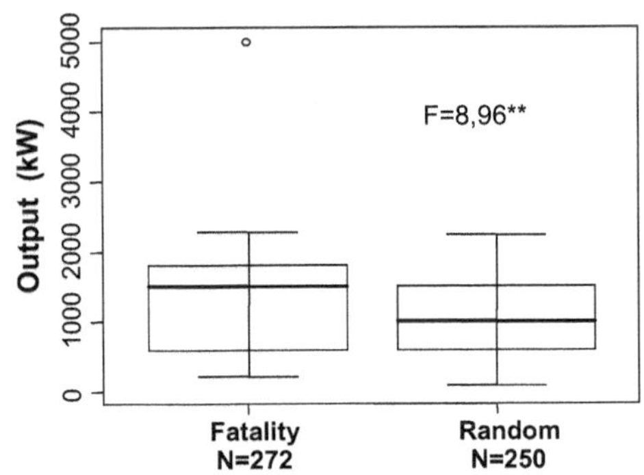

Fig. 12.6 Average nominal capacity of wind turbines. Comparison of wind turbines with reported collisions and randomly selected wind turbines (*horizontal centre line* mean, *boxes* standard error, *whisker* standard deviation)

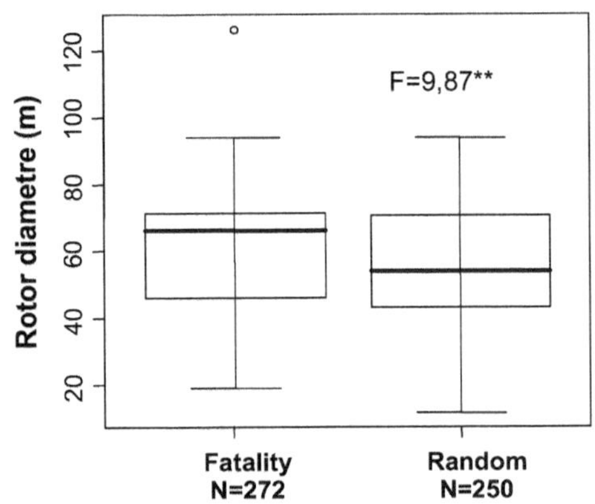

Fig. 12.7 Rotor diameters of wind turbines. Comparison of wind turbines with reported collisions and randomly selected wind turbines (*horizontal centre line* mean, *boxes* standard error, *whisker* standard deviation)

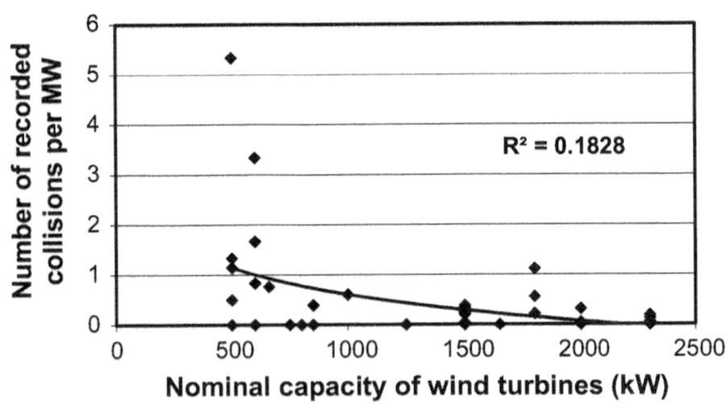

Fig. 12.8 Relation between the capacity of wind turbines and the number of recorded collisions per MW. Regularly searched wind farms consisting of wind turbines of the same nominal capacity were included only (n = 40)

installed in Germany to hub heights at locations with known collisions resulted in a significantly higher collision risk for wind turbines with large hub heights (one-way ANOVA; $F = 20.02$; $p < 0.0001$; Fig. 12.9).

In this case, the increasing space requirement of wind turbines with higher towers will be the decisive factor for the larger number of collision victims even to

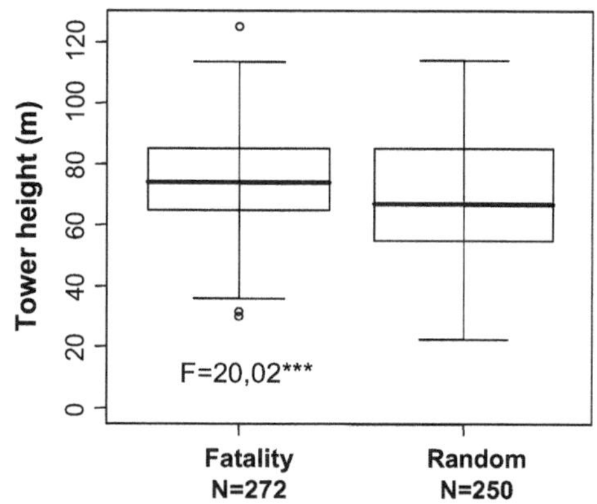

Fig. 12.9 Comparison of wind turbine hub heights with reported collisions and randomly selected wind turbines (*horizontal centre line* mean, *boxes* standard error, *whisker* standard deviation)

a larger extent than for the rotor diameter. Corresponding to the results for nominal capacity, the lack of information on collisions of birds of prey with smaller wind turbines relativizes the statistical power of the results.

12.3.7 Position of Wind Turbines Within the Wind Farm and Wind Farm Size

One of the parameters influencing the probability that a bird collides with a particular wind turbine is, among others, the exposure of the wind turbine in relation to neighbouring turbines, i.e. its position within the wind farm. As this is hard to quantify, a classification based on the number of neighbouring wind turbines was used (see Fig. 12.10).

Figure 12.10 presents a comparison of frequencies of relative positions of turbines with collisions and wind turbines randomly selected from a total of 400 wind turbines in the northern Brandenburg Prignitz and Ostprignitz-Ruppin districts within a wind farm. These two districts were chosen, because many bird of prey collisions were observed in neighbouring areas in Brandenburg, Saxony-Anhalt and Mecklenburg Western-Pomerania, and density and distribution of wind farms is typical for the German inland. All structures from single turbines to linear and areal wind farms could be found within these two districts, respectively.

10% of the collision victims were killed at single turbines, which made up a proportion of only just below 3% of all turbines within the area. Wind turbines with only few neighbouring turbines are likewise overrepresented in reported collisions in relation to the total number of turbines. The distribution of collisions within a wind farm is described in the following using the regularly searched Zitz-Warchau wind farm as an example (7225 searches, Fig. 12.11).

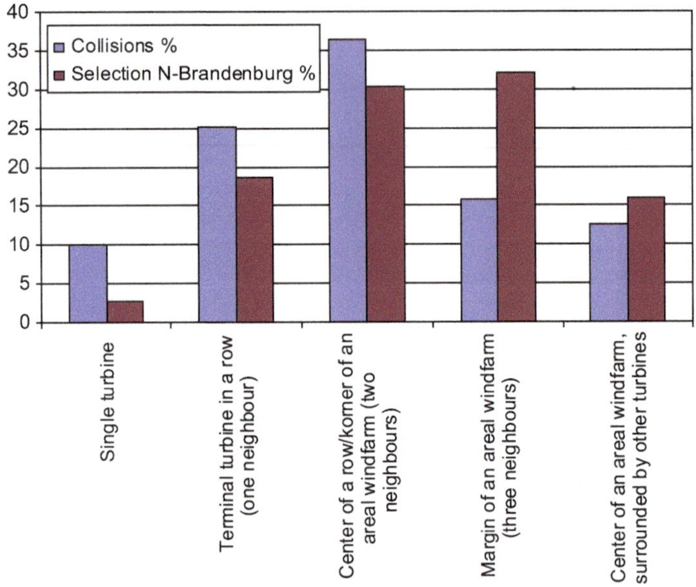

Fig. 12.10 Distribution of locations or positions of wind turbines with recorded collisions in comparison to all turbines of the northern Brandenburg districts within wind farms

Fig. 12.11 Spatial distribution of wind turbines and collision victims found within the Zitz-Warchau wind farm

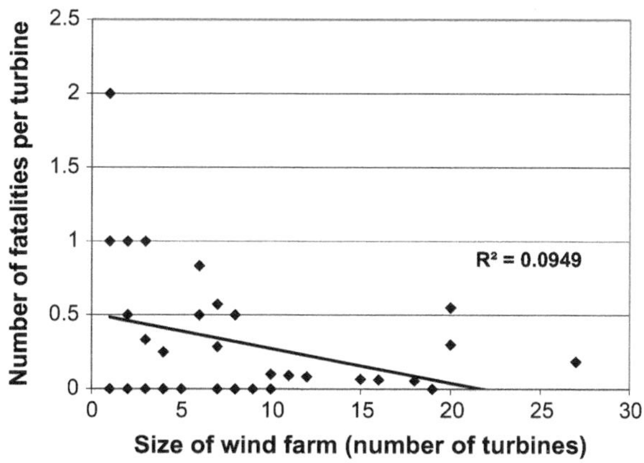

Fig. 12.12 Relation between wind farm size (number of turbines) and number of recorded bird of prey fatalities per turbine. Only wind farms with regular searches were included, also taking account of mixed wind farms, i.e. wind farms with different but spatially connected types and sizes of wind turbines (n = 65)

Eight of the 12 collision victims collided with corner wind turbines (wind turbines 1, 12, 13, 16 and 20), two with middle turbines of the outmost rows (3 and 10) and only two with turbines in the centre of the wind farm (wind turbine 7).

In regularly searched wind farms, the number of birds found dead per turbine was inversely proportional to the size of the wind farm (Fig. 12.12).

12.3.8 Surrounding Area of Wind Farms

The number of collisions depends to a considerable extent on the attractiveness of the site and the surrounding area of a wind farm for birds of prey and the frequency of approaches of wind turbines. Assuming that locations with known collisions present a higher risk of collision than randomly selected wind turbines, landscape structures and characteristics reflecting this difference were looked for.

Land use within a radius of 5 km around the wind turbines was grouped into the following categories: residential areas, roads and infrastructure, dumps and mining, arable land, forest, grassland, water bodies and other. For the smaller radius of 1 km around wind turbines, the categories arable land, grassland and forest were further investigated only (Table 12.2).

Arable land made up the largest proportion in the immediate and more distant surroundings of wind turbines. The proportion of arable land was even significantly larger around wind turbines with collisions than around randomly selected turbines. Wind turbines with a larger proportion of grassland in the immediate vicinity were less risky for birds of prey. The proportion of woodland areas in the surrounding area had no significant influence on the probability of a collision.

The collision risk of single target species was similar with respect to the surroundings of wind turbines (Fig. 12.13).

Buildings, roads, closed forests and water bodies as well as their minimum distance from wind turbines were included as relevant landscape structures. In

Table 12.2 Results of the comparisons between locations of wind turbines with collisions (C; n = 151) and randomly selected wind turbines (R; n = 186) in terms of land use within a radius of 5 km and 1 km and distance from distinctive landscape structures (one-way ANOVA)

Land use (% of the area)	F	*p*	
5 km radius			
Settlements, urban areas	0.01	ns	
Streets, roads	2.83	ns	
Mining areas, waste heaps	0.08	ns	
Arable fields	40.24	***	C > R
Forests	0.31	ns	
Meadows, pastures, heathlands	35.35	***	C < R
Water areas	5.75	*	C < R
Miscellaneous	69.78	***	C < R
1 km radius			
Arable fields	39.14	***	C > R
Forests	0.93	ns	
Meadows, pastures, heathlands	31.2	***	C < R
Distances to structures			
Next grove	0.03	ns	
Next forest	2.13	ns	
Next building	26.62	***	C > R
Next road #	17.29	***	C > R
Next water body (lake, pond, river)	1.76	ns	

Significance levels: * = $p < 0.05$; *** = $p < 0.001$; n. s. = not significant. #—service roads within a wind farm are not considered

addition, the closest woodland area as a potential resting and breeding site for birds of prey was taken into consideration. Avenue trees, hedgerows, copses and similar structures also belong to this category. The distance of wind turbines from the next landscape structure varied between a few metres and more than 0.5 km.

The influence of open woodland structures and forests (potential resting and breeding sites) on the probability of a collision of birds of prey with wind turbines was comparably small. No significant effects could be proved in the comparison to randomly selected wind turbines.

Just as little influence was found for the proximity of water bodies. Paved roads (service roads within wind farms and agricultural roads were excluded) as well as buildings in the vicinity of wind turbines seemed to reduce the collision risk. The distance from these structures was significantly larger for wind turbines with known collisions than for randomly selected wind turbines. The number of collisions depends to a considerable extent on the attractiveness of the site and the surrounding area of a wind farm for birds of prey and the frequency of approaches of wind turbines. Assuming that locations with known collisions present a higher risk of collision than randomly selected wind turbines, landscape structures and characteristics reflecting this difference were looked for.

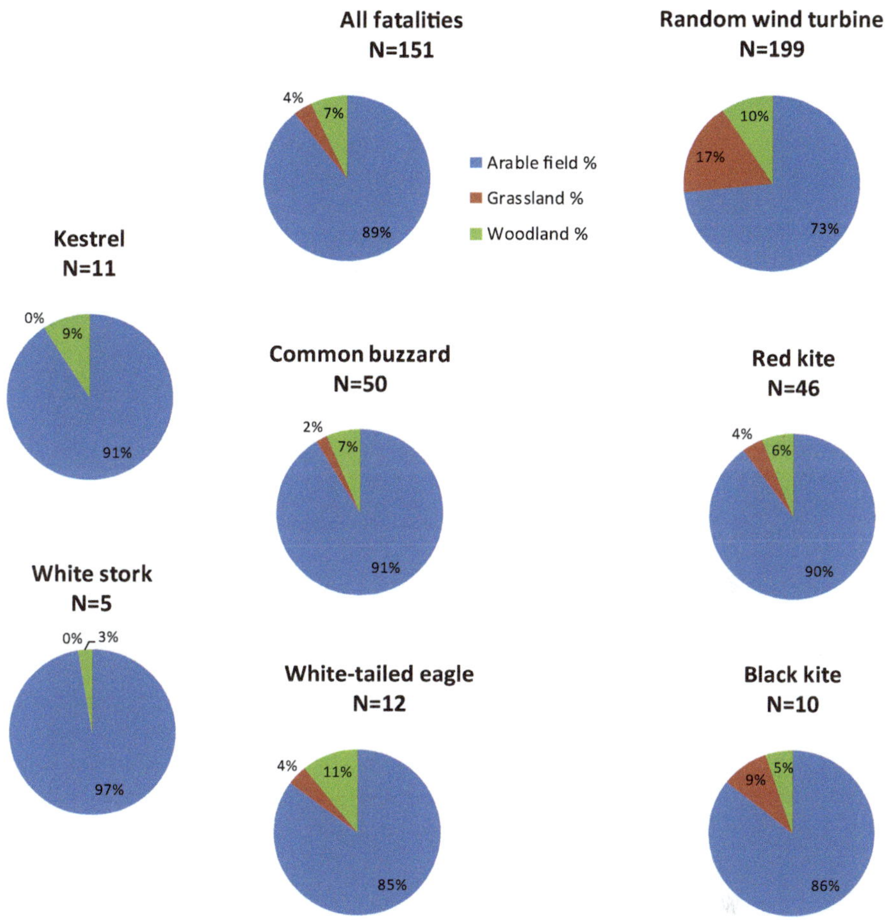

Fig. 12.13 Land use within a radius of 1 km around wind turbines with reported collisions of target species. A random selection of 199 wind turbines from all over Germany was used as a reference

Land use within a radius of 5 km around the wind turbines was grouped into the following categories: residential areas, roads and infrastructure, dumps and mining, arable land, forest, grassland, water bodies and other. For the smaller radius of 1 km around wind turbines, the categories farmland, grassland and forest were further investigated only (Table 12.2; Fig. 12.14).

The importance of the above described parameters for the number of collision victims per wind turbine was determined in step-wise backward multiple regressions. Data for the 5 and 1 km radius were dealt with separately in the regression models. A critical aspect was the selection of the reference group—a random selection of all wind turbines in Germany, used as reference for wind turbines with reported collisions. A lack of comprehensive searches for these wind turbines (due

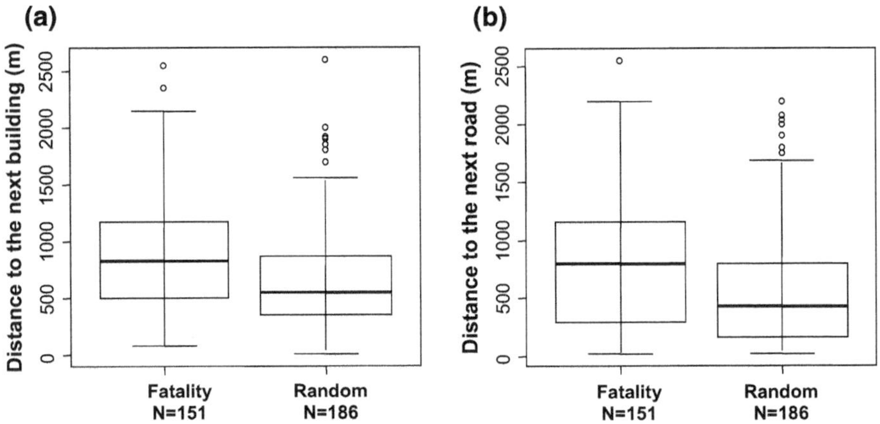

Fig. 12.14 Distance of wind turbines from the next building (**a**) and from the next road (**b**). Comparison of wind turbines with reported collisions and randomly chosen wind turbines (*horizontal centre line* mean, *boxes* standard error, *whisker* standard deviation, *circles* outliers)

to an unmanageable practical effort) does not allow to completely exclude collisions of target species for these turbines. As the target species are eye-catching birds of prey and other large bird species, it can be assumed that the average detection rate will be lower for wind turbines with no reported collision than for turbines with reported collisions. A MANOVA analysis based on several of the investigated location characteristics (land use within a radius of 5 km, position of the wind turbine within the wind farm, capacity) revealed a significant difference between wind turbines with collisions and the reference group (F = 7.88***).

Despite single significant influences of structures in the surrounding area (see Table 12.2), the explanatory value of the regression models was comparably low. For the model looking at the immediate surrounding area, the coefficient of determination R^2 had a corrected value of 0.1155 (F = 10.56***). In descending order of significance, important parameters were the proportion of farmland (t = 4.197; $p < 0.0001$), the distance from buildings (t = 2.755; $p < 0.01$), the position within the wind farm (t = −1.899; p = 0,059) and the distance from roads (t = 1.713; $p = 0.088$). Including land use within a radius of 5 km, the model returned a slightly higher coefficient of determination R^2, corr. = 0.1277 (F = 15.88***). Here, the essential parameters were the distance from buildings (t = 4.296; $p < 0.0001$), the proportion of farmland (t = 4.827; $p < 0.0001$) and the position within the wind farm (t = −2.103; $p < 0.05$). The influence of other parameters was negligible.

The principal component analysis showed that a distinction based on land use is far more important than a distinction based on distances from single structures, the position within the wind farm or technical specifications (Fig. 12.15). In this analysis, wind turbines with collisions and randomly selected wind turbines were distributed within a multidimensional coordinate system based on parameters of the

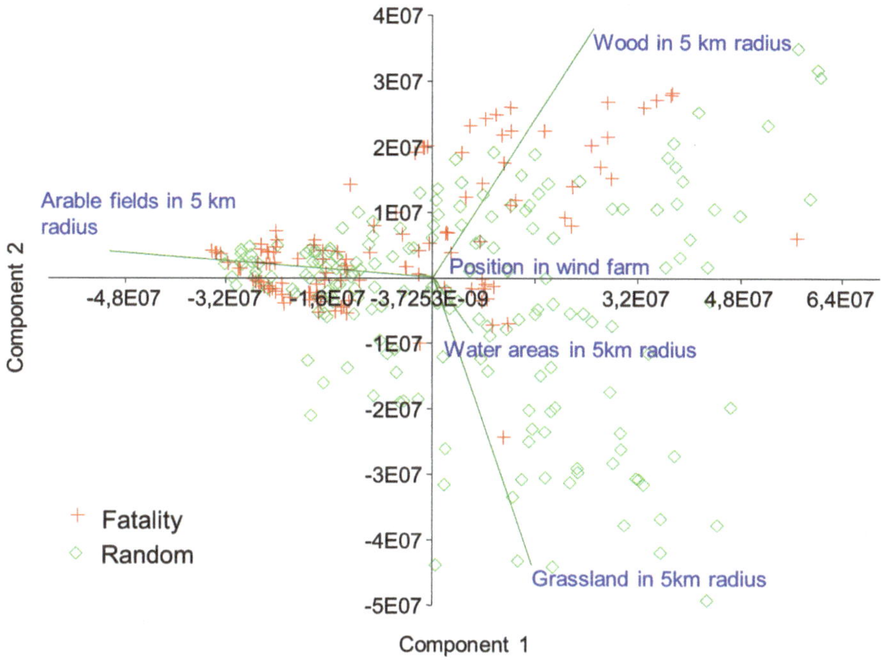

Fig. 12.15 Principal component analysis of the distribution of wind turbines with collisions and randomly selected turbines depending on structures in the surrounding area. *Vectors* indicate the direction and importance of single parameters. The most important single parameters are represented only

surrounding area (proportion of different types of land use, distance from particular structures), the relative position of wind turbines within the wind farm and on technical specifications. The multidimensional system was projected on the hypothetical axes (components) in form of a two-dimensional point cloud for better representation. The parameters of the surrounding area were included as vectors. The length of a vector is proportional to the influence of the parameter on the distribution of locations, whereas the directions depend on the positions of the single points, reflecting particular characteristics of these points. Wind turbines with a large proportion of grassland within the surrounding area, for example, are grouped in the bottom right quarter. Among these, mainly reference turbines and only few wind turbines with collisions can be found (see also Table 12.2).

12.3.9 Nests

It can be assumed that both locally breeding and migrating birds are among the collision victims (see also Chap. 10). In a few (single) cases this assumption is documented by observations or ring recoveries. In a total of 19 cases, a nest was known to be located within a radius of below 6 km from the location of the collision and the victim could be identified as one of the breeding adult birds or one of the young birds of the respective breeding season. The minimum distance between a nest and a wind turbine with the reported collision was 300 m. In six cases (one Eurasian hobby, two red kites, two white-tailed sea eagles, and one white stork) breeding failure as a result of the fatal collision of an adult of a breeding pair with a wind turbine could be documented.

12.4 Discussion

The population of some of the listed species (white-tailed sea eagle, black stork, red kite, peregrine falcon, crane and Eurasian eagle-owl) showed a strong positive trend over the last years, relativizing the threats caused by wind energy use (see also Chap. 7). However, the annually reported fatalities of at least three to four white-tailed sea eagles, representing only a part of the birds actually colliding with wind turbines, cannot be ignored considering the number of about 500 breeding birds. In addition, one rare and in Germany critically endangered species (lesser spotted eagle, about 100 breeding pairs) could be found among collision victims. Generally speaking, soaring bird species, often circling above open landscape areas, appear to be more affected than other bird of prey species such as northern goshawk and Eurasian sparrowhawk, which in Germany are only rarely colliding with wind turbines despite the larger population densities.

The results of this sub-study are afflicted with uncertainties. One of the central criticisms is the actual number of victims in relation to the collision victims recorded in the database. Many of the fatalities were reported by voluntary helpers, hunters or farmers. The circumstances of these discoveries are quite often insufficiently documented. Even for systematic searches of wind farms larger uncertainties remain. For the number of birds found dead, the size of the wind farm (number of wind turbines) is of minor importance in relation to the number of searches, as the number dead birds was recorded per turbine. The direct dependency between the number of recorded collision victims and the search intensity (number of searches) indicates a large number of fatalities, which are not found and confirms the importance of further systematic studies taking the removal of carcasses by scavengers in account. The data collected by the State Bird Observatory and Field Centre of the Brandenburg Environmental Agency is still the most comprehensive database of its kind and is used in the analyses described in the present study.

Collisions of birds with wind turbines are mostly random events, making it difficult to define statistically reliable parameters significantly influencing the frequency of such events. Still some hints can be followed in the identification of parameters, which may increase or reduce the risk of collision. In relation to technical specifications of wind turbines, it seems to be true that increasing size and height of wind turbines will lead to an increased collision risk for birds of prey and other large bird species (see also de Lucas et al. 2008). As direct mechanical impacts are assumed to be the cause of death of the collision victims, the collision risk directly depends on the rotor swept area. One of the implications of higher towers is that the rotor swept area is increasing (e.g. from 30 m for a typical 500 kW turbine of 40 m height to more than 70 m for a 2000 kW turbine of the same manufacturer of 110 m height). For some of the target species, for example Montagu's harriers (Chap. 4), this means that most of the flight activity occurs below the rotor swept area in hunting flights low above the ground, reducing the collision risk considerably (but see Baum and Baum 2011 on an increased collision risk after repowering).

For other species, a rotor swept area shifted upwards to height ranges of 100 or more metres above the ground will result in an increased collision risk. Experiences gained in aviation (Scheller and Küsters 1999) give hints of a particular collision risk for soaring birds of prey (buzzards, kite). These species are especially at risk at heights of 100 m and more (see i.a. Smallwood and Neher 2004). How dangerous small (100–300 kW nominal capacity) and smallest (1–30 kW) wind turbines are for birds of prey could not be investigated so far due to a lack of corresponding observations. Reported collisions of waders, gulls and ducks with small wind turbines (Böttger et al. 1990; Scherner 1999) indicate however that even smaller turbines, which did not attract much attention so far cannot be completely excluded in an estimate of the risks for avian species.

The increased risks caused by larger wind turbines are decreased again with increasing capacity. A number of observations and model calculations lead to the conclusion that the overall collision risk per megawatt generated power is decreasing with increasing turbine size (see also Hötker et al. 2006). Quadruplication of energy output from 500 to 2000 kW slightly more than doubles the number of collision victims. The installation of a few larger instead of many smaller wind turbines with the same total capacity will therefore lead to a lower risk for birds of prey despite a higher collision rate per turbine. This observation is confirmed by other studies on birds in general (Hötker et al. 2006; Barclay et al. 2007).

An improved spatial arrangement of wind farms will add to keep the collision risk at a low level despite the increasing power production. Observations in the Zitz-Warchau wind farm showed a distinct concentration of collision events on peripheral wind turbines. Similar patterns in the distribution of collision victims were for instance observed within the wind farm on the island of Smøla (Norway) (Bevanger et al. 2008), but not for the griffon vulture at rows of wind turbines in Spain (de Lucas et al. 2008). There is evidence that many bird species perceive wind farms as a barrier avoiding large arrays of wind turbines (Drewitt and Langston 2006). A compact layout of wind farms instead of widely scattered single turbines will therefore also help to keep the number of collision victims low. Thus, concentration of wind turbines in large wind farms does not only have logistic and economic advantages. It will at the same time minimise the number of collision victims with no loss of capacity. This should be taken into account in the construction of new wind farms as well as in repowering.

A further essential parameter or rather group of parameters influencing the collision risk are the landscape characteristics in the surrounding area of a wind farm. The number of collisions depends to a considerable extent on the attractiveness of the site and the surrounding area of a wind farm for birds of prey and the frequency of approaches of wind turbines. Assuming that locations with known collisions present a higher risk of collision than randomly selected wind turbines, landscape structures and characteristics reflecting this difference were looked for. It was found that the number of collisions was increasing proportionately to an increasing distance of the wind farm from residential areas and forests. This effect can partly be explained by the typical habitat requirements of the target species. Most activities of the majority of the investigated bird of prey and large bird species take place in open landscape outside residential areas. The informative value of these results should be assessed critically, as calculations of distances and radii was based on the (geometrical) centre of wind farms with a partly larger space requirement, whereas actual collisions quite often occurred at peripheral wind turbines, which were for the most part located far away from this centre. (see also Sect. 12.3.7). More informative results, based on the exact locations of collisions can be found in Sect. 12.3.8.

It is worth noting that birds of prey were particularly often colliding with wind turbines located in arable land and rarely with turbines located in grassland. This will be primarily due to the distribution of small mammals (the main food of most birds of prey) within the differently man-aged locations. In grassland areas, small mammal populations will be reduced by predation after mowing due to the lack of cover, but as no soil cultivation is carried out, the area can still be evenly populated. After mowing and especially soil cultivation, farmland populations of small mammals will take shelter in edge structures and as well the fallows around wind turbine towers, successively repopulating the farmland area with increasing growth of the crop (Volland 2008). At larger distances from edge structures, loss of cover caused by mowing had less influence on the population density of small mammals than to be expected cultivated soils. In areas closer to edge structures, the population density of small mammals dropped considerably with parallel increase of the population in the area around wind turbines (Volland 2008). These variations in numbers of small mammals depending on cultivation will have an essential influence on the spatial distribution and abundance of birds of prey within wind farms and therefore on the collision risk.

In Germany, most wind turbines are concentrated in the northern and eastern lowlands. Locally, large numbers of collision victims for wind farms located in topographical bottlenecks with flyways in a relatively confined area like mountain ridges or passes and isthmi (Smallwood and Thelander 2004, 2008; de Lucas et al. 2008), well documented in literature, do not occur. To a certain extent, the probability of a collision was still found to depend on particular landscape structures (i.e. water bodies) and on land use in the vicinity of the wind farm. Continuing observation of distance recommendations taking account of surrounding structures in the designation of new wind farm sites will remain to be an important means to minimise bird mortality.

References

Baerwald EF, D'Amours G, Klug B, Barclay R (2008) Barotrauma is a significant cause of bat fatalities at wind turbines. Curr Biol 18:695–696
Barclay RMR, Baerwald EF, Gruver JC (2007) Variation in bat and bird fatalities at wind energy facilities: assessing the effects of rotor size and tower height. Can J Zool 85:381–387
Barrios L, Rodriguez A (2004) Behavioural and environmental correlates of soaring-bird mortality at on-shore wind turbines. J Appl Ecol 41:72–81
Baum R, Baum S (2011) Wiesenweihen in der Falle. Beobachtungen in einem ostfriesi-schen Windpark. Falke 58:230–233
Böttger M, Clemens T, Grote G, Hartmann G, Hartwig E, Lammen C, & Vauk-Hentzelt E (1990). Biologisch-ökologische Begleituntersuchungen zum Bau und Betrieb von Windkraftanlagen - Endbericht. NNA-Ber. 3 Jg./Sonderheft, Schneverdingen, pp 124
Bevanger K, Clausen S, Fragstad Ø, Follestad A, Gjershaug JO, Halley D, Hanssen F, Lund Hoel P, Jacobsen K-O, Johnsen L, May R, Nygård T, Pedersen HC, Reitan O, Steinheim Y, Vang R (2008) Pre- and post-construction studies of conflicts between birds and wind turbines in coastal Norway. Progress Report 2008, NINA Report 409, Trondheim, pp 55
Drewitt AL, Langston RHW (2006) Assessing the impacts of wind farms on birds. Ibis 148:29–42
Everaert J (2003) Windturbines en vogels in Vlaanderen: Voorlopige onderzoeksresultaten en aanbevelingen (Wind turbines and birds in Flanders: Preliminary study results and recommendations). Natuur Oriolus 69:145–155
GeoContent International GmbH (2009) http://www.geocontent.de
Hammer O, Harper DAT (2005) Paleontological data analysis. Wiley-Blackwell, Malden, pp 368
Hötker H, Thomsen K-M, Jeromin H (2006) Impacts on biodiversity of exploitation of renewable energy sources: the example of birds and bats – facts, gaps of knowledge, demands for further research, and ornithological guidelines for the development of renewable energy exploitation. Michael-Otto-Institut im NABU, Bergenhusen
Keiler J (2008) Windkraft Betreiber-Datenbasis: Register/Errichtungsdaten deutscher Windanlagen. http://www.btrdb.de/. Assessed 30 June 2010
de Lucas M, Janss GFE, Whitfield DP, Ferrer M (2008) Collision fatality of raptors in wind farms does not depend on raptor abundance. J Appl Ecol 45: 1695–1703
Madders M, Whitfield PD (2006) Upland raptors and the assessment of wind farm im-pacts. Ibis 148:43–56
Möckel R, Wiesner T (2007) Zur Wirkung von Windkraftanlagen auf Brut- und Gastvögel in der Niederlausitz (Land Brandenburg). Otis 15:1–133
Percival SM (2005) Birds and windfarms: what are the real issues? Brit Birds 98:194–204
R development core team (2008): http://www.r-project.org/
Scheller W, Küsters E (1999) Flughöhen von Greifvögeln und Vogelschläge in Deutsch-land. Vogel und Luftverkehr 2/99:76–96
Scherner ER (1999) Windkraftanlagen und „wertgebende Vogelbestände "bei Bremerhaven: Realität oder Realsatire. Beiträge zur Naturkunde Niedersachsens, 52(4):121–156
Smallwood SK, Neher L (2004) Repowering the APWRA: forecasting and minimizing avian mortality without significant loss of powergeneration. Califonia Energy Commission, PIER Energy-Related Environmental Research, CEC-500-2005-005
Smallwood SK, Thelander CG (2004) Developing methods to reduce bird mortality in the Altamont Pass Wind Resource Area. Final Report by BioResource Consultants to the California Energy Commission, Public Interest Energy Research-Environmental Area, Contract No. 500-01-019
Smallwood KS, Thelander CG (2008) Bird Mortality in the Altamont Pass Wild Resource Area, California. J Wildlife Manage 72:215–223
Strickland D, Erickson W, Young D, Johnson G (2007) Selecting study designs to evaluate the effect of windpower on birds. In: de Lucas M, Janss G, Ferrer M. (eds) Birds and Wind Power. Quercus, Madrid, p 117–136

Südbeck P, Bauer H-G, Boschert M, Boye P, Knief W (2007) Rote Liste der Brutvögel Deutschlands. 4. Fassung. Berichte zum Vogelschutz 44:23–81

Volland S (2008) Haben Windenergieanlagen (WKA) einen Einfluss auf Kleinsäugerbestände der Umgebung? Master thesis, Paris Lodron University of Salzburg and Museum für Naturkunde der Humboldt-Universität zu Berlin, pp 83

Chapter 13
Collision Victims and Lattice Towers—Frequency of Collision, Breeding Population and Breeding Success of Birds of Prey in Two Wind Farms in Brandenburg

Tobias Dürr and Leonid Rasran

13.1 Introduction

Wind turbines with lattice towers are assumed to be potentially more dangerous for birds of prey, as they, unlike tubular towers, can be used for resting or perching (Sterner 2002; US Fish and Wildlife Service 2003; but see also Thelander and Rugge 2000). Lattice towers may encourage birds of prey to approach the dangerous rotors. Aim of the present sub-study was to use an example to collect data relating to the question, whether wind turbines with lattice towers cause more bird of prey collisions than wind turbines with tubular towers. To achieve this, two comparable Brandenburg wind farms with lattice and tubular towers were searched for collision victims, correction factors were determined to estimate the number of victims and the behaviour of birds of prey at lattice towers as well as breeding population and breeding success in the vicinity of the wind farms were analysed.

T. Dürr (✉)
Staatliche Vogelschutzwarte (Brandenburg State Bird Conservation Centre) of the Landesamt für Umwelt, Brandenburg (Brandenburg State Environmental Agency), Buckower Dorfstraße 34, 14715 Nennhausen, Germany
e-mail: Tobias.Duerr@LFU.Brandenburg.de

L. Rasran
NABU (Nature and Biodiversity Conservation Union) Michael-Otto-Institute, Goosstroot 1, 24861 Bergenhusen, Germany
e-mail: Leonid.Rasran@boku.ac.at

H. Hötker et al. (eds.), *Birds of Prey and Wind Farms*,
DOI 10.1007/978-3-319-53402-2_13

13.2 Study Areas and Methods

13.2.1 Dretzen and Etzin Wind Farms

Investigations were carried out in the Dretzen wind farm, operational since May 2005, with ten REpower MD77 wind turbines (hub height: 111 m, rotor diameter: 77 m, total height: 149.5 m, height below rotor-swept area: 72.5 m, lattice towers) as well as in wind farms Etzin-I with seven wind turbines (operational since July 2005) and Etzin-II with three wind turbines (operational since November 2006), all Enercon E-70 turbines (hub height: 113.5 m, rotor diameter: 71 m, total height: 149 m, height below rotor swept area: 78 m, tubular steel tower) for reference.

In 34 survey days carried out in both wind farms, all birds of prey observed at a maximum distance of 300 m from wind turbines were recorded. Additionally, the population densities of birds of prey within an area of 2 km around the wind farms were determined.

Field studies to determine the bird of prey territories and the breeding success within both study areas and searches for collision victims of the Etzin-I and Etzin-II wind farms were carried out by Maik Jurke (Ketzin). For the wind farm Dretzen, determination of collision victims was carried out by Sabine Bich (Tuchheim).

Weekly investigations were carried out between calendar week 12 and calendar week 45 including. The complete study period between calendar weeks 12 and 45 could be covered for the Etzin wind farms. The Dretzen wind farm was investigated between calendar weeks 12 and 44 and in calendar week 46, as no investigations were possible in calendar week 45.

13.2.2 Search Efficiency

A search efficiency of 84.6% was determined for the Etzin-I and Etzin-II wind farms between 29 September and 6 October. Laid out birds were marked by angular trimming of flight feathers. 11 of 13 birds laid out the previous day were found. Whether the birds, which could not be found, were removed by predators the previous night, remains unclear to the disadvantage of the search efficiency. For organisational reasons, the observer-related search efficiency could not be determined for the Dretzen wind farm, but was already determined for the same saercher for the nearby Zitz-Warchau wind farm in 2006 and is assumed to be 85.7% for the present study. Within the Dretzen wind farm, searches were carried out with the support of a hunting dog, which was already used in the Zitz-Warchau wind farm between 2003 and 2006.

13.2.3 Search Areas

The proportion of the search area covered per survey day and wind turbine was entered in a daily log and averaged for single turbines and the total wind farm. A circular area with a radius of 100 m around the respective wind turbines was searched. Some of the wind turbines could not be searched on single days due to construction works (construction crane, works on the nacelle). Table 13.1 lists the average area searched per wind turbine. The area, which was not searched, was calculated by subtracting the averaged value from the radius to be searched.

In the Dretzen wind farm, wind turbines 1–7 were surrounded by an intensively used, closely mown golf-like lawn (extremely good visibility). Winter oilseed rape and winter cereals were cultivated around the other wind turbines. This partial area could only be searched to a limited extent between early April and mid-June.

Winter oilseed rape and winter cereals were prevailing in the Etzin-I wind farm. A robinia grove and small-scale maize crops were located in the area of wind turbine 5. Wind turbines 2, 4 and 7 were accompanied by rows of poplars. A completely reeded, water-bearing kettle could be found in the vicinity of wind turbine 2.

Winter oilseed rape and winter cereals were cultivated in the Etzin-II wind farm. One of the wind turbines was surrounded by grassland, which was not mown and ploughed in autumn before oilseed rape was drilled. Wind turbines 2 and 3 were accompanied by rows of poplars. Searches of cultivated areas were at times severely limited due to farming activities in the Etzin-I and Etzin-II wind farms.

Table 13.1 Overview of the average area searched per wind turbine over the study period (percentage of the 100 m radius)

Wind turbine no.	Wind farm Dretzen	Wind farm Etzin-I	Wind farm Etzin-II
1	100	55.4	64.7
2	100	51.3	78.0
3	97	58.0	62.4
4	100	57.9	
5	100	63.1	
6	98	54.4	
7	98	59.4	
8	68		
9	67		
10	73		
Average	90.1	57.1	68.4

13.2.4 Disappearance Rate

A disappearance rate of at least 88.2% within one week was determined between 29 September and 6 October in the weekly search interval using laid out wild bird carcasses (n = 51) for the Etzin-I and Etzin-II wind farms. A disappearance rate of at least 72.2% (n = 18 wild bird carcasses) was found for the Dretzen wind farm between 15 October and 22 October (see Table 13.2).

Two of the birds laid out in Dretzen were plucked. Feathers may provide hints on the carcass, but no sure determination of the cause of death will be possible. The disappearance rate will therefore be between 72.2 and 83.3%.

Six of the birds laid out in the Etzin wind farms were plucked and feathers may have provided hints on the carcass in a survey. On the first day of the survey, a laid out common buzzard was moved by >150 m into the rotor area of a neighbouring wind turbine and was started to feed on. The gorged but complete carcass could still be found one week after the test had ended and could have been assessed as still present. On the second day of the survey, the carcasses could not be checked at four wind turbines due to farming activities. Assuming that the feathers of the six plucked carcasses could still have been found on the 7th day, a removal rate of at least 88.2% can be specified (Table 13.3).

Table 13.2 Results of the tests to determine the disappearance rate for laid out wild bird carcasses in the Dretzen wind farm

Date	15.10.2008	16 Oct.	17 Oct.	19 Oct.	20 Oct.	21 Oct.	22 Oct.
Wind turbine No.	Carcasses	Day 1	Day 2	Day 4	Day 5	Day 6	Day 7
FL318	2	1	1	1	0	0	0
FL319	2	0	0	0	0	0	0
FL320	2	2	2	2	2	1	1
FL321	2	2	2	1	1	1	0 + 1[a]
FL322	1	1	0	0	0	0	0
FL323	2	2	2	1 + 1[a]	0	0	0
FL324	1	0	0	0	0	0	0
FL325	2	2	2	1	1	1	1
FL326	2	2	2	2	0	0	0
FL327	2	2	2	2	2	1	1
Birds found (n)	18	14	13	10 (+1[a])	6	4	3 (+1[a])
Birds found (%)		77.8	72.2	55.6	33.3	22.2	16.7
Birds removed (%)		22.2	27.8	44.4	66.7	77.8	83.3 (72.2)

[a]The bird was plucked and therefore assessed as being removed

Table 13.3 Results of the tests to determine the disappearance rate for laid out wild bird carcasses in the Etzin-I and Etzin-II wind farms

Date	29.09.2008	30.09.2008	01.10.2008	02.10.2008
Wind turbine No.	Carcasses	Day 1	Day 2	Day 3
I/1	6	2	1	1
I/2	6	1 + 2[a]	1[a]	1[a]
I/3	6	2	1	0
I/4	5	4	?	0
I/5	5	1 + 1[a]	?	1[a]
I/6	5	4	?	0
I/7	5	3 + 1[a]	?	1[a]
II/1	5	2	1 + 1[a]	0
II/2	4	3	2	0
II/3	4	1 + 1[a]	1[a]	1[a]
$Birds_{found\ (n)}$	51	23 (+5[a])	5 (+5[a])	(+4[a])
$Birds_{found\ (\%)}$		45.1	9.8	0.0 (7.8)
$Birds_{removed\ (\%)}$		54.9	90.2	100 (88.2)

[a]The bird was plucked and therefore assessed as being removed
? No search possible due to farming activities

13.3 Results

13.3.1 Number of Collision Victims

Four birds and one bat were found dead after collision within the Dretzen wind farm. Without correction factors this equals 0.4 birds and 0.1 bats per wind turbine and year. Eight birds and three bats (see Table 13.4) were found in the Etzin wind farms. This equals 0.8 birds and 0.3 bats per wind turbine and year. Of particular interest for the project were three common buzzards and one red kite, common kestrel and tawny owl each, which were among the collision victims.

In the weekly searches, five of the bird fatalities in the Etzin-I and two in the Etzin-II wind farm (Table 13.5), could not be attributed to the respective wind turbine with absolute certainty. No such victims were recorded in the Dretzen wind farm. These mainly consisted of plucked birds, in one case of a young bird (skylark) killed by mowing during maintenance works around a wind turbine base.

Furthermore, three birds were found, where collisions with the respective wind turbine was not sure.

Table 13.4 Overview of the collision victims found within the single wind farms

	Wind farm Dretzen	Wind farm Etzin-I	Wind farm Etzin-II
	10 WECs	7 WECs	3 WECs
	REpower MD77	Enercon E70	Enercon E70
Common buzzard	1		2
Red kite	1		
Common kestrel	1		
Tawny owl	1		
Hooded crow		2	
Woodpigeon		1	1
Song thrush		1	
Common house martin		1	
Common noctule	1		
Nathusius's pipistrelle		2	
Common pipistrelle			1

Table 13.5 Overview of bird fatalities, which could not be attributed to wind turbine operation with absolute certainty

	Wind farm Dretzen	Wind farm Etzin-I	Wind farm Etzin-II
Skylark			1[a]
Stock dove	1[b]	1	
Woodpigeon		1	1
European goldfinch		1	
Passerine sp.			1

[a]Not yet fledged young bird, killed by mowing during maintenance works around the turbine tower
[b]Ring of a homing pigeon found in fox faeces

13.3.2 Determination of Bird of Prey Fatalities Taking Account of the Retrieval Probability

The number of birds found dead was corrected for search efficiency, size of the searched area and removal rate, as these factors significantly influence the number of birds, which can be retrieved, to estimate the retrieval rates. The following information relates to birds of prey only. The following formula was used:

$$E = F : (1 - O) : (1 - R) : (1 - A)$$

E estimate of the number of victims
F number of bird of prey fatalities found (per wind turbine and year)
O number of birds overlooked (to be determined in tests)
R number of carcasses removed by scavengers (to be determined in tests)
A proportion of the search area, which was not searched.

The formula is based on the assumption that bird of prey fatalities are evenly distributed within the search area. A concentration of collision victims within an area in the vicinity of the tower will lead to overestimation of the number of collision victims, if the searched area is too small. Victims located outside the defined search radius will only be found by chance. In the central database database of wind turbine fatalities, 85.1% of the verified bird of prey collision victims (n = 301) were found within a radius of 50 m from the tower, 8.6% at distances between 51 and 100 m and 6.3% at larger distances. Of the randomly found red kites (n = 49) 13 birds (26.5%) were discovered within distances of >50 m from the turbine tower and ten birds (20.4%) at distances of >100 m (Bellebaum et al. 2012). Collision victims found at the latter distances, were mainly injured birds moving away from the wind turbine or carcasses removed by predators.

Dretzen wind farm (lattice towers)

One bird of prey fatality each was recorded at three of the ten wind turbines within the Dretzen wind farm.

Wind turbines 318, 319, 320, 321, 322, 324, 327: no fatality
Wind turbine 323: 1 Eurasian kestrel
Wind turbine 325: 1 common buzzard
Wind turbine 326: 1 red kite

F(total) = 3 birds of prey
O = 85.7% found → 14.3% overlooked = 0.143
R = 72.2% removed = 0.722
A(average) = 90.1% searched → 9.9% not searched = 0.099

$$\begin{aligned} E(\text{total}) &= 3 : (1 - 0.143) : (1{-}0.722) : (1{-}0.099) \\ &= 3 : 0.857 : 0.278 : 0.901 \\ &= 13.97 \approx 14\,\text{bird of prey fatalities} : 10\,\text{wind turbines} \\ &= 1.4\,\text{bird of prey fatalities per wind turbine and year} \end{aligned}$$

Etzin wind farms (tubular steel towers)

Two bird of prey fatalities were recorded at one of the ten wind turbines within the Etzin wind farms. No fatality was found at any of the other wind turbines.

Wind turbines 1, 2, 3, 4, 5, 6, 7, 8, 10: no fatality
Wind turbine 8: 2 common buzzards

F(total) = 2 birds of prey
O = 84.6% found → 15.4% overlooked = 0.154
R = 88.2% removed = 0.882

A(average) = 60.5% searched → 39.5% not searched = 0.395

$$\begin{aligned} E(\text{total}) &= 2 : (1 - 0.154) : (1-0.882) : (1-0.395) \\ &= 2 : 0.846 : 0.118 : 0.605 \\ &= 33.1 \approx 33\,\text{bird of prey fatalities} : 10\,\text{wind turbines} \\ &= 3.3\,\text{bird of prey fatalities per wind turbine and year} \end{aligned}$$

With 3.3 bird of prey fatalities per wind turbine in 2008, the fatality rate calculated for the Etzin wind farms with tubular steel towers was found to be higher than in the Dretzen wind farm with lattice towers (1.4 bird of prey fatalities per wind turbine in 2008).

13.3.3 Lattice Towers as Perching Point

During all searches in the Dretzen wind farm, all birds sitting on the lattice towers of wind turbines were recorded. The lattice tower was subdivided into three sections (Fig. 13.1):

1. close to the ground, up to the 2nd diagonal brace
2. central, between 3rd and 5th brace (below the rotor swept area)
3. upper third, from the 6th brace upwards (rotor swept part of the tower).

Fig. 13.1 Dretzen wind farm. The *arrows mark* the section boundaries

The observations are summarised in Table 13.6.

Observations showed that diurnal birds mainly use the sections close to ground for perching. Only once a bird of prey (peregrine falcon) could be observed sitting in the section of the rotor swept area.

13.3.4 Bird of Prey Observations and Population Density

To further assess the determined number of collision victims, all birds of prey observed within a radius of 300 m around both wind farms were recorded. Additionally the breeding population was calculated by nest searches within a radius of 2 km. Considerable differences in the number of observed individuals and species were found between the two study areas. During weekly searches, all observed birds of prey were recorded including behaviour and flight height and the vertical and horizontal distance from the tower was estimated (Table 13.7). The rotor area was defined as the horizontal and vertical expansion of the rotor blades.

Within the wind farms with lattice towers, the number of observed birds of prey (n = 74, five species) was by 69.8% lower than within the wind farms with tubular steel towers (n = 245, 11 species). The behaviour of the birds was found to differ considerably as well. With an average of 36.5% of the observed individuals, considerably more birds of prey were found to be sitting within the wind farm with lattice towers as compared to the wind farms with tubular towers (4.5%). Of the 12 observed bird of prey species six (osprey, hen harrier, common buzzard, marsh harrier, red kite, black kite) were recorded at height of the rotor. Three species were found to stay within the immediate rotor area (osprey, hen harrier, black kite). Within the wind farms with tubular towers, a total of 2.7% of the flying birds of prey flying were observed within the rotor area. No such flight movements were observed within the wind farm with lattice towers.

Considerable differences were also found between the population densities within the two study areas. Within a radius of 2 km around the wind farm with lattice towers, six breeding pairs of three species were counted, i.e. 64.7% less bird of prey breeding pairs than in the wind farms with tubular towers (17 breeding pairs of five species).

Table 13.6 Birds recorded sitting on the tower per section

	Section 1	Section 2	Section 3
Common buzzard	5	1	
Red kite	1		
Common kestrel	2	2	
Peregrine falcon			1
Hooded crow	3		
Great grey shrike	1		
Total	12 (75.0%)	3 (18.8%)	1 (6.2%)

Table 13.7 Behaviour of birds of prey in the investigated wind farms

	Dretzen	Etzin
Rotor area	Height: 72.5–149.5 m Radius: 38.5 m	Height: 78–149 m Radius: 35.5 m
Total number of birds of prey (n)	74	245
Flying (n)	47	234
At rotor height	11 (23.4%)	148 (63.2%)
In the rotor area	0	4 (2.7%)
Number of observations per species (1st value in brackets: number of birds flying, 2nd value: proportion (%) flying at rotor height, 3rd value: proportion of flight movement (%) within the rotor area)		
Osprey	0	2 (2; 50.0; 50.0)
Northern goshawk	0	3 (3; 0; 0)
Hen harrier	1 (0)	2 (2; 100; 100)
Common buzzard	40 (20; 35.0; 0)	57 (47; 10.6; 0)
Rough-legged buzzard	0	1 (1; 0; 0)
Marsh harrier	0	20 (20; 15.0; 0)
Red kite	16 (15; 26.7; 0)	11 (11; 27.3; 0)
Black kite	0	138 (138; 97.1; 0.7)
Eurasian sparrowhawk	0	1 (1; 0; 0)
Common kestrel	16 (11; 0; 0)	8 (8; 0; 0)
Peregrine falcon	1 (0; 0; 0)	0
European honey buzzard	0	2 (1; 0; 0)

Information on previous numbers of bird of prey breeding pairs was available for both of the study areas from different avian fauna surveys.

Dretzen wind farm

- 2002: A study carried out by Stadt & Land Planungsgemeinschaft mbH (2002) in the northern part of the study area partly reaching beyond the boundaries of the 2008 area.

Etzin wind farms

- 2002: A study of an area covering about 95% of the 2008 study area. No quantitative information was given on common buzzards (Kolbe and Kolbe 2002).
- 2005: Studies carried out for the neighbouring Bredow-Zeestow II (Göttsche 2005a) and Hoppenrade (Göttsche 2005b) wind farms, located within the eastern and south-eastern 2 km radius around the Etzin wind farms.
- 2007: Survey of the Ketzin II wind farm (Göttsche 2007) on the site of the current Etzin wind farms and within the southern and south-western 2 km radius.

In the surveys of the Bredow-Zeestow-II, Hoppenrade, Ketzin and Etzin-I wind farms, one common buzzard breeding site was recorded twice and one common

Table 13.8 Number of breeding territories determined for the respective wind farms prior to construction of the wind farm and in 2008

	Dretzen 2002	Dretzen 2008	Etzin 2002	Etzin 2008
Proportion of study area in 2008	100%		80%	
Eurasian hobby	–	–	–[a]	–
Common buzzard	4	3	1[b]	9
Red kite	2	1	1	1
Common kestrel	1	2	2	3
European honey buzzard	1	–	–	–
Eurasian sparrowhawk	1	–	–	–
Northern goshawk	1	–	1	–
Marsh harrier	–	–	1	3
Black kite	–	–	1	1

[a]On pair breeding outside the 80%
[b]Incomplete records in the 2008 study area

kestrel breeding site was recorded three times. All other breeding territories were clearly spatially separated. Of the breeding territories recorded in 2008, three common buzzard territories were not recorded in previous surveys. These territories were probably occupied but not recorded in 2002.

A decline of the population of five species (northern goshawk, common buzzard, red kite, Eurasian sparrowhawk, European honey buzzard) and a slight increase in the population of only one species (common kestrel) became apparent after three years of operation of the Dretzen wind farm as compared to the initial population, indicating an overall decline of 40% of all bird of prey breeding pairs (Table 13.8).

Previous surveys in the surrounding area of the Etzin wind farms in 2002 did not cover the entire 2008 study area. Surveys carried out in 2005 (Göttsche 2005a, b, 2007), which are not included in Table 13.8, provide information on further breeding sites of one Eurasian hobby, one common kestrel, one marsh harrier, one European honey buzzard and one red kite within the study area not covered in 2002. A decline of four species (Eurasian hobby, northern goshawk, red kite, European honey buzzard) could be observed after three (seven wind turbines) or two years (three wind turbines) of operation. No estimate of the common buzzard population was possible due to incomplete previous surveys. For two of the species (black kite, common kestrel), the number of occupied territories was found to be the same and in one of the species (marsh harrier) the population even showed a slight increase.

13.4 Discussion

To date, only one further report on results of a survey of collision victims at wind turbines with lattice towers was carried out in Germany (Möckel 2010). The Spremberg wind farm (Slamener Heide), Oberspreewald-Lausitz district, with nine

Fuhrländer FL 2500 wind turbines (hub height: 141 m, rotor diameter: 100 m, total height: 141 m, height below rotor swept area: 91 m) was searched for collision victims on a total of 39 days between early May 2009 and end of April 2010 within a radius of 190 m. One common buzzard was found dead. All retrieved birds (all species) were on average removed after 3.7 days, but none of the carcasses was older than five days. 26.7% of the laid out carcasses (day-old chicks) were overlooked by searchers. Assuming the search radius of the present surveys to be 100%, the average proportion of the area, which was not searched, equalled 39.4%. The removal rate of laid out carcasses (n = 90) determined in two tests was 76% after two nights and reached 98.9% after only four nights of the average 9.4-day search interval. A fatality rate of 23 birds of prey per wind turbine and year was calculated from the search success rate and the determined correction factors using the formula applied in the present study. This value appears to be considerably high and can probably attributed to the unusually high removal rates. With such a high number of victims, at least feathers of some carcasses should have been found, which was not the case.

Several surveys with laid out carcasses carried out by the Brandenburg State Bird Conservation Centre in other wind farms revealed an average removal rate of 44.0% for a ten day search interval and of 80.0% for weekly searches. A rough assessment of a ten day search interval and using the same correction factors will reduce the fatality rate for the Spremberg wind farm to 1.25 collision victims per wind turbine and year, for the Etzin and Dretzen wind farms (with a weekly search interval) to 0.7 bird of prey fatalities per wind turbine and year each and will result in no differences between annual fatalities per wind turbine for lattice and tubular towers.

The present analyses did not provide evidence of a higher risk for birds of prey caused by wind turbines with lattice towers as opposed to wind turbines with tubular towers. Despite the slightly lower number of birds found dead within the wind farm with tubular towers, calculations revealed a slightly higher fatality rate per wind turbine for these wind farms than for the wind farm with lattice towers. As collision victims were found at only three of ten wind turbines with lattice towers and only one of ten turbines with tubular towers, it will be highly probable that other parameters than the tower design had a stronger influence on the collision events.

Furthermore, only one wind farm with lattice towers could be investigated for only one calendar year. Due to the small number of searched wind turbines and the comparably little effort, analyses are additionally liable to be affected by random events while estimating errors (for searching methods see Anderson et al. 1999; Morrison 2002; Grünkorn et al. 2009). The results can therefore certainly not be generalized.

The fact that birds of prey mainly use lower sections of lattice towers for perching, supports the assumption that high lattice towers generally do not present a higher risk for birds of prey than tubular towers. The development of local bird of prey populations in the investigated wind farms does not support this assumption though, as some of the species showed a negative trend after three years of operation or could even not be found within the study area any more.

In both of the investigated wind farms, the total number of bird of prey breeding pairs was declining between 2002 and 2008, within the wind farm with lattice towers by 40% and within the wind farm with tubular towers by 27%. According to Mammen (2012), five species (Eurasian hobby, marsh harrier, black kite, Eurasian sparrowhawk, common kestrel) showed a positive trend in Brandenburg during the study period, whereas the populations of two species (common buzzard, red kite) showed a negative trend and the population of one species (northern goshawk) remained at the same level. The author did not determine a trend for the European honey buzzard and a comparison was therefore not possible. Four species showed a comparable trend (common buzzard, marsh harrier, red kite, common kestrel). For four of the species (Eurasian hobby, northern goshawk, black kite, Eurasian sparrowhawk) the trends within the investigated wind farms differed negatively from the average values determined for Brandenburg. No further conclusion can be drawn due to the comparably small populations and the differing species composition.

The similar number of collision victims calculated for the wind farms with tubular towers and the wind farm with lattice towers does not contradict the assumption of a higher collision risk for wind turbines with lattice towers, as more birds of prey and bird of prey species were recorded within the wind farm with tubular towers with a larger proportion of flying individuals and flights within the rotor area as well as a higher density of birds of prey in the surrounding area of the wind farm.

References

Anderson R, Morrison M, Sinclair K, Strickland D (1999) Studying Wind Energy/Bird Interactions: A Guidance Document. Avian Subcommittee and the National Wind Coordinating Committee, Washington D.C.

Bellebaum J, Korner-Nievergelt F, Mammen U (2012) Rotmilan und Windenergie in Brandenburg – Auswertung vorhandener Daten und Risikoabschätzung. Gutachten im Auftrag des LUGV Brandenburg, Staatliche Vogelschutzwarte, pp 30

Göttsche M (2005a) Untersuchung und Bewertung der Avifauna im geplanten Windpark Bredow-Zeestow II. Unveröffentlichtes Gutachten im Auftrag der MDP & Wiemken GmbH & Co. WP Bredow 2 KG, pp 58

Göttsche M (2005b) Untersuchung und Bewertung der Avifauna im geplanten Windpark Kertzin II. Unveröffentlichtes Gutachten des Büros für ökologische & faunistische Freilandunter-suchung im Auftrag der MDP & Wiemken GmbH & Co

Göttsche M (2007) Untersuchung und Bewertung der Avifauna im geplanten Windpark Hoppenrade. Unveröffentlichtes Gutachten des Büros für ökologische & faunistische Freilanduntersuchung im Auftrag der Ergo Energie-Anlagen GmbH & Co. WP Hoppenrade KG, pp 49

Grünkorn T, Diederichs A, Poszig D, Diederichs B, Nehls G (2009) Wie viele Vögel kollidieren mit Windkraftanlagen? Natur und Landschaft 84:309–314

Kolbe G, Kolbe M (2002) Untersuchungen der Avifauna im Bereich des geplanten Windparkes Etzin. Unveröff. Gutachten für das Büro für Umweltplanung im Auftrag MDP & Wiemken GmbH, pp 9

Mammen U (2012) Monitoring von Greifvogel- und Eulenarten in Brandenburg. Bericht zur Bestandsentwicklung in Brandenburg 1988 bis 2010. Unveröffentlichtes Gutachten im Auftrag LUGV Brandenburg, pp 16

Möckel R (2010) Vogel- und Fledermaus-Monitoring zur Erfassung von Anflugopfern im Windpark Spremberg. Zwischenbericht 1 (2009/10), Unveröffentlichte Studie im Auftrag WSB Projekt GmbH Dresden, pp 34

Morrison M (2002) Searcher bias and scavenging rates in bird/wind energy studies. NREL/SR-500-30876

Stadt & Land Planungsgemeinschaft mbH NL Schenkenberg (2002) Landschaftspflegerischer Begleitplan Windpark Dretzen, unveröffentlichtes Gutachten im Auftrag SeeBa Energiesysteme GmbH, pp 12–36

Sterner D (2002) A roadmap for PIER research on avian collisions with wind turbines in California. California Energy Commission

Thelander CG, Rugge L (2000) Bird risk behaviors and fatalities at the Altamont Wind Resource Area. Proceedings of National Avian - Wind Power Planning Meeting III, San Diego, pp 5–14

US Fish and Wildlife Service (2003) Interim guidelines to avoid and minimize wildlife impacts from wind turbines. United States Department of the Interior, Fish and Wildlife Service, Washington D.C.

Chapter 14
Calculation of the Probability of a Collision of Territorial Birds of Prey with Wind Turbines

Leonid Rasran, Bodo Grajetzky and Ubbo Mammen

14.1 Introduction

A major aim of the project "birds of prey and wind turbines" was the development of realistic and practicable proposals for a coexistence of wind turbines and populations of endangered bird of prey species (Montagu's harrier, red kite, white-tailed sea eagle). Important parameters for planning are recommended distances between newly planned wind farms and breeding or foraging habitats of these birds. Distance recommendations which, on the one hand, do not over-restrict inland areas available for wind energy use but ensure the survival of the populations of endangered bird of prey species on the other hand by keeping the mortality rate caused by wind turbines on a low level, are decisive. Determination or at least realistic estimates of the mortality of birds of prey due to collisions with wind turbines is required to understand the actual impacts of a further development of wind energy use on these birds and to justify necessary protection measures and distance recommendations. In a situation, where comprehensive surveys only provide a rough overview of population dynamics of the species concerned without differentiation of the reasons of mortality (Mammen and Stubbe 2003) and reported fatalities are incomplete with an unknown number of collision victims, which are not found (e.g. see also Sterner et al. 2007; Lekuona and Ursua 2007; Smallwood and Thelander 2008), modelling remains an important instrument in the

L. Rasran (✉)
NABU (Nature and Biodiversity Conservation Union)
Michael-Otto-Institute, Goosstroot 1, 24861 Bergenhusen, Germany
e-mail: Leonid.Rasran@boku.ac.at

B. Grajetzky
BioConsult SH, Schobüller Str. 36, 25813 Husum, Germany

U. Mammen
ÖKOTOP GbR, Willy-Brandt-Straße 44, 06110 Halle (Saale), Germany

H. Hötker et al. (eds.), *Birds of Prey and Wind Farms*,
DOI 10.1007/978-3-319-53402-2_14

determination of reliable, quantitative mortality rates (Strickland et al. 2007). Tucker (1996) and Band et al. (2007) presented detailed proposals for modelling. The basic idea is to determine two components: I—the frequency of flights through the rotor swept area of a wind turbine and II—the probability that a bird collides with the rotor in these flights. A further component (III) was added later to take account for possible avoidance movements of the bird in the vicinity of the rotor (e.g. Urquhart 2010). Based on data collected in other sub-studies of the present study (telemetric monitoring of Montagu's harriers and red kites) the present sub-study aimed to estimate realistic collision probabilities for these birds. These estimates are based on comprehensive model calculations taking account of all three of the above described components.

14.2 Methods

14.2.1 Model Components

Model calculations are based on telemetric data and observations of flight behaviour collected for the three target species of the present study: Montagu's harrier (see Chap. 4), red kite (see Chap. 3) and white-tailed sea eagle (see Chap. 5). The model includes three components (input parameters, see May et al. 2010), calculated as follows:

Factor I (frequency of flights through the rotor swept area) was broken down into individual components:

- Vertical component (V)—Proportion of flights within the rotor swept area.
- Horizontal component (H)—calculated as the footprint of a wind turbine (horizontal projection of the rotor swept area) in relation to the number of flights through this area per season (flight density).
- Temporal component (T)—the time it takes a bird to traverse the area of a wind turbine. This time is divided by the total flight time of an individual determined for the relevant period (usually the breeding season, the period with territorial behaviour, where a bird is regularly present in the vicinity of the nest and possible wind turbines), thus including the maximum number of possible traversing flights in the calculation (hypothetical assumption: the bird exclusively stays in the surrounding area of the wind turbine).

As the activity per area of territorial individuals usually directly depends on the distance from the nest, values were calculated for distance classes around the nest. Different distance classes were chosen due to differing activity patterns and activity ranges of the species (see calculations for the individual species). To estimate mortality rates for wind turbines in the vicinity of the nest, the hypothetical situation that only one wind turbine is present within a distance class around the nest or the activity centre of a bird was assumed. The H value (horizontal component) was calculated as (footprint of a wind turbine/area of the distance segment) * proportion of flight activity within this segment.

Factor II (probability of a collision when flying through the rotor swept area) is calculated using the Band et al. (2007) matrix model simulations based on the theoretical model developed by Tucker (1996). The following variables were used: length (L) and wingspan of the bird (W), average bird speed (BS), nature of the flight movement (flapping or gliding), rotor diameter, number of blades, chord size (maximum chord) and pitch of the wind turbine as well as the average rotation period depending on the prevailing wind speeds. Factor II, as a percentage value, is multiplied as F II by the spatial factor in further calculations. As the model is based on a hypothetical presence of one wind turbine within the investigated distance classes (see above), turbines of different sizes can be included in the calculations. The matrix model simulations developed by Band take account of the technical specifications of a wind turbine. An example of these simulations for red kite can be found in Table 14.2.

Factor III—Avoidance behaviour, i.e. the ability of a bird to actively change the flight path in the vicinity of the rotor to avoid a collision. Displacement from the habitat by the construction of a wind turbine is not included in the model, as this is strictly speaking no parameter directly influencing the mortality.

An avoidance probability is given in percent for each species based on observations in literature.

Montagu's harrier

In the modelling of the collision probabilities, observations of six birds in the North Frisian polders were used. Four of these birds were monitored during one breeding season and two during two breeding season, resulting in a total sample size of eight birds per season. Four male birds and two female birds were monitored (see Table 14.1).

Table 14.1 Telemetric monitoring data of Montagu's harriers in North Frisia used in modelling

Bird/season	Study period	Days recorded	Minutes/hours (used in the present evaluation)	Flight/activity
L1m 2007	21.06–05.08.2007	21	3688/61.5	60.1% (9.6 h/day)
L3w 2007	03.07–21.07.2007	10	4304/71.7	28.1% (4.5 h/day)
FWL2m 2007	21.07–03.08.2007	6	1615/26.9	58.8% (9.4 h/day)
FWL2w 2007	19.07–05.08.2007	7	1654/27.6	39.4% (6.3 h/day)
L1m 2008	25.06–02.08.2008	11	2482/41.4	58.8% (9.4 h/day)
L3m 2008	21.05–22.06.2008	17	4873/81.2	49.5% (7.93 h/day)
FWL1m 2008	05.06–11.07.2008	6	874/14.6	69.97% (11.2 h/day)
FWL2m 2008	04.06–22.07.2008	6	985/16.4	43.1% (6.9 h/day)

The average activities (in flight hours per day) based on the observations were estimated for the duration of an average breeding season (assumption: 100 days).

Based on typical activity profiles of the birds, the following distance classes (distances from the nest) were used: 0–300, 300–650, 650–1000 and >1000 m. For the latter distance class the calculation of the area is based on the segment between 1000 and 2000 m.

As telemetric data was very precise and high-resolution in terms of flight heights, model calculations were carried out for wind turbines of three sizes:

- small—(e.g. E-40), 500 kW, rotor swept height 20–65 m, footprint 1257 m^2
- medium—(e.g. E-66), 1800 kW, rotor swept height 30–100 m, footprint 3421 m^2
- large—(e.g. V90), 3000 kW, rotor swept height 35–130 m, footprint 6362 m^2.

Based on telemetric data, the average flight speed of Montagu's harriers was defined to be on average 4.5 m/s. As birds may approach the rotor at any angle, the rotor area was defined as a sphere with the diameter of the rotor and a fly-through as a flight path of the length of a rotor diameter. This resulted in the following times per fly-through: 9 s for small wind turbine, 15 s for medium-sized wind turbines and 20 s for large wind turbines. These times were the smallest time unit characterising flight activity.

Based on measurements of the tagged birds and literature data (Bauer et al. 2005; Mebs and Schmidt 2006) the following bird sizes were used: male: body length (BL) = 43 cm, wingspan (W) = 105 cm; female: body length (BL) = 47 cm, wingspan (W) = 130 cm.

The avoidance rate of Montagu's harriers was assumed to be 97.5% based on previously published avoidance rates of hen harriers (Urquhart 2010; Whitfield and Band, in preparation). It is therefore assumed that only 2.5% of the flights through the rotor swept area will result in a collision. The fly-through possibility determined by other model components is then multiplied by 0.025 to obtain the actual collision probability.

Red kite

In collision probability modelling, observations of seven birds recorded between 2007 and 2010 in the study area "Querfurter Platte" in southern Saxony-Anhalt were used. Four of these birds were monitored during one breeding season and three during two breeding seasons. 2010 data for three satellite-tagged birds was used in addition, resulting in a total sample size of 13 birds per season for calculation (for more information on telemetric monitoring of red kite see Chap. 3).

As telemetric data, especially for flight heights, was not as high-resolution as the data available for Montagu's harriers, no exact vertical flight profiles in relation to the distance from the nest were created for red kites. For the same reason, no model simulations were performed for different sizes of wind turbines. All calculations for red kite are based on values for medium-sized wind turbines (E-66). The vertical spatial component (V) was set to an average value of 0.25 for all distance classes.

Observations of the red kite sub-study (see Sect. 14.3) showed that about 25% of all observed flights of territorial birds occur within the rotor swept height. An average flight time of 3 h and 47 min was determined in 20 days of continuous monitoring of six birds. The duration of stay in the breeding area is defined as 245 days.

Based on the typical activity profiles of the birds, the following distance classes around the nest were used: <250, 250–500, 500–750, 750–1000, 1000–1250, 1250–1500, 1500–1750 and 1750–2000 m. A sufficient number of recorded locations was available for all distances classes to calculate the probability of use of the respective segment around the nest for the birds during the breeding season. Flights of red kites outside the 2000 m radius were included in the total flight activity, but no collision probability was calculated for these areas due to the low number of flight movements in proportion to the increasing size of the area. The following bird characteristics were used: body length (BL) = 65 cm, wingspan (W) = 160 cm; average flight speed = 6.9 m/s, time per fly-through = 10 s. Individual differences in sizes of the monitored birds were neglected.

The avoidance rate of red kite is assumed to be 98% (Whitfield and Madders 2006; Urquhart 2010).

14.2.2 Calculation and Statistics

Collision probabilities were calculated using Excel. An Excel formula sheet for the calculation of the collision probability for flights through the rotor swept area (factor II) was kindly provided by W. Band. Comparisons between collision probabilities for the different distance classes and sizes of wind turbines (Montagu's harrier) were performed using non-parametric tests (Kruskal-Wallis ANOVAs, Mann-Whitney U test). Statistical tests were carried out using software R 2.11.1 (R development core team 2010).

14.3 Results

14.3.1 Montagu's Harrier

The collision risk per flight through the rotor swept area (factor II) was calculated as 16.4% for large and 17–18% for medium-sized and small wind turbines. Differences in size of male and female birds had only a minor influence on these values. The absolute collision probability calculated based on this factor and on the avoidance rate may well reach around 20% (two fatalities per five breeding pairs and season) in the worst case (wind turbine in the vicinity of the nest). With several wind turbines within the home ranges, the risks need to be multiplied by the number of wind turbines. Independent of the absolute values, which remain an issue for

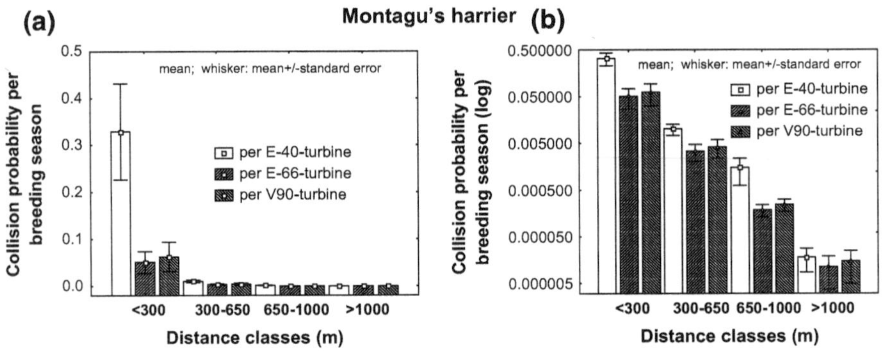

Fig. 14.1 Estimate of the collision probability per breeding season for territorial Montagu's harriers in North Frisia following Band et al. (2007), **a** standard scale, **b** log scale, see text for further details. *Vertical bars* indicate the standard error (n = 8; six birds with eight breeding seasons)

discussion, models revealed a fast and logarithmic reduction of the risk with increasing distance between nest and wind turbine (Fig. 14.1). The high collision risk within a radius of 300 m around the nest for small wind turbines with an average of above 30% is standing particularly out.

The collision probability of Montagu's harriers calculated in the Band model for small wind turbines was significantly higher than for medium-sized and large wind turbines within the same distance class. This tendency was especially pronounced in the vicinity of the nest (<300 m distance class: Kruskal-Wallis test, $Chi^2 = 10.6$; $p < 0.01$). The effect of the wind turbine size decreased with increasing distance from the nest and the differences between small and medium-sized or large turbines were not significant in distance classes of above 1000 m.

14.3.2 *Red Kite*

For red kites, the probability of a collision in flights through the rotor swept area (factor II) was calculated in the Band model as being 13.9% (Table 14.2). Individual differences between the investigated birds were included in the activity profiles (factor I). The absolute probability to collide with wind turbines in the vicinity of the nests was estimated to be considerably high (88% for distances <250 m and about 17% for distances between 250 and 500 m). For distances of below 500 m between wind turbines and nest, this results in one fatality per 4–5 breeding pairs and season (for the first two distance classes together). As expected, the estimated collision probability was continuously decreasing with increasing distance. Particularly notable was the difference between the distance class 1000–1250 m (about 3.5%) and 1250–1500 m (1.2%; Fig. 14.2). The modelled collision risk fell below 1% from distances between 1500 and 1750 m for red kites.

Table 14.2 Excel spreadsheet for the calculation of the probability of a collision for flights through the rotor swept area for the example of red kite and a medium-sized Enercon-E66 wind turbine. Technical specifications (number and length of rotor blades, rotation period, etc.) as well as bird parameters (length, wingspan, speed) are listed in the left-hand column. Collision probabilities for different angles of approach (separately for upwind and downwind) are calculated in the two right-hand columns. The mean is the average collision probability for a flight through the rotor swept area (for more information see Band et al. 2007)

Calculation of alpha and p(collision) as a function of radius

			r/R radius	c/C chord	α alpha	Upwind: collide length	Upwind: p(collision)	Upwind: contribution from radius r	Downwind: collide length	Downwind: p(collision)	Downwind: contribution from radius r
K: [1D or [3D] (0 or 1)	1										
NoBlades	3										
MaxChord	2.5	m									
Pitch (degrees)	30										
BirdLength	0.65	m	0.025	0.575	5.84	13.95	1.00	0.00125	12.51	1.00	0.00125
Wingspan	1.6	m	0.075	0.575	1.95	5.13	0.51	0.00381	3.69	0.37	0.00274
F: Flapping (0) or gliding (+1)	1		0.125	0.702	1.17	3.84	0.38	0.00476	2.09	0.21	0.00259
			0.175	0.860	0.83	3.48	0.34	0.00603	1.33	0.13	0.00230
Bird speed	6.9	m/sec	0.225	0.994	0.65	3.30	0.33	0.00736	0.82	0.08	0.00182
RotorDiam	66	m	0.275	0.947	0.53	2.81	0.28	0.00766	0.64	0.06	0.00173
RotationPeriod	4.39	sec	0.325	0.899	0.45	2.46	0.24	0.00791	0.71	0.07	0.00227
			0.375	0.851	0.39	2.43	0.24	0.00903	1.00	0.10	0.00370
			0.425	0.804	0.34	2.25	0.22	0.00948	1.06	0.10	0.00445
			0.475	0.756	0.31	2.10	0.21	0.00987	1.09	0.11	0.00513
Bird aspect ratioo: β	0.41		0.525	0.708	0.28	1.96	0.19	0.01020	1.11	0.11	0.00576
			0.575	0.660	0.25	1.84	0.18	0.01047	1.11	0.11	0.00633
			0.625	0.613	0.23	1.73	0.17	0.01068	1.11	0.11	0.00685
			0.675	0.565	0.22	1.62	0.16	0.01084	1.09	0.11	0.00730
			0.725	0.517	0.20	1.52	0.15	0.01093	1.07	0.11	0.00769
			0.775	0.470	0.19	1.43	0.14	0.01097	1.05	0.10	0.00802
			0.825	0.422	0.18	1.34	0.13	0.01094	1.02	0.10	0.00830
			0.875	0.374	0.17	1.25	0.12	0.01086	0.98	0.10	0.00851
			0.925	0.327	0.16	1.17	0.12	0.01072	0.95	0.09	0.00867
			0.975	0.279	0.15	1.09	0.11	0.01052	0.91	0.09	0.00877
						Overall p(collision) = Upwind		17.4%		Downwind	10.4%
								Average	13.9%		

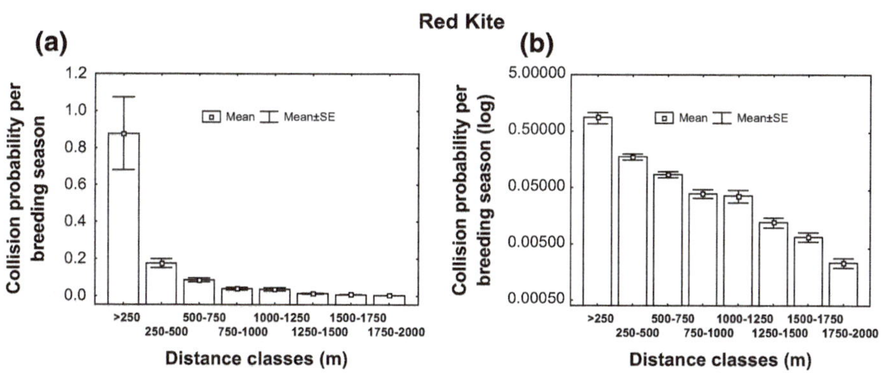

Fig. 14.2 Estimate of the collision probability per breeding season for territorial red kites in Saxony-Anhalt following Band et al. (2007), **a** standard scale, **b** log scale, see text for further details. *Vertical bars* highlight the standard error (n = 13; ten birds with 13 breeding seasons)

Even if the absolute values, which highly depend on the assumed avoidance rate, cannot be clearly defined, the reduction of the risk with increasing distance as well as the significant reduction in risk between the distance classes 1000–1250 and 1250–1500 m remains obvious.

14.4 Discussion

14.4.1 Individual Reaction

The collision risk for birds of prey breeding in the vicinity of wind turbines is present. Observations of red kites prove that even experienced birds, breeding in the same location for several years do collide with wind turbines (Mammen et al. 2009).

The informative value of the models is restricted by several factors. A correlation between the abundance of the birds and the collision risk as proven by Lucas et al. (2008) for the griffon vulture does not exist in all cases but seems to be true for red kites (Chap. 3). The individual behaviour of the birds also plays a major role. As telemetric monitoring of red kite has shown, the tagged birds differed considerably in terms of movement patterns and activity ranges. This variability did not depend on the gender of the birds. Individual characteristics were the main reason for the large range of modelled reactions. The reliability of the model statements was therefore also depending on the number of samples (n) included in the calculations.

Despite the smaller size small wind turbines proved to be more dangerous for Montagu's harriers, mainly due to the small distance between the lower blade tip and the ground. For a species, which is mainly flying low above the ground and therefore below the rotor, the distance between ground and lower blade tip is more important for the probability of a collision than rotor diameter or rotor swept area. A clearance of 20 m was assumed for smaller turbines (as opposed to 30 or 35 m for medium-sized and large turbines) resulting in higher collision values for the vicinity of the nest. With increasing distance this effect was balanced by the smaller space requirement of small wind turbines as compared to larger turbines.

In assessing the estimates for Montagu's harriers, a number of characteristics of the study area, the polders in North Frisia, need to be taken into consideration (very flat relief, special foraging areas within salt marshes on the seaward side of the sea dike), which may influence the patterns of flight movements. It is conceivable that Montagu's harrier will take more advantage of thermals in inland areas with favourable climatic conditions, shifting flight activity towards greater heights. It can still be assumed that the basic pattern of flight activities with the centre of activity in the area around the nest and use of low flight heights in foraging areas can be transferred to different breeding areas. Especially the hunting strategy of these birds is bound to flight heights low above the ground, and area-specific variations of foraging habitats will not be of importance. Thus, the results of the Band model simulations for the collision risk depending on the distance from the nest remain broadly valid, whereas a use of higher flight heights may result in a shift of the risk towards larger wind turbines. A general transferability of the results to other main distribution areas of the Montagu's harrier in Germany (e.g. Mainfranken, Hellweg Börde) seems to be possible. The influences of a changed flight behaviour can be tested in model simulations by adjusting the observed or assumed flight height distribution for different scenarios.

14.4.2 Avoidance Behaviour

The actual avoidance behaviour of the target species, i.e. the reduction of the probability of a collision by birds actively altering the flight path in the immediate vicinity of the rotor blades to avoid a collision, is the most critical point of the model. A video sequence published on the internet (http://www.youtube.com/watch?v=3Cxeaoh55CE; Arbeitskreis Wiesenweihenschutz Ostfriesland ©2010) documents at least one case of such behaviour in Montagu's harriers. Similar observations are generally rare and hardly ever published. This study follows the information given by Urquhart (2010), which is the latest and most comprehensive on this issue. Quantification of the avoidance behaviour, or factor III in the model including further not yet mentioned effects, requires sufficient data to determine collision rates and to compare these to the calculated theoretical risk (based on factors I and II, see May et al. 2010). A correction factor (largely corresponding to the avoidance rate but including further parameters as losses or migration, Bevanger et al. 2009) of between 92.7 and 96.3% depending on rotor speed and season was calculated for white-tailed sea eagles on the island of Smøla, where such a calculation was possible. An avoidance rate of 97.5% was assumed for Montagu's harriers in this study. This value was used by Whitfield and Band (in preparation) as a mean value for the closely related hen harrier showing similar behaviours based on existing results. The actual avoidance rate of these two species may well be even higher (about 99% for the Altamont Pass, California) according to observations in literature (Smallwood and Thelander 2004). Retrieved fatalities of this bird species are still too rare to derive a more exact avoidance rate.

The reduction of the collision probability between the distance classes (Fig. 14.1) is independent of the assumed avoidance rate. Variation of the assumed avoidance rate will only have an influence on the absolute values but not on the decreasing risk with increasing distance between nesting site and wind turbine.

14.4.3 Distance Recommendations

Currently, a minimum distance of 1000 m between nesting habitats of Montagu's harriers and red kites respectively and planned wind turbines is recommended by the Working Group of the German State Bird Observatories (Länderarbeitsgemeinschaft der Vogelschutzwarten LAG VSV 2007). Furthermore, the main flight paths to foraging areas should be kept clear.

The considerably fast decrease in collision probability for Montagu's harriers with increasing distance between wind turbine and nest determined in field studies and model simulations leads to the conclusion that observation a comparably small minimum distance of 300–500 m from the nest would already largely avoid a risk for this species. Today, Montagu's harriers mainly breed in cultivated grass- and farmland areas and therefore do not use the same nesting site over several years.

Even in established breeding pairs, breeding sites may spatially shift considerably between years. Implementation of fixed distance radii will therefore hardly be practicable in agricultural landscapes outside the main breeding areas (e.g. Mainfranken, Hellweg Börde). Quite different activity ranges of red kite during the breeding season are given by different authors (see Nachtigall et al. 2010 and references listed therein). All observations though indicate the existence of an activity centre within a radius of 1–3 km around the nest, where half of the total flight activity occurs. According to model calculations, a 1000 m buffer will decrease the risk of collision. It should however be noted that the reduction of the collision probability between the distance classes 750–1000 and 1000–1250 m is not as significant as the one between 1250–1500 and 1500–1750 m as well as between 1500–1750 and 1750–2000 m. Observations (see Chap. 3) confirm that only 50% of all activities occur within a radius of 100 m around the nest, and an enlargement of the buffer zone to 1500 m will include 70–80% of all activities.

References

Band W, Madders M, Whitfield DP (2007) Developing field and analytical methods to assess avian collision risk at wind farms. In: de Lucas M, Janss G, Ferrer M (eds) Birds and Wind Power. Quercus, Madrid, p 259–275

Bauer H-G, Bezzel E, Fiedler W (2005) Das Kompendium der Vögel Mitteleuropas. Aula Verlag, Wiesbaden

Bevanger K, Berntsen F, Clausen S, Dahl EL, Flagstad Ø, Follestad A, Halley D, Hanssen FE, Hoel PL, Johnsen L, Kvaløy P, May R, Nygård T, Pedersen HC, Reitan O, Steinheim Y, Vang R (2009) Pre- and post-construction studies of conflicts between birds and wind turbines in coastal Norway (BirdWind). Progress Report 2009, NINA Report 505, Trondheim, 70 pp

Länder-Arbeitsgemeinschaft der Vogelschutzwarten, LAG-VSV (2007) Abstandsregelungen für Windenergieanlagen zu bedeutsamen Vogellebensräumen sowie Brutplätzen ausgewählter Vogelarten. Berichte zum Vogelschutz 44:151–153

Lekuona JM, Ursúa C (2007) Avian mortality in wind power plants of Navarra (Northern Spain). In: de Lucas M, Janss G, Ferrer M (eds) Birds and Wind Power. Quercus, Madrid, p 177–192

de Lucas M, Janss GFE, Whitfield DP, Ferrer M (2008) Collision fatality of raptors in wind farms does not depend on raptor abundance. J Appl Ecol 45:1695–1703

Mammen U, Stubbe M (2003) Jahresbericht zum Monitoring Greifvögel und Eulen Europas 15:1–101

Mammen U, Mammen K, Strasser C, Resetaritz A (2009) Rotmilan und Windkraft – eine Fallstudie in der Querfurter Platte. In: Stubbe M, Mammen U (eds) Populationsökologie Greifvogel- und Eulenarten 6, p 223–231

May R, Hoel PL, Langston R, Dahl EL, Bevanger K, Reitan O, Nygård T, Pedersen HC, Røskaft E Stokke BG (2010) Collision risk in white-tailed eagles. Modelling collision risk using vantage point observations in Smøla wind-power plant. NINA Report 639, Trondheim, pp 25

Mebs T, Schmidt D (2006) Die Greifvögel Europas, Nordafrikas und Vorderasiens. Biologie, Kennzeichen und Bestände. Kosmos, Stuttgart, pp 495

Nachtigall W, Stubbe M, Herrmann S (2010) Aktionsraum und Habitatnutzung des Rotmilans (*Milvus milvus*) während der Brutzeit – eine telemetrische Studie im Nordharzvorland. Vogel und Umwelt 18:25–61

R development core team (2010). http://www.r-project.org/

Smallwood KS, Thelander CG (2004) Developing methods to reduce bird mortality in the Altamont Pass Wind Resource Area. Final report to the California Energy Commission, Public Interest Energy Research Environmental Area, Contract 500-01-019, Sacramento

Smallwood KS, Thelander CG (2008) Bird Mortality in the Altamont Pass Wild Resource Area, California. J Wildlife Manage 72:215–223

Sterner D, Orloff S, Spiegel L (2007) Wind turbine collision research in the United States. In: de Lucas M, Janss G, Ferrer M (eds) Birds and Wind Power. Quercus, Madrid, p 81–100

Strickland D, Erickson W, Young D, Johnson G (2007) Selecting study designs to evaluate the effect of windpower on birds. In: de Lucas M, Janss G, Ferrer M (eds) Birds and Wind Power. Quercus, Madrid, p 117–136

Tucker VA (1996) A mathematical model of bird collisions with wind turbine rotors. J Sol Energ-T ASME 118:253–269

Urquhart B (2010) Use of avoidance rates in the SNH wind farm collision risk model. SNH Avoidance Rate Information & Guidance Note, Scottish natural heritage, pp 10

Whitfield DP, Band W (in preparation) Estimates of collision avoidance rates at operational wind farms in the USA

Whitfield DP, Madders M (2006) A review of the impacts of wind farms on hen harriers *Circus cyaneus* and an estimation of collision avoidance rates. Unpublished report, Natural Research, Banchory

Chapter 15
Population Development and Breeding Success of Birds of Prey in Relation to the Development of Wind Energy Use in Germany

Leonid Rasran and Ubbo Mammen

15.1 Introduction

In contrast to the other sub-studies, the present sub-study does not deal with impacts of wind turbines on the behaviour and the risks for individuals, but with the effects of wind turbines within the landscape on the population of relevant bird of prey species. Population size and breeding success of red kites, white-tailed sea eagles and further (more abundant) bird of prey species were investigated in relation to the presence and density of wind turbines within the same area. The Montagu's harrier, playing an important role in the overall study, was not included in the analyses, as the occurrence of this species is limited to only a few areas in Germany only. Furthermore, the question was raised, whether the current density of wind turbines in Germany (as from 2006) has such a strong influence (disturbance during the breeding season, mortality) on birds of prey that the populations of some of the species are threatened with extinction.

L. Rasran (✉)
NABU (Nature and Biodiversity Conservation Union) Michael-Otto-Institute,
Goosstroot 1, 24861 Bergenhusen, Germany
e-mail: Leonid.Rasran@boku.ac.at

U. Mammen
ÖKOTOP GbR and Förderverein Greifvogelmonitoring,
Willy-Brandt-Straße 44, 06110 Halle (Saale), Germany

H. Hötker et al. (eds.), *Birds of Prey and Wind Farms*,
DOI 10.1007/978-3-319-53402-2_15

15.2 Methods

15.2.1 Material

Evaluations of breeding pair densities and breeding success are based on MEROS (Monitoring European Raptors and Owls) results (http://www.greifvogelmonitoring.de, Mammen and Stubbe 2003, 2009). The programme comprises the collection and evaluation of current and historical data on population sizes and reproduction of birds of prey and owls in Europe. Conclusions going beyond local and regional developments, are drawn on the current situation, population dynamics and the development of reproduction rates of these species in Germany and in Europe. 225 of the monitoring areas located within Germany were selected based on the following criteria: the size of the areas (between 30 and 2000 km^2) as well as the availability of data on population densities (number of breeding pairs) and/or the breeding success (number of fledged young birds) of at least one of the following species: red and black kite, white-tailed sea eagle, common buzzard, European honey buzzard, Eurasian hobby, common kestrel, Eurasian sparrowhawk, northern goshawk or marsh harrier. The study period included the years 1991–2006 for population densities and 1991–2002 for breeding success. The selected monitoring areas cover a total of 20% of the German land area but are mainly concentrated on federal states in the eastern part of Germany.

As continuous records over the entire sturdy period were available for only a few of the areas, data with only few years of observation were included and all data were considered un-balanced data in the following. This way, 25 areas or 92 single records from several years were available for the rarest species (white-tailed sea eagle). The sample size available for the more abundant species was considerably larger. The breeding density for red kite was, for example, available for 147 areas with a total of 1005 records (Table 15.1). The statistical power for the smallest sample was determined as 0.81 at a significance level of 0.05 using the GPower 3.0.10 software (Faul et al. 2009).

For all of the selected monitoring areas, the development of wind energy use (number of operational wind turbines, total capacity) was documented per year. Wind turbines constructed and put into operation before April of the respective year (start of the breeding season of most birds of prey) were included. The data given in the operator database were used (Keiler 2008; an example of this data for a monitoring area is given in Table 15.2).

15.2.2 Statistical Methods

The development of bird of prey populations and breeding success was represented in generalised linear models (linear mixed effects models adjusted for maximum likelihood). The time parameter was included in the model to exclude differences

Table 15.1 Distribution of bird of prey monitoring areas in Germany

Federal state	Number of monitoring sites	Area (km^2)	State area	% of monitoring areas in relation to state area
Sachsen-Anhalt	28	7024.6	20,445.7	34.4
Berlin/Brandenburg	38	15,933.9	30,370	52.5
Baden-Würtemberg	11	3303.1	35,751.6	9.2
Bayern	10	6747.6	70,549.4	9.6
Hessen	10	2020	21,114.8	9.6
Mecklenburg-Vorpommern	19	7270.6	23,178.5	31.4
Niedersachsen	16	5821	47,619.6	12.2
Nordrhein-Westfalen	34	8882.5	34,084.1	26.1
Rheinland-Pfalz	3	255	19,853.5	1.3
Schleswig-Holstein	8	2162	15,763.4	13.7
Saarland	2	56	2568.7	2.2
Sachsen	27	6881.4	18,414.7	37.4
Thüringen	19	5172	16,172.1	32.0
Germany	225	71,529.7	357,045.6	20.0

Table 15.2 Example of the operator database (Keiler 2008) data for one of the monitoring areas (No. 78, Altenpleen near Stralsund)

Area Nr.	Year	Size (km^2)	Number of wind turbines [a]	Output (kW)[a]	Number of breeding pairs			
					Red kite	Black kite	Common buzzard	White-tailed sea eagle
G_34	1991	120	0	0	8	0	59	0
G_34	1992	120	0	0	11	2	68	0
G_34	1993	120	0	0	12	2	69	0
G_34	1994	120	1	500	11	2	75	0
G_34	1995	120	1	500	11	2	64	0
G_34	1996	120	6	3200	8	2	62	1
G_34	1997	120	6	3200	10	2	59	0
G_34	1998	120	6	3200	10	1	52	0
G_34	1999	120	6	3200	7	1	57	1
G_34	2000	120	6	3200	7	2	54	1
G_34	2001	120	6	3200	8	1	56	1
G_34	2002	120	6	3200	9	2	55	1
G_34	2003	120	6	3200	11	2	57	1
G_34	2004	120	6	3200	10	3	60	1
G_34	2005	120	6	3200	13	3	53	1
G_34	2006	120	6	3200	12	3	44	1

[a]Only wind turbines which were at work during the breeding season of raptors were taken into account

due to weather conditions between the years. A 0 hypothesis model of the random breeding pair density distribution was compared to a time-dependent distribution per year in an analysis of variance (ANOVA). In a second step, the time-dependent distribution was compared to a model adjusted for wind energy use (number of wind turbines or capacity per area).

The development of wind energy use (wind turbine density, capacity per area) in the monitoring areas over time was separately analysed in linear models and compared to the 0 hypothesis of a random distribution independent of time.

The calculations were carried out using software R 2.8.1 (R development core team 2008).

15.3 Results

15.3.1 Development of Wind Energy Use

Whereas (mostly small) wind turbines were installed in a few of the monitoring areas at the beginning of the study period (1991), wind turbines were in operation in about 2/3 of all areas towards the end of the period (2006) (Table 15.3). For all areas relevant for the target species, a significant increase in density and capacity could be observed over time.

Table 15.3 Number of monitoring areas evaluated per year and areas with wind turbines per year

Year	Number of areas	Areas with turbines
1991	107	7
1992	116	10
1993	124	16
1994	129	24
1995	131	34
1996	134	46
1997	132	54
1998	138	64
1999	145	73
2000	147	75
2001	150	82
2002	141	78
2003	139	80
2004	129	74
2005	110	64
2006	100	62
Total	225	

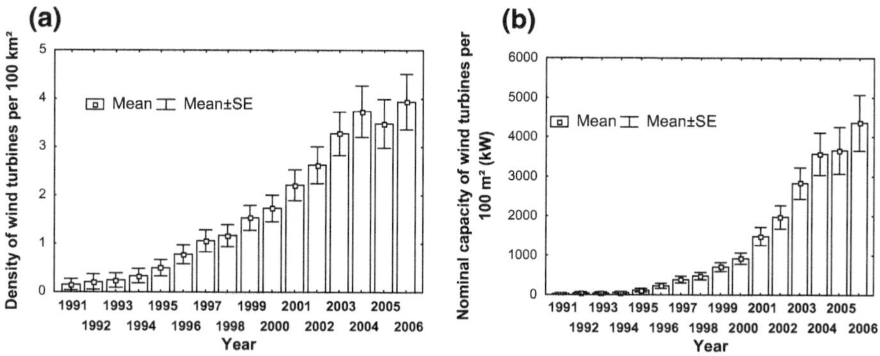

Fig. 15.1 Density (**a**) and nominal capacity (**b**) of wind turbines in monitoring areas per 100 km^2

The average density of wind turbines rose from 0.15 to just under 4 per 100 km^2 (Fig. 15.1). As to be expected, the distribution of wind turbines varied considerably between the areas. Only for 25 of the areas densities of 10 wind turbines and more were reached per 100 km^2. The increase of the capacity per area developed similarly and was even more pronounced due to repowering (replacement of many small wind turbines by a few higher capacity turbines) in some of the areas starting in 2002.

15.3.2 Development of Breeding Populations

Population densities of most of the target species remained relatively stable in most of the monitoring areas. For many of the species (red and black kite, common buzzard, northern goshawk, Eurasian sparrowhawk, marsh harrier) year-to-year fluctuations did not exceed 10% of the population size (see Fig. 15.2a for red kite).

Fluctuations in the European honey buzzard were much more pronounced (see Fig. 15.2j). Significant differences were found between years with no clear trend for the overall population, which remained at a similar level.

A clear increase in population size could be observed for white-tailed sea eagle (Fig. 15.2c), common kestrel (Fig. 15.2g) and Eurasian hobby (Fig. 15.2h) over the study period.

A significant relation between population dynamics and the development of wind energy use could not be established for any of the species. A clear difference in the breeding density distribution between areas with wind turbines and without wind turbines could be determined for red kite (t test; $t = 26.75^{***}$).

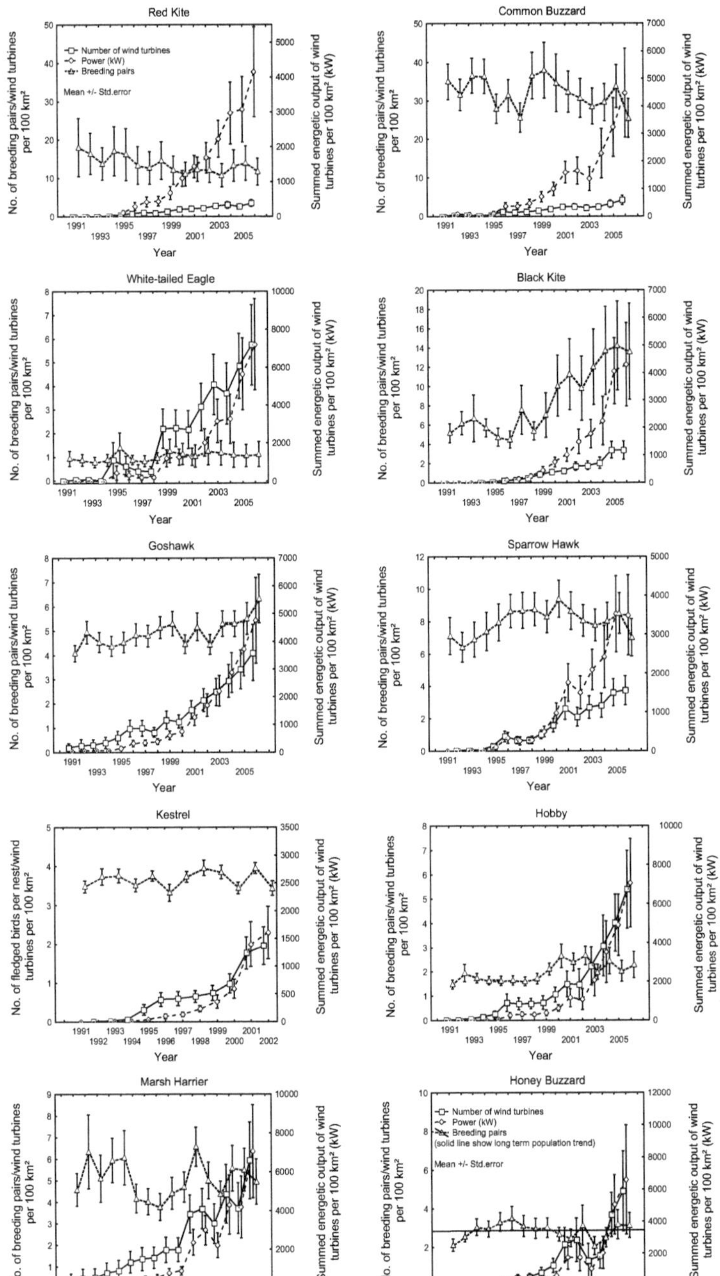

Fig. 15.2 Population densities of bird of prey species in monitoring areas in comparison to the development of wind energy use (number and total capacity of wind turbines)

15.3.3 *Breeding Success*

The number of young birds fledged per breeding pair varied only slightly for all areas and years. For the northern goshawk only, (slightly) significant variations were found between years (Table 15.4).

The positive trend of the total population size found for the white-tailed sea eagle in recent years occurred parallel to the increase in wind turbine density. An insignificant interaction value proves though that these developments were independent of each other and no mutual influence can be assumed.

A significant relation between breeding success and density of wind turbines was found for the European honey buzzard and black kite. For both species the general trend was comparably stable since 1991. In the last 3 or 4 years of the study (1998–2002) however, the breeding success was clearly decreasing. The rapid increase in wind turbine density also occurred during this period (negative correlation of breeding success and wind turbine density for the period between 1998 and 2002, $r = -0.87$ for black kite and $r = -0.83$ for the European honey buzzard, see Fig. 15.3d, j).

15.4 Discussion

15.4.1 *Methods*

The mathematical soundness and validity of the applied statistical models was verified and confirmed by an external expert for statistics (Dr. Michael Breuer, University of Kiel). However, a number of limitations, considerably reducing the informative value of the results, need to be taken into account:

- Data on birds of prey was collected by numerous voluntary helps at different times. For many of the monitoring areas records were only available for single years and usually the accuracy of the data was decreasing with increasing size of the area to be monitored.
- Monitoring areas consisted of smaller areas, covering only or at least to a major proportion habitats of relevant bird of prey species but also of larger areas with a large proportion of forest or residential areas within the monitored area.
- The exact locations of nests were not included in the available data. Therefore, no conclusions on the actual distance between nests and wind turbines could be drawn and the analyses are based on presence within partly extensive areas. Directly adjacent wind turbines outside monitoring areas were not included.
- Data were generally inhomogeneous and not normally distributed. The data included many 0 values and strong outliers.
- Data collection ended in 2006, for breeding success already in 2002. For this reason, the rapid increase in wind turbine density especially over the last years of the study period could not be analysed. About 19,000 wind turbines were in

Table 15.4 Influence of the development of wind energy use on the size of the population (number of breeding pairs) and the breeding success of birds of prey

		Species	Red Kite *Milvus milvus*	Common Buzzard *Buteo buteo*	White-tailed Eagle *Haliaeetus albicilla*	Black Kite *Milvus migrans*	Goshawk *Accipiter gentilis*
	Model	Test					
Wind Energy							
N of observations			1005	843	210	630	1031
N of areas			147	130	37	98	135
Number of wind turbines:			F	F	F	F	F
0-Hypothesis	1						
Time	2	1 vs 2	**63,89*** **	**40,62*** **	**33,22*** **	**50,74*** **	**93,23*** **
Energetic output:							
0-Hypothesis	1						
Time	2	1 vs 2	**53,18*** **	**35,36*** **	**29,10*** **	**38,37*** **	**86,36*** **
Population							
N of observations			1005	843	210	630	1031
N of areas			147	130	37	98	135
			F	F	F	F	F
0-Hypothesis	1						
Time	2	1 vs 2	3.04	0.02	**10,80*** **	1.14	0.9
Number of wind turbines	3	2 vs 3	0.01	0.59	0.13	0.19	2.73
Time x Number	4	3 vs 4	0.12	0.26	1.7	0.02	0.68
Energetic output	5	2 vs 5	0.02	0.03	0.56	0.06	1.64
Time x output	6	5 vs 6	0.03	0.11	0.62	0.01	2.26
Breeding success							
N of observations			673	605	92	739	653
N of areas			122	113	25	74	119

(continued)

Table 15.4 (continued)

			F	F	F	F	F
0-Hypothesis	1						
Time	2	1 vs 2	1.05	1.78	3.58	0.83	**4,71***
Number of wind turbines	3	2 vs 3	0.01	2.48	0.02	**4,96***	1.05
Time x number	4	3 vs 4	1.23	0.38	**5,65***	2.11	0.15
Energetic output	5	2 vs 5	0.01	0.01	0.74	**3,97***	0.3
Time x output	6	5 vs 6	2.13	0.79	**5,99***	0.83	0.23
		Species	Sparrow Hawk *Accipiter nisus*	Kestrel *Falco tinnunculus*	Hobby *Falco subbuteo*	Marsh Harrier *Circus aeruginosus*	Honey Buzzard *Pernis apivorus*
	Model	Test					
Wind Energy							
N of observations			668	732	629	515	454
N of areas			115	117	107	83	91
Number of wind turbines:			F	F	F	F	F
0-Hypothesis	1						
Time	2	1 vs 2	**70,53*** **	**60,81*** **	**50,43*** **	**42,54*** **	**71,02*** **
Energetic output:							
0-Hypothesis	1						
Time	2	1 vs 2	**43,02*** **	**48,61*** **	**39,42*** **	**36,9*** **	**59,35*** **
Population							
N of observations			668	732	629	515	454
N of areas			115	117	107	83	91

(continued)

Table 15.4 (continued)

			F	F	F	F	F
0-Hypothesis	1						
Time	2	1 vs 2	0.16	**4,92***	**6,32***	1.19	**30,82*****
Number of wind turbines	3	2 vs 3	0.82	1.49	0.77	0.22	0.79
Time x Number	4	3 vs 4	0.62	0.01	0.01	3.36	0.08
Energetic output	5	2 vs 5	0.01	1.14	0.34	0.01	0.48
Time x output	6	5 vs 6	0.75	0.04	0.46	2.42	0.02
Breeding success							
N of observations			436	485	292	254	174
N of areas			83	96	70	57	52
			F	F	F	F	F
0-Hypothesis	1						
Time	2	1 vs 2	2.46	0.47	0.46	0.13	0.01
Number of wind turbines	3	2 vs 3	2.85	0.62	1.75	0.01	**10,64****
Time x number	4	3 vs 4	0.97	2.43	0.25	0.01	**6,72****
Energetic output	5	2 vs 5	2.02	0.13	2.31	0.03	**8,03****
Time x output	6	5 vs 6	2.35	0.63	0.01	0.08	**11,83*****

Analyses of variance of linear models. $*p < 0.05$; $**p < 0.01$; $***p < 0.001$

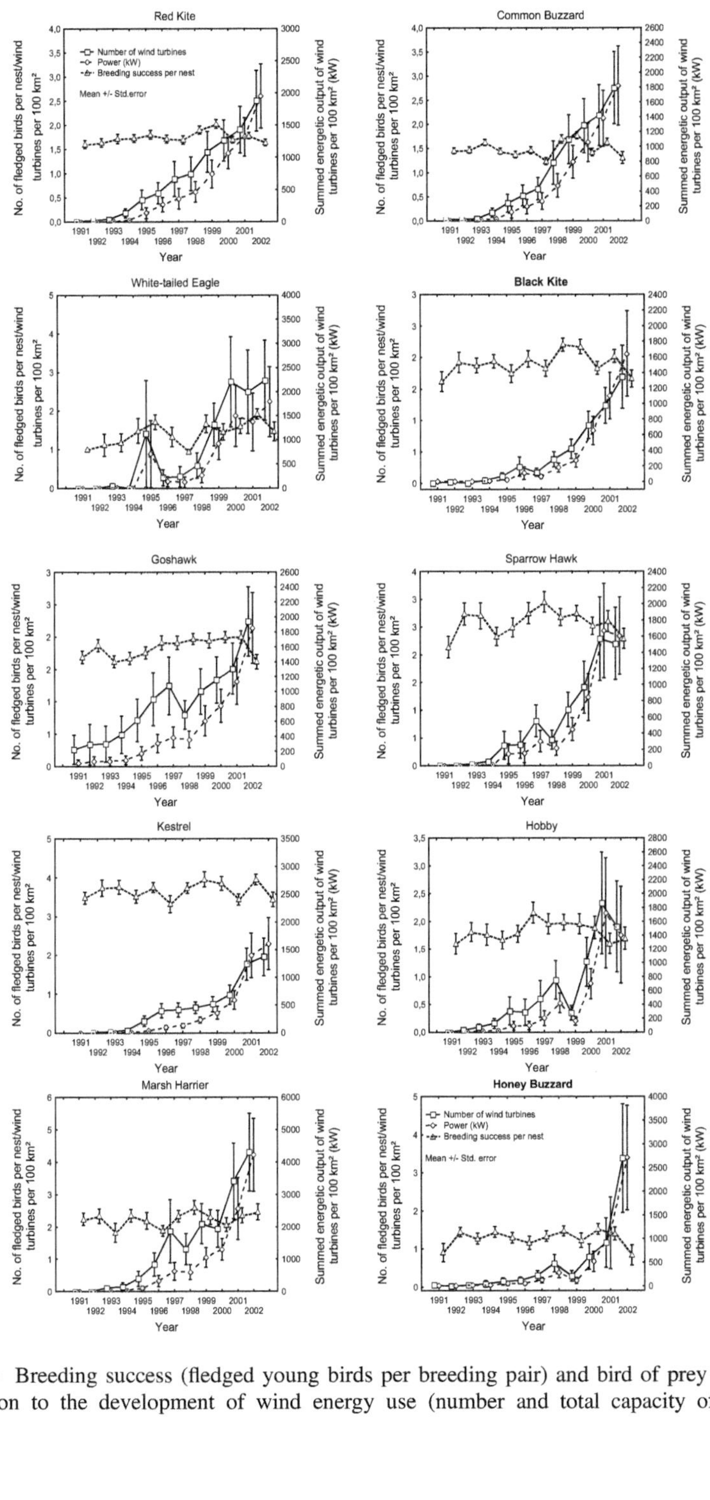

Fig. 15.3 Breeding success (fledged young birds per breeding pair) and bird of prey species in comparison to the development of wind energy use (number and total capacity of all wind turbines)

operation in Germany towards the end of 2006 (overall capacity about 20.6 GW). With 21,000 wind turbines and a capacity of 27.8 GW these numbers were considerably higher in 2010 (Sources: DEWI 2011, EnergyMap.info 2011). In comparison to 2002 (about 14,000 wind turbines, 11.8 GW nominal capacity) the difference is even more obvious.

The key statements derived from model calculations need to be critically reviewed, as already small changes in the sample resulted in strong variations in data. The β error could only be estimated instead of for the existing data. Therefore, the possibility that non-existing correlations may have been found cannot be excluded.

15.4.2 Results

The results of the statistical analyses could not prove an influence of the present densities of wind turbines within the monitoring areas (as from 2006) on the investigated bird of prey species. The total white-tailed sea eagle and red kite populations did not seem to be at risk from wind energy use despite the fact that these species were among collision victims on a regular basis. The assumption that wind energy use did play a comparably minor role among the anthropogenic mortality factors for birds of prey so far (e.g. power lines, traffic, hunting, i.e. see Bevanger 1998; Kalpakis et al. 2009) was already suggested by other authors (e.g. Erickson et al. 2001; de Lucas et al. 2004).

Furthermore it is important to note that for some of the species (e.g. white-tailed sea eagle) distance recommendations were broadly observed and wind turbines were rarely located in the immediate vicinity of nests. The lack of a relation between the development of populations and wind energy development will have been influenced by these precautionary measures.

It should be kept in mind, that the lack of a relation between both developments found in the analyses needs still to be qualified for the above mentioned reasons. A clear reduction of the population density of a bird of prey species due to the development of wind energy use would imply that the previous development (up until 2006) of wind energy use in Germany already had serious consequences for species protection, severely threatening the population of the respective species.

High fatality rates putting a population at risk of extinction usually are reported for wind farms in the immediate vicinity of topographical bottlenecks (Lekuona 2001; Telleria 2009) used by large numbers of migrating birds (mountain passes, straits). Birds in breeding areas seem to be much less affected (Madders and Whitfield 2006). Hunt (2002) assessed the breeding population of golden eagles at a wind farm in California to be stable despite the high fatality rate. The study did not consider regulation of losses by immigration of individuals from more distant areas (Smallwood and Thelander 2008). Evidence of an entire population at risk of extinction due to wind energy use would be provided, if collision fatalities together with mortality for other reasons balance or exceed the reproduction rate as proven for the burrowing owl (*Athene cunicularia*) in the Altamont Pass by Smallwood et al. (2007). Wind turbines however did not present the only risk in this study, and due to the small size of the population and the low reproduction rate, a minor additional increase in mortality had a strong impact.

The lack of a direct impact of wind farms on the monitoring areas analysed in the present study cannot be understood as a complete all-clear. It merely means that the current (as from 2006) and Germany-wide impacts of wind turbines on birds of prey did not exceed the threshold of an immediate risk.

Displacement of birds of prey from breeding areas due to avoidance of wind turbines proven for other bird species of the open landscape (Hötker et al. 2006) will not have played a role, as species investigated in the present study did not show avoidance behaviour (i.e. see Chap. 3).

As higher densities of wind turbines mainly occurred from 2002, population development under the intensified influence of wind turbines could only be analysed for a comparably short period of time. No clear trends are usually apparent for such short period. Furthermore, data on breeding success was only available until 2002 and will therefore be even less informative (e.g. see Reichenbach and Steinborn 2007).

Particularly high population densities of for example red kite were only recorded in areas free from wind turbines with significantly higher densities than found for areas with wind turbines. These mainly small monitoring areas consisted of especially valuable areas already designated as a nature conservation area where the construction of wind farms is not allowed. A significant relation between the higher breeding pair densities and the lack of wind turbines in an area may suggest that at least locations with larger accumulations of breeding birds (colonies) may be impaired by the construction of wind farms. No wind farms should be constructed in these areas in the future (see Percival 2005).

The impacts of wind farms on European honey buzzards and black kites should be further investigated in the future. A slightly negative trend of the overall breeding population was found for European honey buzzards (Mammen and Stubbe 2009). Data relating to wind energy use is comparably incomprehensive for this species, and statistical results indicating an influence of wind energy use on the breeding success may as well be interpreted as non-existing correlation. For this species, no immediate risk caused by wind farms can be derived from the results to date.

References

Bevanger K (1998) Biological and conservation aspects of bird mortality caused by electricity power lines: a review. Biol Conserv 86:67–76

de Lucas M, Janss GFE, Ferrer M (2004) The effects of a wind farm on birds in a migration point: the Strait of Gibraltar. Biodiv Conserv 13:395–407

DEWI (2011) DEWI GmbH - Deutsches Windenergie Institut. http://www.dewi.de/dewi/index.php?id=1&L=1. Assessed 30 June 2010

EnergyMap.info (2011) Deutsche Gesellschaft für Sonnenenergie e.V. http://www.energymap.info . Assessed 30 June 2010

Erickson WP, Johnson GD, Strickland MD, Young DP, Sernka KJ, Good RE (2001) Avian collisions with wind turbines: a summary of existing studies and comparison to other sources of avian collision mortality in the United States. National Wind Coordinating Committee (NWCC), Western Ecosystems Technology Inc. Washington D.C.

Faul F, Erdfelder E, Buchner A, Lang A-G (2009) Statistical power analyses using G*Power 3.1: Tests for correlation and regression analyses. Behav Res Methods 41:1149–1160

Hötker H, Thomsen K-M, Jeromin H (2006) Impacts on biodiversity of exploitation of renewable energy sources: the example of birds and bats – facts, gaps of knowledge, demands for further research, and ornithological guidelines for the development of renewable energy exploitation. Michael-Otto-Institut im NABU, Bergenhusen

Hunt G (2002) Golden eagles in a perilous landscape: predicting the effects of mitigation for wind turbine blade-strike mortality. Consultant Report to California Energy Commission, Sacramento

Kalpakis S, Mazaris AD, Mamakis Y, Poulopoulos Y (2009) A retrospective study of mortality and morbidity factors for Common Buzzards *Buteo buteo* and Long-legged Buzzards *Buteo rufinus* in Greece: 1996–2005. Bird Conserv Int 19:15–21

Keiler J (2008) Windkraft Betreiber-Datenbasis: Register/Errichtungsdaten deutscher Windanlagen. http://www.btrdb.de/

Lekuona JM (2001) Uso del espacio por la avifauna y control de la mortalidad de aves y murciélagos en los parques eólicos de Navarra durante un ciclo anual. Direccion General de Medio Ambiente, Gobierno de Navarra, Pamplona

Madders M, Whitfield PD (2006) Upland raptors and the assessment of wind farm impacts. Ibis 148:43–56

Mammen U, Stubbe M (2003) Jahresbericht 2002 zum Monitoring Greifvögel und Eulen Europas 15:1–101

Mammen U, Stubbe M (2009) Jahresbericht 2003 und 2004 zum Monitoring Greifvögel und Eulen Europas 16/17:1–118

Percival SM (2005) Birds and windfarms: what are the real issues? British Birds 98:194–204

R development core team (2008) http://www.r-project.org/

Reichenbach M, Steinborn H (2007) Langzeituntersuchungen zum Konfliktthema "Windkraft und Vögel", 6. Zwischenbericht, Gutachten im Auftrag von MMJ GmbH, http://arsu.de/de/media/fiebing_gutachten_2007.pdf. Assessed 30 June 2010

Smallwood KS, Thelander CG, Morrison ML, Rugge LM (2007) Burrowing owl mortality in the Altamont Pass Wild Resource Area. J Wildlife Manage 71:1513–1524

Smallwood KS, Thelander CG (2008) Bird Mortality in the Altamont Pass Wild Resource Area, California. J Wildlife Manage 72:215–223

Tellería JL (2009) Potential impacts of wind farms on migratory birds crossing Spain. Bird Conserv Int, doi:10.1017/S0959270908008137

Chapter 16
Conclusions, Risk Assessment, Conflict Minimisation, Practical Recommendations, Need for Further Research

Hermann Hötker, Tobias Dürr, Bodo Grajetzky, Thomas Grünkorn, Ralf Joest, Oliver Krone, Kerstin Mammen, Ubbo Mammen, Georg Nehls, Leonid Rasran, Alexander Resetaritz and Gabriele Treu

16.1 Risk Assessment

The investigations carried out for this project confirmed the assumption expressed for other bird of prey species that birds of prey are not or hardly disturbed or displaced by wind turbines. Instead they collide disproportionately often with wind turbines in comparison to their abundance. This is not only true for Germany but is

H. Hötker (✉) · K. Mammen · L. Rasran · A. Resetaritz
NABU (Nature and Biodiversity Conservation Union) Michael-Otto-Institute,
Goosstroot 1, 24861 Bergenhusen, Germany
e-mail: Hermann.Hoetker@NABU.de

T. Dürr
Staatliche Vogelschutzwarte (State Bird Observatory and Field Centre)
of Landesamt für Umwelt, Gesundheit und Verbraucherschutz Brandenburg
(Brandenburg Environmental Agency), Buckower Dorfstr. 34,
14715 Nennhausen-Buckow, Germany

B. Grajetzky · T. Grünkorn · G. Nehls
BioConsult SH, Schobüller Str. 36, 25813 Husum, Germany

R. Joest
Arbeitsgemeinschaft Biologischer Umweltschutz im Kreis Soest
(Soest District Working Group for Nature Conservation
and Landscape Management), Teichstraße 19,
59505 Bad Sassendorf-Lohne, Germany

O. Krone · G. Treu
Leibniz Institute for Zoo and Wildlife Research (IZW),
Alfred-Kowalke-Str. 17, 10315 Berlin, Germany

K. Mammen · U. Mammen · A. Resetaritz
ÖKOTOP GbR, Willy-Brandt-Straße 44, 06110 Halle (Saale), Germany

H. Hötker et al. (eds.), *Birds of Prey and Wind Farms*,
DOI 10.1007/978-3-319-53402-2_16

confirmed by many international studies (Smallwood and Thelander 2004; Drewitt and Langston 2006; Madders and Whitfield 2006; de Lucas et al. 2004; Dahl et al. 2012).

16.1.1 Red Kite

No evidence of on avoidance of wind turbines or spatial displacement by wind farms could be found for red kites in the present study. Where red kites forage is apparently above all determined by the availability of food in the breeding area independent of wind farms, which may be present. Certain structures created in the construction of a wind farm (service roads, fallows around turbines, parking and turning areas) will even attract red kites, if populations of small mammals and thereby important sources of food develop around these structures. Further attractive foraging areas may be available to red kites within wind farms. Freshly harvested and/or ploughed agricultural land may provide access to food which is not as easily accessible in not yet harvested surrounding areas due to high growing vegetation. Manure heaps are a further source of food. Even if the respective sources of food are located just outside a wind farm red kites will visit the investigated wind farm more often.

Red kites mostly use the area around the nest with high flight intensity. The closer the wind farm to the nest, the more likely did red kites enter the wind farm.

Red kites staying in a wind farm were almost inevitably at risk, as they were often flying at height of the rotor (46.6% of the observed flight time was spent in the potentially dangerous height band of between 26 and 150 m) approaching wind turbines without hesitation and even flying between rotating turbines (observations of U. Mammen).

Band model calculations confirmed the results of telemetric monitoring for several scenarios. The collision risk was considerably decreasing with increasing distance from the nest (Chap. 14). In the close-up area (<250 m) the collision risk was ten times as high and still twice as high at distances up to 750 m as compared to the far range (>1250 m).

16.1.2 White-Tailed Sea Eagle

White-tailed sea eagles did not show distinct avoidance behaviour towards wind turbines. Among all large bird species investigated in the present study, the collision risk was highest for white-tailed sea eagles measured by the number of reported collisions in relation to the size of the population in Germany. One reason may be found in the size of the eagles. The statistical likelihood that an object moving through the rotor swept area is hit by a rotor blade also depends on its size

(Band et al. 2007). The probability to find a collision victim under a wind turbine will also be higher for this large eagle species in comparison to e.g. Montagu's harriers.

Additionally, the surrounding area of the nest is an area with especially intense flight activity for white-tailed sea eagles and therefore an area with a particularly high risk of collision with present wind turbines. Collisions of white-tailed sea eagles with wind turbines may as well occur at larger distances from the breeding territory. An analysis of the so far documented collision victims did not indicate an influence of the distance from existing breeding sites. Depending on habitat conditions and quality white-tailed sea eagles may travel large distances on their way to foraging grounds crossing wind farms on the way or visiting attractive structures within wind farms. The results of the study lead to the following assessment: Wind farms within the home range of white-tailed sea eagles will always imply a risk of collision. This risk is not limited to the close vicinity of breeding sites and should therefore be taken into consideration for the entire home range of white-tailed sea eagles.

16.1.3 Montagu's Harrier

While the present study was still ongoing, the first collisions of Montagu's harriers with wind turbines were reported in 2010 (Dürr 2011) so that the species can now be found among the bird of prey species where birds found dead confirm the collisions. The very small number of retrieved victims of collisions with wind turbines in Germany in comparison to the other two species does not allow assuming a low collision risk, especially as further collisions are known from Spain. Several factors indicate that the proportion of birds found or reported dead is relatively small (Langgemach and Dürr 2011). Due to the rarity of the species in Germany (410–470 breeding pairs, Südbeck et al. 2007), the comparably small body size and the fact that Montagu's harriers are in Germany mainly present between April and August when vegetation is highest it is relatively unlikely that a harrier will be found and reported after a collision. The number of unreported Montagu's harrier collisions may therefore be relatively high.

The results for flight behaviour (Sect. 4.1.8) have shown that flights in the risk area (courtship, food passes, thermal soaring, interaction) made up a relative small proportion of the total flight time. These potentially risky behaviours are concentrated on the surrounding area of the nest and constantly repeated throughout the breeding season. Outside a radius of about 300 m around the nest, the intensity of flight activity and use of higher flight heights is clearly decreasing with distinct individual variability. As Montagu's harriers show no avoidance behaviour towards wind turbines according to the results for flight behaviour, an increased risk of collision can therefore be assumed for the area of the nest.

A merely sporadic use of wind farms was found in the relatively richly structured territories within the study areas in Schleswig-Holstein, where the presence of attractive hunting grounds (salt marshes, richly structured grassland) was able to keep Montagu's harriers away from wind farms. It could be demonstrated that attractive habitats in or in the vicinity of wind farms will instead lead to a concentration of Montagu's harriers in the impact zone of wind turbines. In general it can be assumed that especially in poorly structured agricultural landscapes with a lack of suitable structures (edge structures, etc.) or habitats (extensively farmed grassland, fallows, etc.) wind farms will be used for resting and foraging.

In summary it can be concluded that the collision risk for Montagu's harriers depends first of all on the distance between nest site and wind farms and the accompanying structures. This assessment is confirmed by results of collision modelling following Band et al. (2007) for different radii around the nest. Model simulations show that the distance between nest and wind turbines has a considerably stronger influence than other factors like wind turbine model or dimensions (see Chap. 14). The results indicate that a further increase in wind turbine density in Montagu's harriers breeding areas will intensify conflicts.

Nest sites with distances of <100 m from wind turbines were found in Schleswig-Holstein. The breeding distribution shows that habitat selection strongly depends on the preference for winter barley crops as a breeding site in this area. Where these crops were located in the vicinity of wind turbines, partly small distances were found between nest and wind turbine. Displacement effects of wind turbines could not be found.

The GIS-based analyses in the Hellweg Börde resulted in partly considerably larger distances between nest sites and wind turbines (median: 500 m, minimum distance: 170 m). The breeding distribution there was influenced by further disturbing habitat factors (e.g. slope gradient, height above sea level, vertical structures). No influence of existing wind turbines alone could be proven for this region, but all point vertical objects together, i.e. wind turbines together with power pylons, had a significantly negative effect on the breeding distribution. These results indicate that displacement effects in breeding areas will occur due to additive effects of already present disturbing structures (settlements, forest edges, power pylons), which will result in increased avoidance distances, if the amount of available open landscape is too small (Griesenbrock 2006, see also Sect. 5.5.5).

16.1.4 Further Bird of Prey Species

For birds of prey in general other characteristics of wind turbines than distance from the nest and preferred foraging grounds had considerably less influence on the collision risk. Wind turbines on bare mountain ridges or fells constitute a particular danger for birds of prey and especially vultures, as the birds will often take advantage of favourable thermals or may hunt in these places (Hötker et al. 2004).

Such phenomena were especially observed in wind farms in Spain and the American south-west (de Lucas et al. 2004; Smallwood and Thelander 2004) but not so much in Germany due to a lack of corresponding habitats. Apparently, solitary wind turbines lead to higher collision risks than wind turbines within wind farms. Furthermore, wind turbines located in crops will apparently lead to more collisions than wind turbines in other habitats, but the effects, i.e. the differences in collision risk, will probably be rather small.

16.2 Practical Recommendations

16.2.1 Red Kite

A reduction in collision risk can be achieved by observing a sufficiently large distance between wind turbines and red kite nests. The larger the distance between wind turbine and red kite nest, the lower the risk for red kites to collide with a wind turbine rotor. According to present data, the collision risk can be considerably reduced by observing a distance of 1000 m, as a major proportion of red kite flight activity takes place within this radius.

Apart from a sufficiently large distance from the nest, the collision risk can be reduced by keeping the wind farm and its immediate surroundings as unattractive as possible for foraging red kites. This can be achieved, if no field forage or other crops, to be mown during the breeding season of red kite (up until mid-July), is cultivated within a wind farm. Harvest within the wind farm should preferably start after harvesting other fields within the region. In large parts of Germany, this applies to winter barley and winter oilseed rape, as these are the first crops to be harvested. No additional sources of food like for example manure heaps which may attract red kites should be created within a wind farm. Fallows around wind turbine bases should be as small as possible and designed as unattractive for red kites as possible. They should not be mown or ploughed before end of July. Wind farms should be constructed at a sufficiently large distance from obvious points of attraction for red kites.

Attractive foraging locations created outside wind farms may possibly prevent red kites from visiting wind farms (e.g. regularly cut alfalfa crops, see Sect. 3.6). It needs to be taken into consideration that these crops may be used by several red kite breeding pairs in densely populated areas, which may in turn be attracted to the wind farm depending on the spatial constellation. Furthermore, the presence of other bird of prey species and non-breeding red kites needs to be taken into consideration.

The presented measures will not completely resolve the problem of red kite colliding with wind turbines, but will reduce the problem, if implemented consistently.

16.2.2 White-Tailed Sea Eagle

Due to missing avoidance behaviour of white-tailed sea eagles towards wind turbines, a reduction of the collision risk can especially be achieved by excluding the construction of new wind turbines in the vicinity of known nest sites and on preferred flight paths between nests and foraging grounds. To minimise the collision risk for white-tailed sea eagles, the fact that certain structures on the site of a wind farm may attract the birds (perching points for foraging flights, grazing grounds for geese, kettle holes and ponds as habitat for waterbirds, etc.) needs to be taken into consideration in the planning and evaluation of wind farms, as these structures may lead to an increased risk. Furthermore, specific, regularly used flight corridors of white-tailed sea eagles to foraging grounds further away should be taken into account in the planning of wind turbines.

The results of the present studies lead to the following recommendations: Sites at a radius of 6 km around known nest sites should as rule be checked for the presence of eagles (flight movements) in wind farm planning. Location and structure of potential wind farm sites should be checked with respect to preferred flight paths of white-tailed sea eagles between breeding site and foraging grounds. Regions with a high density of white-tailed sea eagles bear a high collision risk. This should be taken into account in the choice of potential areas for wind energy use.

16.2.3 Montagu's Harrier

The risk of collision with wind turbines for Montagu's harriers can be considerably reduced by spatial separation of breeding areas and wind farm sites. Primary objective of protection and avoidance measures for Montagu's harriers is to enhance the attractiveness and quality of habitats outside wind farms to reduce or avoid usage of wind farm sites by harriers. To achieve this, a network of areas with suitable vegetation or agricultural use should be created within the home ranges to serve as breeding and foraging habitats. Due to the dependence on farmed grassland, nutritious fallows etc. as well as the clear preference for linear edge structures as feeding habitat, habitat use of the Montagu's harrier can easily be redirected.

Protection measures should primarily concentrate on established traditional main breeding areas. The risk of collision can be considerably reduced by spatial separation of breeding areas and wind farm sites. It needs to be taken into consideration though that breeding sites within an area are often changed partly involving large-scale spatial relocation. It is therefore crucial to define and preserve sufficiently large, undeveloped breeding areas with suitable structures within the core areas of occurrence of this species. These areas should be identified by monitoring the distribution of breeding sites over a sufficiently long period and to protect these through conservation and landscape planning measures.

Potential breeding areas outside wind farms may be enhanced by the creation or conservation of attractive breeding habitats with suitable vegetation, i.e. cultivation or increased cultivation of winter barley (breeding habitat), extensively farmed grassland (pastures and hay meadows as hunting habitat) as well as the creation and expansion of edge of field buffer strips, edge strips of water bodies and edge fallows as linear hunting habitats together with areal foraging habitats within the main foraging areas.

No attractive foraging or breeding areas should be created within wind farms or relocated to areas outside wind farms. This especially applies to the cultivation of winter barley as the preferred breeding habitat as well as extensively farmed hay meadows and pastures within wind farms. Existing grassland areas in close proximity to wind turbines should not be cut or grazed during the breeding season to render the soil surface less accessible while hunting (see also red kite).

In conflict areas, the above measures should be part of large-scale habitat management, linking specific measures for rotating crops (breeding habitats) and grassland extensification programmes (foraging habitats). Measures should be spatially planned as to achieve a relocation of breeding sites and guidance of the birds along linear hunting habitats (edge of field buffer strips, ditch systems, etc.) to areas outside wind farms.

Like red kites Montagu's harriers intensively use freshly mown grassland. Probably due to rapid decrease in small mammal abundance, the attractiveness of these areas is limited to a few days. In grassland areas, hay cutting within the framework of habitat management programmes may lead to effective enhancement of foraging areas. Similar to red kites such measures may be effectively implemented in areas with alfalfa cultivation by cutting alternating partial areas several times per season.

16.3 Need for Further Research

A review of the comprehensive bird of prey monitoring data in comparison to data on wind energy development did not indicate an influence of wind turbines on bird of prey populations up until 2006 (Chap. 15). From end of 2005 until end of 2010 a further 4033 considerably larger and higher performance wind turbines were put into operation (German Wind Energy Association 2011) and plans are to further increase the number of wind turbines substantially to meet the climate protection goals of the federal government. Investigations of red kites clearly show that the collision risk increases with decreasing distance between wind turbine and bird of prey nest. As to weather and, if so, to what extent a further development of wind energy use will influence of bird of prey populations cannot be judged on the basis of so far collected data. The central database of collision victims of the State Bird Observatory and Field Centre Brandenburg can only provide a first impression of the frequency of birds found dead, as the probability that a collision victim is retrieved and reported is unknown. Therefore, the number of birds of prey colliding with wind turbines in Germany or the collision rate of birds of prey is still unknown. Based on extrapolations, Bellebaum et al. (2013) assume that about 3.1% of the post-breeding population of red kites in Brandenburg collide with wind turbines. Reliable estimates of collision rates are methodically complex (Smallwood and Thelander 2004; Grünkorn et al. 2009; Niermann et al. 2011) and therefore time consuming but after all indispensable to assess the role of wind energy development in the development of bird of prey populations.

Concerning this matter, the Federal Ministry for the Environment, Nature Conservation, Building and Nuclear Safety (BMU) is funding a research project to determine the collision rates of birds (of prey) and to establish planning-related recommendations to estimate and assess bird collision risk at wind turbines (project period: 01.11.2011 to 31.10.2014, FKZ 0325300A-D). Aim of the project is to collect representative data on collision rates of birds with onshore wind turbines in several North German federal states in a systematic field research project. Line transect searches for collision victims are carried out on a regular basis. The actual number of collision victims is calculated with inclusion of experimentally determined correction factors. Furthermore, behavioural observations of birds are included in the assessment of a risk of collision with existing wind turbines. Furthermore, modelling of the influence of additional mortality is used to clarify significance on population level. The project is intended to help in a final assessment of this issue, which is very present in public perception.

The impacts of mortality caused by wind turbines on the bird of prey populations in Germany can be assessed using population modelling (Carrete et al. 2009; Krüger et al. 2010) and should be carried out for the most important species.

Habitat use of white-tailed sea eagles could not be sufficiently investigated within the project and questions on the behaviour of white-tailed sea eagles within wind farms remain unanswered. Further investigations of the flight behaviour of white-tailed sea eagles by means of spatially and temporally high resolution transmitters (GPS) are therefore recommended.

As far as the Montagu's harrier is concerned, testing, control and further development of the above described habitat management measures can be considered to be crucial to achieve the intended reduction in collision risk. Apart from the collection of further basic data (e.g. flight behaviour, habitat use of special locations, which have not been investigated to date), future research projects should always include practical implementation of habitat management measures. To achieve this, partly long-term monitoring programmes need to be included.

References

Band W, Madders M, Whitfield DP (2007) Developing field and analytical methods to assess avian collision risk at wind farms. In: de Lucas M, Janss G Ferrer M (eds) Birds and Wind Power. Lynx Edicions, Barcelona

Bellebaum J, Korner-Nievergelt F, Dürr T, Mammen U. (2013) Wind turbine fatalities approach a level of concern in a raptor population. J Nat Conserv 21:394–400

Bundesverband Windenergie (2011) http://www.wind-energie.de/infocenter/statistiken/deutschland/windenergieanlagen-deutschland. Assessed 30 June 2011

Carrete M, Sanchez-Zapata JA, Benitz JR, Lobin M, Donazar JA (2009) Large scale risk-assessment of wind-farms on population viability of a globally endangered long-lived raptor. Biol Conserv 142:2954–2961

Dahl EL, Bevanger K, Nygård T, Røskaft E, Stokke BG (2012) Reduced breeding success in white-tailed eagles at Smøla windfarm, western Norway, is caused by mortality and displacement. Biol Conserv 145:79–85

Drewitt AL, Langston RHW (2006) Assessing the impacts of wind farms on birds. Ibis 148, Suppl 1:29–42

Dürr T (2011) Zur Gefährdung von Wiesenweihen durch Windenergieanlagen (WEA). http://www.mugv.brandenburg.de/cms/detail.php/bb1.c.206060.de. Assessed 30 June 2011

Griesenbrock B (2006) Habitat und Nistplatzwahl der Wiesenweihe *Circus pygargus* L. in der Hellwegbörde. Diploma thesis, Westfälische-Wilhelms-University of Münster

Grünkorn T, Diederichs A, Poszig D, Diederichs B, Nehls G (2009) Wie viele Vögel kollidieren mit Windenergieanlagen? Natur und Landschaft 84:309–314

Hötker H, Thomsen K-M, Köster H (2004) Auswirkungen regenerativer Energiegewinnung auf die biologische Vielfalt am Beispiel der Vögel und der Fledermäuse - Fakten, Wissenslücken, Anforderungen an die Forschung, ornithologische Kriterien zum Ausbau von regenerativen Energiegewinnungsformen. Endbericht für das Bundesamt für Naturschutz, Michael-Otto-Institut im NABU, Bergenhusen, pp 1–80

Krüger O, Grünkorn T, Struwe-Juhl B (2010) The return of the white-tailed eagle (*Haliaeetus albicilla*) to northern Germany: Modelling the past to predict the future. Biol Conserv 143:710–721

Langgemach T, Dürr T (2011) Informationen über Einflüsse der Windenergienutzung auf Vögel. Stand 13. Mai 2011. Pp 33. http://www.mugv.brandenburg.de/cms/media.php/lbm1.a.2334.de/vsw_tak_2011.pdf

de Lucas M, Janss GFE, Ferrer M (2004) The effects of a wind farm on birds in a migration point: the Strait of Gibraltar. Biodiv Conserv 13:395–407

Madders M, Whitfield DP (2006) Upland raptors and the assessment of wind farm impacts. Ibis 148, Supplementum 1:43–56

Niermann I, Brinkmann R, Korner-Nivergelt F, Behr O (2011) Systematische Schlagopfersuche - Methodische Rahmenbedingungen, statistische Analyseverfahren und Ergebnisse. In: Brinkmann R, Behr O, Niermann I, Reich M (eds): Entwicklung von Methoden zur Untersuchung und Reduktion des Kollisionsrisikos von Fledermäusen an Onshore-Windenergieanlagen. Cuvillier, Göttingen, p 40–115

Smallwood KS, Thelander CG (2004) Developing methods to reduce bird mortality in the Altamont Pass Wind Ressource Area. Final report by BioResource Consultants to the California Energy Commission, pp 363

Südbeck P, Bauer H-G, Boschert M, Boye P, Knief W (2007) Rote Liste der Brutvögel Deutschlands, 4. Fassung, 30. November 2007. Berichte zum Vogelschutz 44:23–81

FSC
www.fsc.org
MIX
Papier | Fördert
gute Waldnutzung
FSC® C083411

Kolibri 360 GmbH
Ferdinand-Jühlke-Straße 7
99095 Erfurt, Deutschland
produktsicherheit@kolibri360.de